CONFRONTING DEATH

CONFRONTING DEATH

⌢

Values, Institutions, and Human Mortality

David Wendell Moller

New York   Oxford
OXFORD UNIVERSITY PRESS
1996

Oxford University Press

Oxford   New York
Athens   Auckland   Bangkok   Bombay
Calcutta   Cape Town   Dar es Salaam   Delhi
Florence   Hong Kong   Istanbul   Karachi
Kuala Lumpur   Madras   Madrid   Melbourne
Mexico City   Nairobi   Paris   Singapore
Taipei   Tokyo   Toronto

and associated companies in
Berlin   Ibadan

Copyright © 1996 by Oxford University Press, Inc.

Published by Oxford University Press, Inc.
198 Madison Avenue, New York, NY 10016

Oxford is a registered trademark of Oxford University Press

Library of Congress Cataloging-in-Publication Data
Moller, David Wendell.
Confronting death : values, institutions, and human mortality /
David Wendell Moller.
p. cm.   Includes bibliographical references and index.
ISBN 0-19-504295-6 (clothbound).—ISBN
0-19-504296-4 (paperbound)
1. Death—Social aspects.   2. Thanatology.   I. Title.
HQ1073.M6   1996
306.9—dc20   94-30042

1 3 5 7 9 8 6 4 2

Printed in the United States of America
on acid-free paper

For Mike Vogel—My Hero

PREFACE

When I called the American Cancer Society to request permission to include some of their materials in this book, their representative responded: *"Absolutely not. In no way do we want to be associated with a book on death. We want to emphasize the positive aspects of cancer only."* Although misguided, such avoidance of death and dying should not be interpreted as being intentionally callous; rather, it reflects the prevailing view of our time. Thus, the response of the American Cancer Society expresses a culturally grounded fear that association with death will tarnish its image and hinder its ability to fulfill many of the functions it performs so well. Of course, people do die from cancer. Cancer is a leading cause of death in our nation, and individuals who die from cancer often experience slow, lingering, and turbulent deaths. The emotional, psychological, and social needs of such individuals are great. Yet, American society's attitude and response to these needs is too often one of avoidance and denial. Unfortunately, death cannot be averted by denial. Death is universal and inescapable. Additionally, the needs of dying individuals and their loved ones inevitably persist and become all the more urgent when we as individuals and as a society fail to respond meaningfully to them.

It is my intention in this book to advocate for all individuals whose lives are touched by dying and death and, in this way, to affirm the value of life and living. If we can replace our attitudes of denial and avoidance with understanding and compassion, we will be able to provide greater solace for dying and grieving persons. In this way, the value of life, even during the extremity of suffering, will be respected and proclaimed. Thus, the goodness and value of life in its wide range of styles are heartily affirmed by the study of the ways and styles of death. Especially affirmed are the living styles of individuals facing their own mortality and the mortality of their loved ones.

Over the years I have become more and more grateful for the lasting impact that Richard Connors has had on shaping my personal and scholarly life. His influence has continually led me to challenge the existence of evil in the world, the inhumanity of humanity, and the relationship between personal growth and openly confronting the problems of human suffering and mortality. William Moller, Jr. has shown me that enormous physical debility does not have to lead to despair of life. In fact, in many ways, his courage is a profound inspiration for all who believe in the inalienable value of human life.

I would like to thank my colleagues at Indiana University for their support during the writing of this book. I am particularly grateful for the collegiality of Bob White, Bill Gronfein, John Liell, Morris Weinberger, and Colin Williams. I am indebted to John Barlow and Bill Plater. I'd like to thank the Indiana University President's Council on the Social Sciences for their support of this project. I especially thank Barbara Bogue for her friendship and hard work during the preparation of this manuscript. I thank Mary for her continuing nurturance and love. I am grateful to the anonymous reviewers who carefully studied the manuscript in earlier versions and made important and helpful comments. I am grateful to Stan Stackhouse for his sensitivity and diligence in helping me gather some of the illustrations for this book. I thank Mike Vogel for touching my life so deeply and for showing me that the holocaust must be remembered in any serious study of life and death. I appreciate the faith Susan Rabiner had in this project and thank Valerie Aubry for her encouragement and support. I am grateful to my students for all they teach me and for continually reaffirming my belief that the study of death is an important means of affirming the value and love of life.

Indianapolis   D.W.M.
May 1994

CONTENTS

CONFRONTING DEATH

The world has become a world "given over to death," and this is the more terrible because there exists in the world a basic despair that seems impossible to remedy.

JACQUES CHORON

1

Dying and Historical Context

I'm not afraid of death; I just don't like the hours.

WOODY ALLEN

The idea of death, whether it evokes images of peace, fellowship, beauty, fear, or loneliness, is a product of culture and society. The environment surrounding human death inevitably shapes the nature of the death experience. Consider, for example, a person dying at home surrounded by a loving family, and concerned friends and community. Contrast this situation with that of a person dying in the austere and impersonal surroundings of a medical center, attended by professional staff and equipment. Assuredly, we would expect that the experience of dying at home would be qualitatively different from that of dying in a hospital. On a broader scale, we would also anticipate that human dying and death would differ in societies that fear and deny death and in those that accept this process as a natural part of life.

To begin with, let us ask ourselves how society and culture affect living life. From birth to death, human beings are social creatures. Many of the things that people regularly do and believe have been taught to them by parents, teachers, peers, and institutions. Thus, going to the hospital to give birth, choosing blue wallpaper for little Johnny's room, learning the value of democracy, speaking and reading the English language, being buried in a coffin in a single grave, having death certified by a physician, and brushing one's teeth every morning are all behaviors that are approved and taught to us by American society. Not all cultures and societies, however, behave as Americans do. It is important to understand that human behavior is largely a product of cultural values, social structures, and interaction with other members of a particular society. Likewise, human dying is defined by the time and place in which dying and death occur. Specifically, the meaning of the dying experience, the place where death occurs, and the response of other people to dying and death will, on balance, be shaped by the values and institutions of the society.

3

The purpose of this chapter is to illustrate how dying and death are influenced by historical and social contexts. One of the major themes of this book is that *the way one dies is a reflection of the way one lives*. This is not just applicable for the lives of individuals, but is also reflective of the patterns of death and dying in the society as a whole. In promoting this theme, this chapter will explore distinct and changing views of death throughout European and American history. We may find these unfamiliar patterns of human death bizarre in that they differ from the patterns we are familiar with and take for granted. They are, however, when considered in their appropriate cultural and historical contexts, quite normal and understandable.

Traditional Patterns of Death and Dying

Historically, death and dying held a deeply meaningful place in human culture. From the fifth to the nineteenth century in European civilization, people openly confronted death on a regular basis. As we shall see, there were unique and interesting variations in Europeans' attempt to deal with dying and death during these centuries. People commonly defended themselves against the threat of death through community, spirituality, and ritual. These traditional social forces shaped popular responses to death and helped provide a sense of comfort to dying people and their families. Philippe Ariès, in *The Hour of Our Death*,[1] provides a comprehensive, scholarly analysis of the evolution of European and American attitudes and behaviors concerning dying and death. Let us consider his account.

The Tame Death

During the early Middle Ages in the West, beginning in the fifth century, human dying and death were characterized by an attitude of acceptance and tranquility. Death during this time was a communal and public act whose approach could be anticipated and met by ritualized expression of mourning around the deathbed. During this period of the Tame Death an unexpected death was a source of social shame, for people perceived it as a sign of God's displeasure. A good death was one that could be anticipated, allowing the dying person to prepare spiritually for life's end and the community to gather and offer support to the dying person and itself, as well. People considered a sudden death disruptive and frightening, for it imperiled the community's ability to adjust to its loss. The only exception that was allowed to the requirement that people be forewarned of death was for the knight killed in combat. In a society prizing chivalry and military conquest, a knight's death in battle was seen as saintly. Since most fallen knights did not die immediately from their wounds, rituals of death could be carried out in an abridged form behind the lines of battle. Additionally, as a knight's occupation was inherently dangerous and life-threatening, he was viewed as being in a perpetual state of anticipating and preparing for the possibility of death. Anyone other than a knight who died suddenly was cast to the outskirts of society and quickly and quietly forgotten.

Traditionally, ritual was an important way to defend oneself against fears of dying and death. Ritual acts in the shadow of death offered a sense of familiarity, meaning, and reasonableness to an otherwise unexplainable event. During the Tame Death the dying person was surrounded by the family and community. Goodbyes were exchanged, prayers said, and final instructions given to family members. After these farewells had been completed, if nothing remained to be said, the dying person quietly and peacefully waited for death. This process of dying took on a mandatory and ceremonial quality; a person would be horrified to think of death without the presence of others and a ceremony of farewells. Consider for a moment how much death and dying have changed over the years. During the Tame Death, up to the final moment of death, one was never alone; in modern society, one has a very good chance of dying alone someday, perhaps in a hospital, nursing home, or other chronic care facility.

During the Tame Death, it was believed that upon death the individual entered a sleep-like state to peacefully await salvation. It was believed that the dead person slept tranquilly during this period in a garden of flowers. (It is interesting to note that this image of the dead peacefully at rest, surrounded by flowers, persists today. Sending flowers to the funeral home and draping the casket with a spray of flowers is common practice in contemporary American funerals as is commenting on how peaceful the deceased looks as he or she lies in the casket.) Awakening and salvation would take place at the end of the world, when the body would be resurrected to join the soul. In this view, there was an image of collective destiny for all humanity, as it was believed that everyone would be saved. Of course, if one died suddenly and privately, one would not be saved. But, from a societal standpoint, one who died shamefully in this way was quickly forgotten—as if he or she had never existed at all.

Burial practices during this time also differed radically from those of modern individuals. Ironically, despite regular familiarity with death and dying, people during the early era of the Tame Death were frightened in the presence of a corpse. As a result, cemeteries were initially built far away from towns to protect the living from contamination by the dead and by the evil spirits that were thought to ooze from them.

People also sought protection from the wicked spirits of the dead by appealing to martyrs and saints. It was felt that the presence of a martyr's grave would drive away harmful spirits that could jeopardize the living. The belief that martyrs protected the dead and the living became so widespread that the fear of being close to a corpse subsided and the cemetery itself began to be defined as a sacred place. Sometime during the eighth century, the living and the dead were brought together as the cemetery was moved from the outskirts into town, specifically into the town churchyard.

As cemeteries moved from the countryside into towns, the dead were buried either inside the church or in its courtyard. The honored and preferred place of burial was within the church: under altars or entombed in a wall or under the floor. Since burial in the church was a sacred honor, only those whose lives were recognized by the church as being virtuous and holy (martyrs, saints) could be buried there. Others had to be buried in the outside cemetery, the exception being those who were sufficiently affluent to make sizable donations to the church and thereby "purchase" the church's recognition of their holy worth. It would seem, then, that buying oneself a place close to God is neither a new idea nor a new practice.

The cemetery during the Tame Death was replete with community graves that were largely unmarked. While burial in the church was in an individually identifiable tomb, the dead in the courtyard were piled into mass graves containing from 30 to 1,500 bodies. This practice became common around the thirteenth century as local towns were forced to deal with the hoards of dead bodies resulting from the plague.

The most striking characteristic of the traditional cemetery was the digging up of the dead, the removal of their bones, and the reusing of the grave for new burials. As the European population began to multiply, space in churchyard cemeteries became increasingly limited. In order to deal with the problem of scarce burial space, graves were "turned over" and the bones removed and placed in a charnel. A charnel is literally a "house of bones," and was designed specifically as a storage gallery—a final resting place—for bones that had been exhumed. Initially, the charnels were no more than storage bins for bones and skulls, but as they became more widespread, they began to take on aesthetic meaning. In fact, they became places where bones and skulls were arranged with artistic flair in order to create a sensual museum of the dead that the living would regularly visit.

By the turn of the twelfth century, the cemetery had become what the suburban shopping mall is to people of modern society. It was the center of social life. Residences sprang up surrounding the cemetery. It became the central place of public interaction—of making speeches, of courting, of picnicking, of baking bread in communal ovens, and so on. Indeed, the cemetery served as a forum and public square where all community members could stroll, socialize, assemble, and make merry.[2] It is important to remember that this activity often took place in the presence of mass graves that were partially full and fully open, in the midst of funerals that were taking place and, of course, in full view of the charnels.

Death's Surprise: The transi, as prevailing image of death, selects its victim.
© *Bale, Kunstmuseum*

Museum of the Dead: Artistically arranged, these bones in the charnel are exhibited to the visual delight of the community. © *D.R.*

Bottom right: **Cemetery of Mummies:** The almost theatrical setting created in the Capuchin Charnel in Rome illuminates the intimate connection between the world of the living and of the dead. © *Alimari-Giraudon*

House of Bones: Skull boxes are the centerpiece of this charnel which is designed to artfully and efficiently store exhumed bones. © *Jean-Robert Masson*

Common Graves: Mass graves reflected the openness toward death and the prevalence of a community orientation during the Tame Death. They will, in forms indicative of racial hatred and hi-tech destruction, reappear in the twentieth century during the holocaust. © *Marc Durand*

An important social function of the bustling public atmosphere of the cemetery was to regularize and normalize the presence of death. An ironic relationship of indifference and intimacy between life and death was created. Living could proceed untroubled in the shadow of death itself, as its terrifying and frightening aspects had been stilled, that is to say, death was tamed.

The Death of the Self

The Tame Death as a general and prevailing image of death persisted until the eighteenth century. However, several variations of the Tame Death emerged as the years and centuries passed. The Death of the Self is the first variation in the generalized pattern of Tame Death.

As noted earlier, the Tame Death was a collective phenomenon; every person's death was the same. All were saved and there was no fear of punishment, although there was a fear of quick and unexpected death. At the beginning of the eleventh century, the idea of a universal, collective destiny disappeared. Growing societal images of courts, justice, and judgment supported the developing idea of a societal distinction between good and evil. In this context, the individual self began to take on meaning and the concept of biography emerged. The acts of individuals could now be evaluated as being good or bad, and it was felt that people could choose between good and evil. It was believed that one's biography, or account of one's life, consisted of the composite picture of the choices made between good and evil.

This growing idea of biography and self diminished the notion that the fate of the dying person was bound to the collective destiny of humanity. The concept of biography brought with it the belief that all of life's activities could be itemized and summarized. The devil, it was thought, kept a record of all of one's evil choices, and the guardian angel recorded all of a person's good deeds. At the moment of death, a person's whole history was compressed into a singular statement. If good outweighed evil, then a person was judged as being wholly good and worthy of salvation. If the scale of accounts was tipped in favor of evil, one was judged as wholly evil and sent to hell.

During the fifteenth century, a new idea of the final judgment appears. Salvation or nonsalvation would no longer be determined by record keeping of life's acts, but rather by an active spirtual battle between good and evil, a battle that would take place at the bedside of a dying person.

The salvation of a person's soul was now to be determined by the act of death rather than the acts of life. Dying a good death became the key to salvation, and books on the art of dying a good death became as popularized by the printing press in the fifteenth century as books on how to have good sex are today.

A good death was a religious death. It involved praying, an act of contrition, and a deathbed confession. During the period of dying, it was thought that the

person would be tempted by the devil. If the devil's temptations were successful, all would be lost and the person would be damned. However, it was believed that God was also present at the deathbed, offering spiritual support to
help the dying person reject the enticements of the devil. Indeed, the prevailing belief of the time held that God and the devil actively competed with each
other for the soul of the dying person and that the person actually witnessed
their supernatural battle. Naturally, the dying person was surrounded by people,
but prevailing custom indicated that he became unaware of their presence as
he attended to the spiritual battle over his soul and concentrated on rejecting
the devil's temptations.

The important difference between the Tame Death and the Death of the Self
is the movement from universal salvation to individual judgment. According
to Ariès,[3] because an individual "bad death" was now possible, the fear of
punishment began to haunt humanity. Ariès argues that people during this time
were still generally without fear of death, as dying itself was not frightening,
but damnation and eternal punishment became pronounced sources of terror.

In addition, pungent physical images of death and of putrid rotting flesh,
termed the "macabre," began to become quite popular during the fourteenth to
sixteenth centuries. Anyone who has spent a night camping in the mountains

---

Thanatology Snapshot 1.1

The Common Graves

This was where they buried the poor, those who could not afford the high price
of burial in the church or under the charnels. They were piled in huge common
graves, veritable pits thirty feet deep and fifteen by eighteen feet in area, which
contained between twelve hundred and fifteen hundred bodies; the smaller ones
between six hundred and seven hundred. There was always one open, and sometimes two. After a few years (or months), when they were full, they were covered over and other pits dug nearby. . . . These pits were barely covered with
earth when they were closed, and it was said that during cold winters, the wolves
had no trouble digging up the bodies—nor did the thieves who supplied ama-
teur anatomists in the eighteenth century.

Philippe Ariès
The Hour of Our Death
(New York: Alfred A. Knopf, 1981), p. 56

---

can empathize with John Denver's analogy: "You fill up my senses like night in the forest." To be sure, one's senses of sight, hearing, and smell are swelled and accentuated by the forces of nature in the forest, especially in the darkness of the night. Similarly, the images of death in its naked and "vulgar" forms filled the senses of Europeans of the fourteenth to sixteenth centuries. For example, skeleton men dominated the art of this period, much as sexy ladies now predominate on magazine covers.[4] Poetry, paintings, woodcut art, drama, and stories of the time were characterized by themes and images of disfigured corpses, rotting skin, and dead people covered with shreds of worm-eaten flesh.

The Danse Macabre, also known as the "death play" of the fifteenth century, enacted scenes whereby gruesome dancing skeltons would survey and suddenly select people from the living to join them in the eternal dance of death. The social meaning of the macabre and the Danse Macabre was to illustrate that death is everywhere, that no one knows when death will choose him or her, and that, ultimately, all are equal in the face of death.

As we progress into the sixteenth century, we see that spiritual images are replaced by physical images of decomposition. The appeal of sudden death is also reflected in the Danse Macabre. These changes are especially understandable in light of the increasing social value placed on material possessions in this period. The pervasive threat of having one's life and possessions torn away by plague or other infectious diseases helped render images of sudden death and the physical dimensions of death quite common.

Upon first glance this emphasis on sudden death and physical decomposition seems rather bizarre. Yet, even in our contemporary death-avoiding lifestyle an ironic fascination exists with sudden, unanticipated death and grotesque physical images thereof. To be sure, the crimes and personality of Jeffrey Dahmer captured the attention of many Americans, and the brutal murders of Nicole Simpson and Ron Goldman received unprecedented coverage by both mainstream and tabloid media. In fact, it is not pressing too far to suggest that those brutal deaths, and their associated celebrity defendant, became a source of ordinary conversation and entertainment for many people. In addition, the novels and films of Stephen King are enormously popular because of, not in spite of, their ability to titilate their audience with themes of the macabre. In fact, the entire genre of horror films does this as well. Similarly, the body of films labeled as "thrillers" captivates its audience by the use and attraction of furious and sudden death. And, let us not forget that an entire population of readers and moviegoers were delightfully terrorized by the savagery of a great white shark swimming and feeding off of the shores of Long Island, New York. Thus, although the forms of the macabre have changed, the appeal of sudden, violent death, and iconic representations thereupon, are a salient part of the contemporary American cultural experience.

Remote and Imminent Death

The crucial moments of dying took place in the bedroom during the Tame Death and the Death of the Self. As we move through the sixteenth and on to the seventeenth century, the days of dying and the moment of death lose their magical and mystical appeal. Death is no longer seen as an elusive spiritual phenomenon but becomes a natural part of, and merely the cessation of, living. Thus, as Ariès describes it, the art of living replaced the art of dying. Life became a preparation for death, and living as full and virtuous a life as possible emerged as the key to salvation. The devastating experience of the plague and the macabre orientation that characterized the sixteenth century propagated the idea that death could happen at any time. Since death was always imminent, the ideal that emerged during the seventeenth century was to be always prepared for death by living as good a life as possible. Thus, when a famous monk of the time was asked during a soccer game what he would do if he had only five minutes left to live, he replied that he would finish playing soccer. Apparently he felt he had lived as he should and was ready to die whenever God called.

As living a good life became more important than dying a good death, death became less immediately consequential. The result was that the spiritual drama of the deathbed was replaced by a calmer understanding that life would someday lead to death and that as long as one lived a good life, death would be sweet and peaceful. It is also intriguing to note how the fear and drama of sudden death, so prevalent earlier on, had been abolished.

While concern about death was becoming more remote, the idea of physical death continued to be of importance. Death, or, more precisely, the dead body was seen during this time to have two important values: (l) as a source of macabre eroticism and (2) as a means of furthering scientific knowledge about life.

Necrophilia, or love of the dead, became a prevalent theme in the seventeenth and eighteenth centuries. Unlike in the era of the macabre, the idea of death was no longer associated with images of decompositon and decay, but rather elicited themes of sensuality and desire. In much of the drama and literature of the period, dying as a human process heightened the intensity and expression of love and passion. There were also widespread tales of the living making love to the dead in the literature, drama, and especially the oral tradition of this period.

It is difficult to tell whether many of the tales of necrophilia were more fantasy than reality; nevertheless, the idea of loving the dead was a popular theme in everyday life. Let us listen to the words of Ariès:

> Were these fantasies inspired by actual events? Sade tells us, "I often saw a man who bought the cadavers of young girls and boys who had died violent deaths

and had been recently placed in the ground. He paid their weight in gold. He had brought them to his home where he committed an infinity of abominations on these fresh bones." The marquis is not a reliable witness. However, his remark is strangely confirmed by a memorandum on the indecency of burials addressed to the procurator general of Paris in 1781. "The bodies lowered into this common pit are every day exposed to the most infamous violations. On the pretext of study, certain persons . . . steal dead bodies from cemeteries and commit on them everything that impiety and debauchery might inspire."[5]

The theme of loving the dead undoubtedly seems perverse to even the most enlightened of modern readers, but before we simply dismiss these necrophilic themes as deranged and disgusting, let us reflect on the presence of such themes in everyday American life and culture. Dracula and his undead women continue to delight modern readers and theater- and moviegoers with their erotic seduction of the living. Many of the above-mentioned Hollywood produced thrillers that delight and terrify contemporary audiences also have strong and vivid erotic impulses connected with murder, blood, and dismemberment. The recent bestseller, *The Alienist*, explicitly connects unorthodox sexuality with murder and the macabre, and Anne Rice's enormously popular trilogy of novels about vampires distinctly blends the sensual with the macabre. "The Addams Family," with their unabashed love of the dead, Mel Brooks' humorous version of Dr. Frankenstein's corpse making love to Madelyn Kahn in a haystack while she sings "Sweet Mystery of Life," the presence of Elvira on late night television and her appearance in national commercials, and Alice Cooper's songs "Cold Ethyl" and "I Love the Dead," to name just a few, suggest that one doesn't have to look far to find some indicators of necrophilia in everyday popular culture.

During Remote and Imminent Death, dead bodies were not only a source of erotic inspiration, they were an important source of knowledge. Hence, the cadaver became a focus of societal attention. Given that society was on the edge of the Industrial Revolution and every day becoming more secularized, it is not difficult to understand the growing fascination with gross physical anatomy. People felt that the cadaver held the secrets to life, and that knowledge of anatomy was a basic requirement for understanding life and the ways of God. Anatomy was so popular in the seventeenth century that the anatomy lesson was often a social "happening" that included good-natured joking, refreshments, and people wearing gay, masquerade-like apparel. Dissection had become an ironically fashionable activity, an ancient version of the modern cocktail-theme party.

The study of anatomy became so celebrated that more and more dissections were being performed outside of the medical arena in private homes. Indeed, it was not unusual for an affluent person to have a private anatomy lab in his home, where he would diligently dissect and study the physical features of the human cadaver. The growing societal notion that anatomy was for everyone of culture and sophistication led to a shortage of cadavers during the seventeenth century. As Ariès comments, people complained that young surgeons could not find enough bodies because of the intense competition for cadavers from private dissectors.[6] Indeed, the cadaver shortage had such an impact on the times that cemeteries were raided, black-market cadaver enterprises emerged, and as a result, cemeteries had to be staffed with watchmen in order to control the pilfering of the dead. Ironically, as concern for living a good and spiritual life made death more remote, an emphasis on the immediate physicality of death (necrophilia, dissection) blossomed.

The Death of the Other

The dawning of the nineteenth century brought a romanticization and a heightening of the beautification of death. The deathbed scene became a prevailing theme in the literature of the Victorian period. The Victorian frame of mind sought to build a society based upon restraint, politeness, proper etiquette, and decorum. The everyday emphasis on these ideal behaviors gave rise to a sweeping cultural fear of sex and pleasure. The passions and irrationality of sexuality threatened to disrupt the order and civility the Victorians were trying to achieve. Hence, in this context, death was removed from the sphere of the erotic and became defined as sentimental rather than sensual. The typical deathbed scene of Victorian society embraced suffering and dying with idyllic romanticism and excessive emotionalism. Death became beautified.

The Victorian deathbed differed from other deathbeds of the traditional death in that a smaller fellowship surrounded the dying person. The nuclear family had become the normal living arrangement, and the Industrial Revolution had contributed to the growth of cities, with their corresponding sense of isolation and alienation. In this changing social setting, the dying person was surrounded by a community generally limited to immediate family and loved ones.

In an important way, the emotional closeness of the fellowship with the dying person intensified the pain and separation of death. The material and emotional excesses of grief, which were characteristic of this era, may be seen as a way of compensating intensely grieving individuals for their loss. Thus, the flowery consolation literature of the times, the fancy embossed memorial cards, the extended wearing of mourning clothes, and the extensive adornment of specially designed mourning jewelry established required rituals of mourning that helped to fill the void created by the death of a loved one.

The Victorian person was, indeed, fascinated by dying and death. Despite this fascination, the peaceful resignation and tranquility that characterized the deathbed of preceding eras were gone forever. The extravagant statements that the Victorians made about death, grief, and mourning signal that death was no longer calmly being accepted. Although ritual, religion, and fellowship remained strong sources of support at the deathbed, one can almost hear the howling of death and the scurrying of Victorian people as they tried to quiet its highly emotional and disturbing presence. No longer tranquilly accepted, death was becoming unleashed—that is to say, untamed.

Across the Ocean: Death and the American Puritan

The Puritans who landed at Plymouth Rock and settled New England brought with them an ambivalent attitude toward death and dying. The Puritans valued death as a vehicle for attaining salvation. In this sense, death was a release of the soul from its earthbound imprisonment. At the same time, however, the Puritan was possessed by an unrelenting fear of death and eternal damnation. In order to explain the contradictory relationship of both valuing and fearing death, we must examine the religious and social contexts in which the Puritans lived.

Puritan religion had strong roots in Calvinism. The ideas of John Calvin stressed that all human beings were predestined before birth to salvation or nonsalvation. In addition, these ideas emphasized that the motives of the Almighty were beyond human comprehension. According to Calvinism, if humans could understand God and influence his judgment by their earthly acts, God would be ordinary and weak. Since God was considered neither, and since humans were seen as not only weak but also sinful and depraved, human beings were seen as incapable of affecting God's judgment. Because it was believed that humanity was unable to influence the process of salvation or damnation, Puritans lived in ubiquitous and helpless fear of nonsalvation.

As death drew near for the Puritan, tension naturally grew, and the question of whether one's sin-ridden soul would be admitted to Heaven or cast into the fiery pit of Hell haunted every conscious moment.[7] In this way, tension and anxiety in the face of death were built into Calvinist doctrine. The idea of death as a welcome part of the journey to Heaven became overshadowed by the intense fear of nonsalvation.

The dread of eternal damnation was softened by the presence of religious ritual and human fellowship. In the event that a Puritan was predestined to Heaven, he or she wanted to be well prepared for salvation. In order to encourage and provide support for this preparation, people gathered around the deathbed. Prayers were said, songs were sung, and the Bible was read—all of which were intended to urge acceptance of the will of God. From a sociological perspective, this communal religious activity offered comfort and support to those who were inevitably at the mercy of God's will. Anyone who has participated in an intense religious ceremony of group prayer and song will easily recognize the feeling of euphoria and well-being that can result from a communal religious experience. This sense of elation was instrumental in aiding Puritans to cope with their own dying and in assisting the family, loved ones, and the community to cope with their loss.

Since the Puritan believed that death freed the soul from earthly bondage, the dead person was considered to be nothing more than an empty, bodily shell. Consequently, funeral and grieving practices were supposed to be nonelaborate and generally understated. The body was washed, wrapped in a plain white cloth, and placed in a simple wood coffin. Friends and neighbors would then come to the home, or sometimes to the deceased's church, where the body would be laid out. Generally, the family would remain awake during the night, with the body in their presence. This gave rise to what we know today as going to a wake. Prayers would be said during the viewing of the body—not for the soul of the deceased, but rather for the consolation of the living. After a few days, the body would be buried. After burial, the mourners would return to the home of the deceased for festivity, food, drink, prayer, and words of condolence. Again, the focus of attention was the grieving survivors, not the deceased.

Burial of the dead was to be a simple affair, without elaborate ritual and ceremony. As the appropriate guidelines pointed out: "when a person departeth this life, let the dead body upon the day of Buriall, be directly attended from the house to the place appointed for the publique Buriall, and there immediately be interred, without any ceremony."[8] Extended periods of mourning were also discouraged, as they were seen as an unwillingness or inability to accept the will of God.

The Puritan imperfectly lived up to the ideal of simple commemoration of the dead. As the seventeenth century came to a close, we find that funeral sermons were delivered on the day of burial at the gravesite. An increasing air of formality began to be associated with the funeral, and symbols of grief (mourning gloves, ribbons, scarves, etc.) were typically worn. Funeral verses were often inscribed on coffins and on the sides of the horse-drawn hearses that transported the body to the cemetery. Elegant mourning rings—"marked with black enamel death's-heads, skeletons, coffins, and other reminders of the frailty of life"[9]— were often given to those who attended the funeral procession and the subsequent celebration and festivities. Ironically, many Puritans became proud collectors of these artifacts of death.

The Puritan was once more caught in a contradiction between theology and social reality. The doctrines of Puritan religion emphasized the need for simplicity in funerals and restraint in rituals. The reality of Puritan funeral practices, however, increasingly gravitated toward the elaborate and extravagant.

This appearance of more complex and ornate rituals is understandable from a sociological perspective, as the aggrandizement of funeral ceremonies helped the Puritan to live with the overwhelming fear of death.

The nineteenth century brought with it a rise in individualism. Evolving societal forces in America were diminishing the tightly woven sense of community that had been present in the 1700s. The nuclear family was becoming more prominent and increasingly isolated from community and kin networks. As David Stannard comments in his study of the Puritan way of death: "it was into this sort of world that there emerged an attitude towards death and dying that was characterized by self-indulgence, sentimentalization and ostentation—a world diversifying and compartmentalizing . . . a world that had lost something central to the cohesiveness of Puritan culture: a meaningful and functioning sense of community."[10] The traditional Puritan way of life, deeply rooted in community ritual and cohesiveness, was rapidly being dissipated by the social forces of modernization.

As in Victorian England, the movement away from a broad base of communal solidarity brought to nineteenth-century America images of death that were romanticized and beautified. Elegant cemeteries were built and proudly displayed to visitors. Tombstones were artistically and dramatically etched. Dying became accentuated by elaborate deathbed rituals and was popularly characterized as life's final and most edifying experience. The death of children became transformed into a sentimental deliverance to eternal salvation. Mourning apparel flourished, and the time period of mourning was extended to a year or more. This inclination toward a lavish response to death, as already mentioned, was logically consistent with the growth of individualism and the corresponding intensification of grief. Unlike in prior times, the armor of expansive community no longer shielded the vulnerable, grieving person from intense pain. Elaborate sentimentalization of death helped to provide a strengthening sense of comfort to fill some of the emotional vacuum that was created by the age of declining community.

The Disappearance of Death

The twentieth century crystalized a new attitude toward dying and death. In this modern era, dying and death were no longer considered to be important experiences that would absorb the attention and energies of humanity. To the contrary, death and dying became something to be shunned, avoided, denied, and, if possible, conquered. Contrary to earlier times, dying and death in the twentieth century have become devoid of meaning, ritual support, and cultural approval. What has been termed the "age of death denial" has arrived.

Leo Tolstoy, in his most powerful and perhaps greatest short story, "The Death of Ivan Ilych," portrays the human implications of the meaninglessness of dying and death in modern society. Written in 1859 and clearly a prophetic look at things to come, the story begins with the announcement that Ivan Ilych has died. The news of his death elicits responses from others based upon self-concern, business implications, and a general sense of unease. As Tolstoy writes:

> Besides considerations as to the private transfer and promotions likely to result from Ivan Ilych's death . . . each one thought or felt "Well, he's dead, but I'm alive!"[11]

And, as an acquaintance of Ivan Ilych reflected when fulfilling the "tiresome demands of attending the funeral service":

> The dead man lay as dead men always lie, in a specially heavy way, his rigid limbs sunk in the soft cushions of the coffin, with the head forever bowed on the pillow. . . . Besides that there was in [his] expression a reproach and a warning to the living. This warning seemed to Peter Ivanovich . . . not applicable to him. He felt a certain discomfort and he hurriedly turned and went out the door. . . .[12]

The story then shifts to a retrospective account of Ivan Ilych's life and his dying. The point that Tolstoy makes about the life of Ivan Ilych is that he lived in a most normal way; he worked hard and achieved professional respect, social status, and affluence. His life "flowed with pleasantries" but was without family solidarity, the support of community, tradition, and spiritual meaning. It is for this reason that Tolstoy observes that Ivan Ilych's life was ordinary in that it was steeped in a culture of materialism, superficiality, and self-interest. Yet, it was precisely these qualities that made his life most terrible as they clearly were unable to provide for meaning, human comfort, and the alleviation of his sufferings.

Thus, the life of Ivan Ilych, while socially proper and successful, ultimately proved to be hollow and empty. The qualities of his life, with its absence of support systems, affected his dying in most painful and horrible ways. On his deathbed, he found himself deceived by the medical profession, lied to by friends and family, alone, and terrified—without the support of ritual, community, and tradition to sustain him. Ivan Ilych reflects upon the horrors of this modern way of dying in his final days:

> Anger choked him and he was agonizingly, unbearably miserable. "It is impossible that all men have been doomed to suffer this awful horror!" . . .[13]

> Ivan Ilych saw that he was dying, and he was in continued despair . . . [his wife became] impatient with illness, suffering and death because they interfered with her happiness. . . .[14]

He suffered the same and unceasing agonies and in his loneliness pondered the same inevitable question, "What is this? Can it be that this is Death?" And the inner voice answered: "Yes, it is Death."[15]

"Why these sufferings?" And the voice answered, "For no reason—they just are so." . . .[16]

Tolstoy's short story marvelously and effectively characterizes the meaninglessness of the modern experience of dying. In addition, it illustrates dramatically the theme: The way one dies is a reflection of the way one lives. Living in a framework of individualism, materialism, absence of ritual and weakening of religion, Ivan Ilych's death process was filled with loneliness, was without meaning, and was full of terror. The relevant point to be extracted from Tolstoy is that the values and institutional arrangements of modern society do not provide regular patterns of comfort and support to dying people and their grieving survivors.

Dying, no longer supported by deathbed ritual and community, has become an unbearable and intolerable intrusion into the order of everyday life. Edwin Schneidman[17] speaks of death as a social disease, something not appropriate for American life and, hence, something to be hidden away. Ariès[18] comments on how most people come to a modern American funeral with emotional restraint and little sadness or mourning. Indeed, the funeral home is generally a very calm, stoic environment. The physical focus of attention is the deceased, who has been made into a beautiful-sleeping-memory picture by the embalm-

---

Thanatology Snapshot 1.2

Sex and Death Reversed

At present, death and mourning are treated with much the same prudery as sexual impulses were a century ago. Today it would seem to be believed, quite sincerely, that sensible, rational men and women can keep their mourning under complete control by strength of will and character, so that it need be given no public expression, and indulged, if at all, in private, as furtively as if it were an analogue of masturbation.

Geoffrey Gorer,
Death, Grief, and Mourning
(New York: Doubleday and Co., 1965), p. 128.

ing and funerary art of the undertaker. According to Mitford,[19] this process of beautification of the corpse aids in denying the realities of death.

In addition, American society goes out of its way to hide death and to dissociate it from the flow of everyday life.[20] Dying people are often hidden away in hospitals or nursing homes, and upon death they are swept away from "the corridors of the living" to the basement morgue as quickly and discreetly as possible. In the aftermath of the death of a loved one, approximately one week, sometimes less, is allowed for public expression of grief. Indeed, many unions have been successful in negotiating for five paid workdays off in the event of the death of an immediate family member. After this "grace period," one is expected to get on with one's life, and any remaining remnants of grief are supposed to be handled privately. Doctors, as we will see in the next chapter and as highlighted in Thanatology Snapshot 1.3, are uncomfortable with the idea of death and often declare war against death and dying. Autopsies are performed out of public view, and identification of dead bodies in coroners' offices is increasingly being done through video screens. Material artifacts of death are also generally excluded from everyday life. When was the last time you went to a local department store to price the selection of caskets? Items for virtually every other form of human activity—sex, eating, sleeping, entertainment, learning—are present. Why are items related to death excluded from places where people ordinarily congregate?

Geoffrey Gorer writes that death in modern society has become taboo, unmentionable.[21] Think of the words "dying," "death," "dead." There's nothing especially difficult about pronouncing these words, but actually saying them is a different story. The words do not flow easily in everyday conversation, and when they are expressed, they evoke an uneasy feeling in both the speaker and listener. Consider, for example, the many linguistic devices that are used to circumvent the realities associated with these words. Euphemisms, or ways of speaking about death without directly doing so, are a significant part of everyday language and the vocabulary of the medical profession. Phrases such as "the patient expired," "respirations have ceased," "she passed away," "he kicked the bucket," "there's nothing more that we can do," and "the patient has gone sour" help to minimize the direct involvement of people in the death of another person. In addition, Hollywood's presentation of death and dying, with its dramatic and violent flair, promotes a popular cultural image that makes death and dying unreal. The objective reporting of "body counts" by the news media also tends to make death unreal and fantasy-like. Familiar statements such as "the holiday death toll has reached 61," "200 were killed in the plane crash," or "140 American soldiers were killed today" become empty statements, statistical abstractions that have little or no personalized meaning to the viewer.[22]

Thanatology Snapshot 1.3

Do Doctors Know the Real Enemy?

His name was Eli Kahn. He was admitted to the hospital because of abdominal pain and vomiting. X-rays taken on admission suggested a small bowel obstruction. Having reviewed his films, I walked over to the division to work Kahn up.

He was a thin, frail old man with a weathered face and marvelously bright eyes. When I entered the room, his attention was fixed on Kovanich in the next bed, an old man recently operated on for colonic cancer. Kovanich had not done well, and now he lay entwined in a tangle of drains and tubes, breathing laboriously.

I introduced myself. Kahn wrenched his gaze from his neighbor and looked up at me. "I'm dying," he said.

"Don't be silly."

"What's silly about dying?"

"Nothing, but it's not allowed. You are in a hospital, a university hospital, equipped with all the latest technology. Here you must get well."

"My time has come."

"Time is measured differently here."

"Wait until you are 78 years old and tired and alone and have a pain in your belly."

There was no arguing with him.

Physical examination revealed an erratic heartbeat, a few crackles in the lungs, a tender distended abdomen, an enlarged prostate and arthritic changes in the joints.

"You see," said Kahn, "the engine is broken down; it is time for the engineer to abandon it."

We discussed the case with our attending and elected to decompress the bowel for a few days before attempting surgery. When I went into Kahn's room to pass a Miller-Abbott tube I saw him again staring at the patient in the next bed. Kovanich was comatose.

"We have to pass a tube into your stomach, Mr. Kahn."

"Like that?" He gestured toward the tube protruding from Kovanich's nose.

"Something like that."

"Listen doctor, I don't want to die with tubes sticking out all over me . . ."

"Now I'm dying, okay, I'm not complaining. I'm old and tired and have seen enough of life, believe me. But I still want to be a man, not a vegetable that someone comes and waters everyday—not like him." He looked over at Kovanich. "Not like him . . ."

(*continued*)

19

Thanatology Snapshot 1.3 (*continued*)

Early the next morning I heard the hospital page issue the code for cardiac arrest. I raced up to the division to find nurses dashing in and out of Kahn's room. Inside I saw Kovanich lying naked on his bed in a pool of excretions with the house officers laboring over him—pounding on his chest, squeezing air into his lungs, injecting one medication after another, trying to thread a pacemaker down a jugular vein. The whole thing lasted about an hour. Kovanich would not come back, and finally all labors ceased.

The nurses began clearing the resuscitation equipment out of the room, while we filed out to begin the round of post-mortem debates.

"Doctor, wait a minute," Kahn was signaling me. I went over to his bed.

"What is it, Mr. Kahn?"

His eyes were frantic. "Don't ever do that to me. I want you should promise you'll never do that to me."

"Mr. Kahn, I know that this has been very upsetting . . ."

"Promise!" He was leaning forward in bed and his eyes were boring through me. There was an interminable silence.

"All right, Mr. Kahn, I promise . . ."

On the fourth hospital day, Kahn went into congestive heart failure. I found him cyanotic and wheezing on morning rounds. Swiftly, the house staff swung into the practiced and coordinated action of acute care: morphine, oxygen, IPPB, tourniquets, digitalis, diuretics. But despite our skilled efforts, Kahn responded poorly. "He's exhausting himself trying to breathe, and he's still hypoxic," our attending said. "I think he ought to be intubated, it will give him a rest and will help us oxygenate him and get at his secretions."

When the anesthesiologist arrived to intubate him, Kahn was gasping. I explained to him about the endotracheal tube. His breathing became more labored as he struggled for words. "You promised . . ." was all he could say.

"But this is different, Mr. Kahn. This tube is just for a short while, maybe just a day. It's to help you breathe."

He stared off in another direction. The anesthesiologist intubated him without difficulty, and we hooked him up to the ventilator.

"I think he ought to be monitored also," our attending said.

So we brought in the cardiac monitor and pasted the leads onto Kahn's chest while he looked on, not stirring, his face expressionless, his eyes dull.

Kahn was asleep that night when I stopped in for an evening check. The room was still, save for the beep-beep of the monitor, the rhythmic whoosh of the ventilator and the hum of the nasogastric suction apparatus. And Kahn looked suddenly so very old and frail, lost among tubes and wires and enormous imposing machines . . .

Thanatology Snapshot 1.3 (*continued*)

Sometime late that night, Kahn woke up, reached over and switched off his ventilator. The nurses didn't find him for several hours. They called me to pronounce him dead. The room was silent when I entered. The ventilator issued no rush of air, the monitor tracked a straight line, the suction machine was shut off. Kahn lay absolutely still.

I mechanically reached for the pulseless wrist, then flashed my light into the widened unmoving pupils, and nodded to the nurses to begin their ritual over the body.

On the bedside table I found a note scrawled in Kahn's uneven hand: "Death is not the enemy, doctor, inhumanity is."

Nancy Caroline,
"Do Doctors Know the Real Enemy?"
The New Physician, October 1985, pp. 18–19.

In large part, as Gorer and Becker[23] suggest, the American approach to dying, death, and the dead depersonalizes and isolates these phenomena from everyday life. There is in American society today a collective tendency to maintain death in a deep freeze of silence. And when it oozes to the surface, there is a similar thrust toward its privatization or trivialization.

Curiously, however, death is neither totally denied nor hidden from everyday life. Cemeteries are open to the public and can easily be seen from roadways. There are more than a few individuals who avidly read the daily obituaries, and sympathy cards are publicly displayed and sold in all greeting card stores. In addition, a surge of interest in death and dying has emerged during the past fifteen years. Many books have been published, numerous films produced, and countless professional meetings organized around the themes of death and dying. Thanatology, the study of death and dying, has become institutionalized on college campuses. It is also true that many people in America today are familiar with the phrase "death-with-dignity." This term has become the rallying call of professional thanatology activity in the past decade and has found its way into the popular media: television, magazines, radio, and so on. The growth of the American hospice, to be discussed in a later chapter, is a direct descendant of the growing public awareness of the "death-with-dignity" movement and reflects a trend toward greater acceptance of dying and death in some sectors of our society.[24]

Some thanatologists have argued that the idea of death denial in American society is overstated. Kellehear[25] argues that death is not denied as much as it is "organized" and managed. He suggests that dying is not produced out of existence but rather is devalued. Like the skid row bum, who exists physically but whose human worth is devalued and disregarded, the phenomenon of dying is not obliterated. Instead, the social status of dying persons is diminished. Often the dying person is put at the mercy of technology—for example, Kovanich and Eli Kahn—and their worth as human beings becomes less and less important. Death in this sense is not wholly denied. It is organized and contained within a technological system by the profession of medicine.

It has also been argued that death is not necessarily feared by everyone. A person who has been suffering from a serious and debilitating illness for a long time may see death as a relief from suffering. Families who have witnessed the pain and suffering of a loved one for an extended period of time may likewise see death as a relief for themselves and for the deceased. Studies have also shown that people may not fear death as much as they fear being dependent on others as they become sicker and more disabled. Kellehear and others argue that fear of death may more appropriately be fear of dying or, more precisely, of those qualities associated with modern dying: loneliness, meaninglessness, physical deterioration, a growing sense of helplessness, and so on.

There is, then, at least some ambivalence in the relationship between modern people and the processes of death and dying. There is growing interest in studying death and dying and in promoting the idea of death-with-dignity. There is also, however, a widespread reluctance to speak of death in everyday interaction and to show grief openly. Dying, with its physical deterioration and social isolation, is a frightening experience. We find this to be true not only in the fictional world of Ivan Ilych but also in the real world of American life. Despite this ambivalence, one thing remains clear: The traditional orientation to death, with its essential patterns of religion, ritual, and community, has been replaced by the denial, confusion, contradiction, and meaninglessness of the modern styles of dying and death.

Summary

The Tame Death represented a general attitude toward death and dying that lasted from the fifth to the eighteenth centuries. During the early era of the Tame Death, prevailing images included fellowship at the deathbed, dying as a public activity, peace and tranquility in the face of death, and the idea of collective salvation for all who died. As society moved into the eleventh century, some changes in the Tame Death took place. Popular views of life after death shifted from salvation for everyone to individual judgment. During this period, characterized by the Death of the Self, the idea of biography took on meaning, and images of a deathbed struggle between good and evil prevailed. The theme of the macabre, with its emphasis on physical decomposition, was also very significant during this time.

Dying a good death was replaced by living a good life during the period of Remote and Imminent Death. Although the particular form changed, physical images of death continued to be important as death became linked with eroticism (necrophilia) and science (dissection of the cadaver). Death became sentimental and beautiful during the nineteenth century, when extended fellowship and community were replaced by an intimate gathering around the deathbed during the Death of the Other.

American Puritans were deeply ambivalent toward death, alternating between acceptance and intense fear, the religious ideal of simple death and the growing reality of overstated and elaborate ritual. Death in Puritan culture became idealized and romanticized as society evolved toward individualism and commercialization.

As the twentieth century emerged, growing secularization, loss of community, and increased reliance on science and technology radically changed humanity's relationship to death and dying. Prevailing images of death became medicalized as technology and machines replaced human ritual at the deathbed. This move toward reliance on medical technology in dealing with dying has made death and dying increasingly invisible. In this sense, the submerging of death in technology has become the modern way of "taming" the inescapable human problem of mortality. The nature and implications of the medicalization and technologization of dying and death will be discussed in the following chapters.

2

The Modern Organization of Death

I didn't want her to die. I just wanted to put her back in her natural state
and leave her to the Lord. If the Lord wants her to live in a natural state,
she'll live. If He wants her to die, she'll die.

JOSEPH QUINLAN
(Father of Karen Ann)

As the discussion of premodern dying indicated, traditional patterns of death
and dying were closely connected to the hub of everyday social life. Dying
and death were ritualized through ceremony and the presence of the human
community. However, a rather abrupt and sweeping reversal of humanity's
response to death and dying has emerged in contemporary times. Premodern
societies tamed death through ritual and fellowship, whereas modern societies
hide it from public view through a bureaucratic and technological system of
care for the dying.

Bureaucracy is an arrangement of human activities that values the following:
specialization; rationalization; the development of power through expert and
specialized knowledge, knowledge secretly protected; and, perhaps above all,
depersonalization. As Max Weber observes, the more the nature of bureaucracy
is perfectly developed, the more the bureaucracy is dehumanized.[1] The more
it succeeds in eliminating love, hatred, and all purely personal and emotional
elements from its daily operation, the closer the bureaucracy comes to perfection. In his classic discussion, Weber convincingly argues that modern societies
are inherently bureaucratic societies. Indeed, one does not have to look far to
see that bureaucratic coordination of activities and institutions is a salient characteristic of contemporary American life. In our political life, in a hospital or
outpatient clinic, in the vast array of social service activities, and in the process of registering for university classes, the features of bureaucracy are readily
visible. And, as we shall see, care of dying people in the modern American
setting is very much fashioned around bureaucratic principles.

In addition to principles of bureaucratic coordination, American society
values science and technology. The benefits and virtues of scientific discov-

24

ery and technological progress are continually extolled by politicians, the media, and corporate leaders. Americans continue to place great faith in science, believing that life will be made better through scientific breakthroughs and technological accomplishments.[2] As we show no signs of lessening the pace with which we introduce technology into our society—and into our offices and homes[3]—the task before us is to create a sense of human purpose as we venture into our technological future.

A society's response to dying people is largely consistent with the values and structures that shape the society as a whole. Thus, the evolutionary movement away from ritual and community to bureaucratic management and medical treatment of dying patients is logically consistent with broader patterns of social living. The modern image of a dying person connected by tubes and wires to life-sustaining equipment—the image represented by Eli Kahn in Thanatology Snapshot 1.3—is a logical derivative of America's bureaucratic and technological orientations. This chapter will explore the ways in which the human experience of dying is organized around bureaucratic and technical principles, as well as the recent social reactions to and against the overtechnologized treatment of dying individuals.

The Medicalization of Dying

Weber describes how the bureaucratization of society removes many social functions from family and fellowship networks and places them in autonomous institutions that are independent of emotional ties. As dying becomes bureaucratized, at least two things happen: Dying takes place in specialized institutions—namely, hospitals, nursing homes, and hospices—and the social role of "formal caretakers of the dying" emerges. In previous eras, it was the responsibility of the community and family to care for the dying. In the contemporary setting, the control of dying has been largely placed in the hands of medical professionals. Dying has become medicalized.

In describing the modern medicalized-dying scenario, Lofland identifies six societal factors that give the dying process its present shape: (1) a high level of medical technology; (2) early disease detection; (3) a complex definition of death; (4) a high prevalence of chronic disease; (5) a low incidence of fatal injuries; and (6) active intervention into the dying process.[4] In *The Craft of Dying*, she argues that these factors combine to ensure that the typical experience of dying will not be brief. Thus, prolongation of the dying process is a basic feature of death in modern society. When one speaks of a "patient's lingering," one forms a newly developed image shaped by the above factors that is far removed from the Tame Death.

As dying is prolonged by an active process of medical treatment, an irreconcilable hostility is established between medical technology and death. In the scenario of medicalized death, death becomes transformed into an enemy to be defeated. Studies have shown that physicians often see death as a sign of failure and clearly as something that is never to be an accepted outcome of medical intervention.[5] This medical definition of death as an enemy is also reflected by our societal commitment to sophisticated medical technologies and by how quickly extraordinary medical techniques, such as transplant and by-pass surgery, have become commonplace.

In addition to establishing death as an adversary and prolonging the dying process, the technologization and bureaucratization of death promote depersonalization. Physicians, who are largely responsible for defining how a patient will die, are socialized to remain emotionally neutral and undisturbed in the presence of dying and death. From the beginning of their medical training, student physicians quickly learn to remain emotionally aloof when facing suffering, tragedy, and death. In the bureaucratic organization of medical care, death

---

Thanatology Snapshot 2.1

Bureaucratic-Technological Dying

Mr. V. awoke in room 5143 of the Metropolitan Hospital, remembering only his sharp chest pain at the family barbeque, the dizziness and the sirens. After two weeks of struggling, the sixty-year-old factory worker died, alone in his hospital room. The only sounds had been the "bleep bleep" of the heart monitoring machine. A clear plastic tube had connected his windpipe to a respirator. Another machine had aided his badly infected kidneys. A glucose tube was inserted in his left forearm. The nurse disconnected the machines and the tubes, wrapped the body in a sheet, and with the help of an orderly, lifted it onto a special stretcher with a false bottom. With the cover in place, the rolling cart appeared to be empty, and none of the patients seeing it in the corridors would suspect that the orderly was pushing a dead man to the basement vault. The nurse returned to her desk to fill out the necessary papers and notify the physician who would contact the relatives.

James Nelson,
Human Medicine,
(Minneapolis: Augsburg Publishing Co., 1982), p. 123

and dying cannot become emotionally disturbing experiences. For this reason, the student physician must be desensitized to dying, death and their symbols—blood, bones, corpses, urine, stench, and so on—symbols that are disturbing to ordinary people but that become virtually unnoticed by the trained medical professional.

Consider, for example, the impact of anatomy lab and the introductory pathology courses on first-year and second-year medical students. Generally in their first semester of study, twenty-two- and twenty-three-year-olds walk into an anatomy lab, where a roomful of cadavers awaits their attention and dissection. Clearly, the initial contact with an assigned cadaver is a traumatic emotional experience. However, the scientific requirements of a student's training direct him or her to find ways to submerge the shock and overcome the fears. Through a variety of adaptive mechanisms, successful student physicians become effectively desensitized to death so that they can go about their business in an emotionally unaffected way. So dramatic is this process that it is not long until students become so coarsened that they can eat their lunches in the presence of the corpses.[6] It is important to recognize that this ability to remain cool and aloof in the presence of disturbing stimuli is a highly valued quality for both physicians at work and bureaucratic organizations in general.

Renée Fox, in an important essay, explores the impact of the autopsy experience on the development of the professional self-image of medical students. Fox argues that the autopsy experience is even more dramatic for the student physician than the anatomy lab, as the corpse waiting to be autopsied bears a greater connection to human life than the preserved and mummified cadaver of the anatomy lab. The body to be autopsied is not mummified, will bleed when cut into, is often still warm, and is being grieved for by loved ones, who are perhaps still on the hospital grounds. The way in which the autopsy is tied to human life is voiced by a student: "When you see the initial incision and first bleeding, that's a point at which you realize you are very aware of the whole person . . . you realize that this is someone who has died, and that you are going to look inside that person."[7]

The formal, manifest function of the autopsy is to acquaint second-year students with the study of pathology. However, as Fox observes, the experience also has an underlying latent function: It promotes the quality of depersonalization or detached concern in the developing role image of the student physician. As her study points out, medical students promptly learn to suspend their emotional fears and turbulence in deference to emotional neutrality and focus their attention on scientific and technical matters. As one student comments: "Most of the fellows are so eager to be good doctors that they force themselves to look at things in a scientific way. I was noticing that at the autopsy. Nobody was hanging back from the table. Everyone was leaning forward trying to see everything they could. Every guy was interested in the facts . . . asking questions and wanting to learn about what had happened."[8]

The point to be made, which is illustrated by these two examples, is that the professional training and socialization of physicians plant the seed for a depersonalized response to dying and death. Medical doctors quickly learn, through their training, that the proper response to threatening stimuli is to contain and nullify rather than ventilate their emotions. It is for this reason that a physician looks at the enemy of death with professional composure and emotional disinterest, both of which are essential components of the bureaucratic way of death.

In addition to and as a by-product of emotional neutrality and depersonalization, rationality is an important feature of the bureaucratic management of dying and death. David Sudnow, in a classic study, shows how medical staff respond to death in a standardized, routinized manner. Medical doctors, according to Sudnow, are not likely to be present when a dying patient nears the threshold of death. Whatever (custodial) care is required by a dying patient is generally provided by nurses and other health care personnel. And when the patient does die, a preestablished and well-coordinated ritual of disposing of the body is swiftly brought into action.

> Wrapping a body is a well organized routine having . . . a clear beginning, sequence of steps and closure; it is done collectively by two or more persons and is automatically carried off . . . the procedure essentially involves the complete removal of the deceased's clothing, including all jewelry, and the folding of a heavy gauge muslin sheet completely around the body, pinning it down the front with large safety pins, in mummy style. . . . A diaper-like sheet is wrapped around the genital area; the hands and feet are crossed and bound together with a special cotton-covered string. Two pre-cut gauze pads are placed over the eyes, after the lids have been closed. Before the body is finally wrapped in the outside sheet . . . all I-V tubes are removed, nasal suctioning equipment detached, catheters taken out, etc.[9]

Thus, the process of wrapping a body is a well-conditioned secular routine—a step-by-step process that, while less than pleasant, is not emotionally absorbing. It is a rational, efficient medical technique akin to setting a cast, preparing someone for surgery, or readying a woman to deliver a baby.

Other processes related to death are also approached with a rational, nonemotional orientation. Informing relatives, transferring the body to the morgue, filling out the death certificate, requesting consent for organ donation, and getting permission for an autopsy are accomplished without emotional involvement of the medical professional in the private, personal sphere of loss and grief. Ultimately, death work becomes so routinized that personnel can move in and out of death-related activities without the standardization of their everyday work routine being jeopardized.

The issue we have been considering is how society adjusts to the fact of human dying and death. Death, as many have argued, may be absurd, terrifying, and senseless.[10] At the very least, it is a disruptive societal and personal force that is dramatic in its impact. In many ways death is also incomprehensible, for it means the ending of an individual.[11] And one of the paradoxical facts of being human is the inescapable coupling of the knowledge that each of us will someday be dead with the impossibility of conceiving that the world will continue without us. Indeed, the thought of death, and of the world continuing to exist without us, is unimaginable and preposterous to the human mind.

While the incomprehensibility of death may have always haunted humanity, the contemporary response to dying is radically different from responses found in the eras of traditional death patterns. Modern society copes with dying and death largely through institutionalized depersonalization and bureaucratic routine. In this way, society is protected from the ravages of death and dying because the generalized societal response is emotionally neutral and the grieving process becomes the domain of a small group of people, namely, the immediate family and the dying person, who are isolated from the flow of everyday life. Dying as a human process becomes managed, as it is regulated by a group of emotionally detached medical professionals. In this way, the ending of an individual's life becomes the province of medical managers, and the potentially disruptive impact of death, which is so terrifying and incomprehensible, is softened and contained.

The Caretaker's Role: Interactions with the Dying

The physician is the primary enforcer of the bureaucratic code. Individually and collectively, physicians set the stage for dying. Their decisions will affect life prolongation, technological aggressiveness in responding to dying, and often the place of death itself. In this way, as Thanatology Snapshots 1.3, 2.2, and 2.3 illustrate, responsibility for control of dying has been removed from the individual and the community and placed largely under the jurisdiction of the medical professional. Illich compares this phenomenon of medical dominance to a game:

> The chief function of the physician becomes that of an umpire. He is the agent or representative of the body social, with the duty to make sure that everyone plays the game according to the rules. The rules of course forbid leaving the

game and dying in any fashion that has not been specified by the umpire. Death
no longer occurs except as the self-fulfilling prophecy of the medicine man.[12]

As mentioned earlier, physicians, often threatened by the prospects of a
patient's dying, define death as an enemy to be defeated. Feifel and his colleagues, interested in the question of why physicians are so uncomfortable with
dying and death, sought to determine if physicians have a heightened fear of
death. They compared doctors with two groups, healthy people and sick people,
and found that physicians were significantly more fearful of death than either
group of nonphysicians. The conclusion drawn by Feifel and his coauthors is
that this fear of death is a significant factor in physicians' choice of a medical
career and their defining of death as an enemy.[13] After all, where else does one
get to fight and defeat the grim reaper on a regular basis? Of course, it could
also be that they fear death because they witness it much more often as physicians. In this way a self-perpetuating cycle of fear is created: Exposure to death
spawns a fear of death that is both unrelieved and intensified by the absence of
meaningful support systems to assist physicians in responding to dying and
death with openness and compassion. And, as mentioned earlier, most programs
of medical education are notoriously inept at preparing student physicians to
deal with the psychosocial dimensions of patient care.

Technology is the primary weapon that physicians use to fight death. Although there are several different ways in which medical doctors respond normatively to dying patients, technological activity is the common denominator.
The first way is the save-life-at-all-costs orientation. This reliance on technology to rescue the dying reflects our society's shift from a moral and social order
to a technical order.[14] Thus, in our technologically oriented society, a common
approach to dying is to strive officiously to keep the patient alive.[15]

The attempt to prolong life through the active use of technology is central to
a physician's role definition and to his or her battle against death. The more
successful a doctor is in warding off death, the more successful he or she is as
a physician:

> When his self-esteem is involved it is imperative that he prevent death from
> occurring, for death makes him feel vulnerable. But if he can keep the "corpse"
> alive for a few more days or weeks, his mastery over death is demonstrated. "He's
> not about to allow his omnipotence to be challenged," said one, "but if you talk
> to the family you'll find they are usually sorry that the patient had to suffer those
> extra days and that the hospital bill ran up so high."[16]

In the save-life-at-all-costs framework, the physician defines death as failure and hence seeks to delay the inevitable and to hide behind a technical veil
of postponement.[17] Technology is brought into action in a no-holds-barred
fashion, and frenetic attempts are made to outwit death. Remember when

M*A*S*H*'s Hawkeye was feverishly attempting to save a patient and exclaimed, "Don't let the bastard win?" The bastard, of course, was and is death, the ultimate threat to the magical powers of modern medical technology.

Some physicians are more frightened about death than others. Preliminary research into varying degrees of the fear of death among physicians indicates that doctors with higher death anxiety are more likely to use technology aggressively to prolong dying. In fact, the patients of physicians with high death anxiety are in the hospital an average of five days longer before dying than patients treated by physicians with medium and lower death anxiety.[18]

The medical justification for save-life-at-all-costs behavior is that the use of technology to save a life is always a reasonable goal. But, as Annas notes, if this attitude is taken to its logical extreme, anything goes—that is, any technique or even any human experiment becomes justified.[19] In addition, there are serious questions as to whether the save-life-at-all-costs approach to the dying person is motivated by concern for the patient and family or by concern for the demands of the physician's professional role definition and ego.[20] As Jay Katz notes in his book, *The Silent World of Doctor and Patient*:

> At such time, all kinds of senseless interventions are tried in an unconscious effort to cure the incurable magically through a "wonder drug," a novel surgical procedure, or a penetrating psychological interpretation. The doctors' heroic attempts to try anything . . . may turn out to be a projection of their own needs onto patients.[21]

Another medical pattern of interaction with the dying is avoidance-neglect.[22] This occurs when a physician responds to his or her unease and battle with death by avoiding contact with the dying. Studies of terminal care have reported that the sicker terminal patients get, the more physicians withdraw their presence, contributing to the establishment of an "emotional quarantine" of the dying.[23] Sudnow explains this avoidance-neglect pattern in terms of the ending of a physician's responsibility to the patient.[24] As physicians are in the business of saving life and preventing death, when death becomes inevitable, they may define their job as being over. In this case, dying patients are "abandoned" as physicians direct much more of their time and attention to nonterminal patients who are physically able to respond to the healing powers of medical technology. Strategies of avoidance include spending less time with dying patients, specialists resisting consulting in terminal cases, physicians timing their rounds to avoid interacting with dying patients (e.g., when the patient may be sleeping or scheduled for tests), and selectively trivializing many of the patients' psychosocial and palliative needs.

The third pattern of relating to the dying is detached-sympathetic support.[25] This occurs when the physician recognizes the importance of a dying patient's

Thanatology Snapshot 2.2

The Phoenix Heart Implant

On Tuesday morning, March 5, 1985, Dr. Jack Copeland, Chief of University Medical Center's Transplant Team in Tucson, Arizona, performed a heart transplant on Thomas Creighton, a thirty-three-year-old divorced father of two. Copeland later explained that the donor, an accident victim who had been hospitalized for several days, "wasn't what we'd call an excellent donor candidate, but in view of the urgency, we elected to proceed with the transplant" (*Arizona Daily Star*, March 7, 1985, p. 2). The procedure was not a success because of rejection of the heart. At 3:00 A.M. Wednesday, a search for another human heart began; Mr. Creighton was placed on a heart-lung machine.

At 5:30 A.M., a call was placed to Dr. Cecil Vaughn of Phoenix asking if he had an artificial heart ready for human use. Dr. Vaughn was scheduled to implant an experimental model developed by dentist Kevin Cheng into a calf later that day, but had never considered use of the device in a human. Nonetheless, he called Dr. Cheng, who told him, "It's designed for a calf and not ready for a human yet." Asked to think about it for ten minutes, Dr. Cheng recalls, "I knelt and prayed." When Vaughn called him back, he said, "The pump is sterile, ready to go" (*The New York Times*, March 19, 1985, pp. C1–C2).

The two flew by helicopter from the hospital to the airport, chartered a jet to Tucson, and then took another helicopter to the Tucson hospital. They arrived at 9:30 A.M., Wednesday. The implant procedure began at noon. Designed for a calf, the device was too large and the chest could not be closed around it. The implant maintained circulation until 11:00 that night when, in preparation for a second heart transplant, it was turned off, and Mr. Creighton died.

The press treated the story like a modern American melodrama. *USA Today* called the implantation of Dr. Cheng's device "the fulfillment of an American dream" (March 8, 1985, p. 1A). *The New York Times* editorialized that "the artificial heart has at last proved it has a useful role" (March 9, 1985, p. 22). *Newsweek* faulted the FDA, noting, "It's hardly fair to doctors, or their patients, to make them break the law to save a life" (March 18, 1985, pp. 86–88). The FDA initially termed the unauthorized experiment a violation of the law, but by week's end had done an about-face and was flailing itself as "part of the problem" (*The New York Times*, March 17, 1985, p. 87).

George J. Annas,
"The Phoenix Heart: What We Have to Lose,"
The Hastings Center Report,
Vol. 15(3), 1985, p. 15.

needs. Such a physician seeks to incorporate a sense of responsiveness to the patient's emotional and social needs as a regular part of the program of patient care. Detached-sympathetic support physicians are generally willing to spend time talking to dying patients and their families. However, their concern for a dying patient remains professionally detached and does not move toward emotional involvement. The major interest of this physician is technical treatment of the patient's disease and symptoms. Although less likely to employ heroics until the last moment, this physician considers the patient's comfort and the management of pain important priorities. As we shall see later, the orientation of detached-sympathetic support is partially consistent with the hospice philosophy of caring for the dying.

A key point to be made regarding each mode of interaction is that technology and its application are the primary focus of the physician's activity. Why some physicians adopt a save-at-all-costs approach, or an avoidance-neglect approach, or a detached-sympathetic-support approach to their interactions with their dying patients is not clear given the present state of thanatological research. Researchers are beginning to look at some possibly relevant factors, such as choice of specialty, the doctor's age, his or her personality and death fears, place of training, and the place in which the physician practices medicine. If the factors that determine why a physician pursues the path of heroics, neglect, or support could be identified, it would then become feasible to screen or train physicians on the basis of these factors. The results of such screening and training would provide greater support for a patient's psychological and social needs.

While doctors are the dominant actors in shaping a patient's dying experience, the nurse is the primary giver of care.[26] Unfortunately, the role of the nurse as primary caretaker is an underdeveloped area of study within the sociomedical sciences and thanatology.[27] Only lately are sociomedical scientists and thanatologists seriously beginning to investigate this role.

The nurse's role brings her in contact with dying patients more closely and continuously than other members of the medical team. Due to their regularity of contact with a patient who is dying, nurses are more exposed to the human side of dying than physicians. They have more informal contact with the dying, often serve as intermediaries between doctors and patients, serve as patient comforters, and are likely to see patients when they are least poised and most vulnerable.[28] For these reasons, it is not difficult to understand that while doctors have an intellectual-scientific relation to the dying process, the nurse is regularly enmeshed in a web of both scientific and personal dimensions of the dying experience.

Although most nurses say they are reasonably comfortable working with dying patients,[29] they too regularly employ strategies of avoidance. LeShan found that nurses respond more slowly to the signal calls of terminal patients than to those of patients who are not critically ill.[30] As we will see in greater depth in the following section, nurses often use diversionary tactics when a conversation with a patient turns toward the prognosis. Research has shown that while nurses report a general sense of comfort with dying patients, they are likely to feel uncomfortable in unstructured interactions. When feelings of unease are evoked by unstructured, free-flowing conversation, nurses customarily respond with avoidance strategies and techniques (tending to life-support equipment, fixing the patient's bed and surroundings, scurrying off to complete paperwork, etc.).[31] In effect, nurses normatively erect what LeShan calls "a protective glass curtain" between themselves and their dying patients.[32]

It is also fascinating to note that although nurses have more contact with the human dimensions of dying and regularly employ avoidance strategies, they have a more positive attitude toward death than physicians. Medical doctors tend to see death as the enemy and describe it in stark, negative terms ("unsafe," "alone," "cold"), whereas nurses see death in a more positive light and describe it as "safe," a "rebirth," and a "form of victory." This tendency of physicians to see death in more negative terms than nurses see it is a consistent attitude among the two groups.[33]

It thus follows that nurses are more likely to provide emotional support for the dying patient and are less likely to support the use of heroics to prolong terminal existence.[34] But perhaps the crucial factor that ultimately defines the ways in which both physicians and nurses interact with the dying is the inadequacy of their professional training. Neither physicians nor nurses have been well prepared to deal with the human, social, and emotional implications of the dying process as it affects patients and their families.[35]

Awareness Contexts: Staff, Patients, and Patterns of Dying

Barney Glaser and Anselm Strauss are pioneers of the sociological study of dying. In their book *Awareness of Dying*, they report studies of the patterns of interaction between staff and patients, considering what each interacting party knows of a patient's terminality and how people relate to each other within differing "contexts of awareness." In their study of dying in metropolitan hospitals, they identify four different types of awareness: closed awareness, suspected awareness, mutual pretense, and open awareness.[36]

Closed awareness occurs when the medical staff is aware that a patient is dying, but the patient is not aware of that fact. Several factors contribute to the creation of this context. The first has to do with the alien nature of death itself. Most Americans are not afforded a dress rehearsal for their own dying and are often not present for long periods during the dying process of others. Hence, most patients are unfamiliar with, and many have difficulty recognizing, the signs of impending death. A second factor is that physicians contribute to patients' ignorance by not telling them the truth or by actively diverting attention away from considerations of dying. Some studies, especially during the 1960s and 1970s, showed that physicians are often reluctant to tell patients the truth about their dying and regularly find medical rationalizations ("he'll go to pieces" or "she'll give up hope") to justify their deceit.[37] Although it is true that more physicians today than in the past believe that patients have a right to hear the truth, physicians are neither trained in nor highly adept at the art of communicating successfully with patients about the dying process. Additionally, the truth is often told in ways that are medically correct but simultaneously serve to obscure the truth about the preeminent question of living and dying. In this way, modern physicians have become adept at dancing around the truth of the truth.[38] A third factor that gives rise to closed awareness is that families aware of the patient's terminal diagnosis tend to guard the secret in order to "protect" their loved one. Two other contributory factors are the bureaucratic management and guarding of information and the fact that patients are isolated in the hospital and have no allies to help them find information about their condition. In this way then, patients, doctors, families, and bureaucratic organizations conspire together in building the closed awareness context.

In closed awareness, the medical staff seeks to keep terminal realities from surfacing. However, the maintenance of closed awareness is often easily imperiled by unwitting disclosure of information, the worsening of symptoms, and the patient's increased worry that something may be seriously wrong. When this situation emerges, an atmosphere of suspicion awareness is created. Suspicion awareness occurs when the staff knows that the patient is terminal and the patient begins to suspect it. As the patient's suspicions are aroused, like a man who suspects his wife of having an affair, the patient seeks to uncover information relevant to those suspicions. A contest for information emerges: a cat-and-mouse game of patients seeking and staff hiding. The following passage from the work of Glaser and Strauss illustrates some of the strategies employed by nurses in this interactional game:

First Nurse:	A stern face. You don't have to communicate very much verbally, you put things short and formal . . . yes, very much the nurse.
Second Nurse:	Be tender but don't . . .
First Nurse:	Sort of distant, sort of sweet.
Second Nurse:	Talk about everything but the condition of the patient.

First Nurse: And if you do communicate with them when you are not too
                much the nurse, you could talk about all kinds of other things;
                you know, carefully circling the question of death.[39]

If and when patients do discover the truth about their dying, their own typically modern unease with dying and death may contribute to a mutual mas-
querade of denial between staff and patient. This third awareness context, in
which staff and patient are aware of the situation but pretend otherwise in their
interactions with each other, is termed the "ritual drama of mutual pretense."
The ultimate task for both patient and staff, in this context, is to devise strategies that enable them to carry out the charade of denial in their daily interactions.

One may view these three awareness contexts as forms of evasive interaction between staff and patient. The staff's contribution to these "conspiracies
of silence" is understandable, as they are distinctively functional for staff and
their work habits. An aware dying patient can become a source of distress and
disturbance that impedes the flow of a standardized work routine. Patients may
be managed more easily if their dying takes place within the context of evasive interaction. The reasons of the patient for continual evasive responses to
dying are a bit more complicated. Undoubtedly the general cultural avoidance
of dying is a contributing factor. In a society where dying is largely avoided,
denied, managed, and so on, it is understandable that an individual's response
to the incomprehensible reality of his or her own death will entail mechanisms
of denial and evasion. In addition, however, as Charmaz suggests, the evasive
interaction patterns of the medical staff may give patients the impression that
interactions about dying and death are off limits.[40] Patients already tending
toward denial may have this tendency reinforced as they learn that "proper"
patient behavior is to keep quiet about their suffering and dying. In this way,
patients are socialized to accept and participate in a depersonalized definition
of their own dying. In a fashion ideally suited to the bureaucratic ethos of modern death, they become, in terms of the public presentation of self, nonemotional spectators of their own demise.

The fourth interaction context described by Glaser and Strauss is open awareness. This occurs when both staff and patient are aware that the patient is dying
and openly acknowledge this in their interactions with each other. Although
frankness and candor are the foundation of open awareness, this does not imply
that the context is free from complexity and confusion. To the contrary, patients
who are aware of their own terminality often lack adequate information. When
can they realistically expect to die? How will they die (in great pain?, quickly?,
etc.)? Often there is ambiguity about where they will be when they die (at home,
in a hospital, nursing home, or hospice?). But despite these uncertainties, the
idea of open awareness has been rigorously promoted by many advocates of

Thanatology Snapshot 2.3

Prolonging Life

It was an ordinary day for Dr. David Finley, the chief of intensive care at a Manhattan hospital. There were seven terminally ill patients in his ward, and he had to decide how long to keep them alive.

He called a lawyer.

Dr. Finley oversees seven doctors in an 18-bed world of tubes, capsules, wires, pumps and pins that allow modern medicine to keep the heart beating, lungs breathing, kidneys pumping and immunological system fighting—long after the body has given out.

It is the most expensive unit of Roosevelt Hospital, a place that sparkles with large metal monitors whose screens blink green squiggles and lines all day. Buzzers, bells and gongs punctuate the otherwise soothing sound of running oxygen. It is the place where day after day, hour after hour, the most difficult issues of dying are played out.

Of the 143 patients who died at St. Luke's–Roosevelt during one typical month—last June—nearly 40 percent received this highly specialized care.

The intensive care unit is a trying place to work, a place where doctors speak in euphemisms. They talk of "levels of commitment"—a phrase that measures whether a patient is worth the effort of keeping alive. They speak of "aggressive" care, reserved for those patients with hope, and care that is "supportive," for those without.

At a time when technology has made heroic, life-sustaining procedures routine, doctors and nurses in this unit are increasingly finding themselves professionally and emotionally ill prepared to undertake what amounts to a new addendum to the Hippocratic Oath to "do no harm."

They are now being asked not merely to preserve life at all costs, but sometimes to decide when the cost of preserving life is too high—and thus when to shut off respirators, to withhold dialysis, to deny resuscitation. In short, they are asked to decide when life should end.

"More than ever," Dr. Finley said, "the house staff is lost." You don't have to do everything for everybody. The question is, 'Where do you stop?' At what point do we say, 'What are we doing? Do we care what we are doing?'"

Dena Kleinman,
"In the Intensive Care Unit, Doctors Find
the Hippocratic Oath Is Redefined,"
New York Times, January 16, 1985, Section Y, p. 11.

the death-with-dignity movement.[41] The value of openness in interaction has also become a cornerstone of two currents of thanatological and social change: the hospice and negotiated death. The hospice as an emerging form of the modern organization of death will be considered in the next section. Let us now turn to the emerging concept of negotiated death.

If the patient and the medical staff approach the patient's dying within a framework of openness, the patient is provided with a greater opportunity to be involved in his or her own dying process. In the open context, it becomes possible for the patient to negotiate in order to obtain information regarding dying, to influence the ways in which pain will be managed, to influence the extent of the role that technology will play in shaping the dying process, and to negotiate on the place of death itself. Although the process of negotiating one's own dying may be difficult, especially with save-life-at-all-costs and avoidance-neglect physicians, it is only within the context of open awareness that the possibility for meaningful patient input into the dying experience emerges.

On a personal level, negotiated death offers the prospect of dying a death that is carved out through active participation by patient, family, nurses, doctors, and even lawyers. The consequence for society is that the more negotiated an impending death becomes, the more death in that society will become individualized; hence, the concept of pluralistic death is created by the modern organization of dying. However, while open-awareness, negotiated death seems straightforward and represents a preferable alternative to physician-defined dying, it, too, brings a new complexity and intricacy to the dying scenario. As Thanatology Snapshot 2.3 illustrates, not only is it becoming increasingly difficult to identify the correct course to follow with dying people, the interests of the concerned parties often are inherently contradictory. The popular play and movie *Whose Life Is It Anyway?* highlights some of the complexities involved in modern dying, as does the case of Karen Ann Quinlan. On a less dramatic plane, negotiations and conflict over how people should die are being waged in private, isolated circles on a daily basis in hospitals, and even in courtrooms, all over the country. These negotiations can typically involve physicians, family members, hospital administrators, and lawyers for both sides, and may concern ending life-sustaining treatment for terminally ill or comatose patients, who may or may not have left directions for their own care. These ad hoc, sometimes formal, negotiations take numerous forms: families and doctors against administrators, family members against patient, patient against doctor, family member against family member with doctors as anxious onlookers, and doctors against doctors. The possible presence of attorneys may worsen an already confused and messy scenario.

As the complexities of dying increase without the development of social, legal, and sociomedical norms to help stabilize the pattern of pluralistic death, the inevitable results are uncertainty and anxiety. The irony of this is that the recent societal emphasis on open awareness and truth telling is bringing its own set of problematic consequences to the dying process. But, as pointed out in Chapter 1, this process of individualizing the dying experience, the growing normlessness, the declining traditional patterns of dying, and the resultant confusion are characteristic features of present-day social change and modernity.

The Hospice Alternative

The technical orientation of patient care found in the hospital setting serves to alienate the dying patient and to facilitate personal isolation and powerlessness.[42] At this point, it should not be surprising to find that studies have shown that the emotional needs of dying patients are generally not met by the social organization of hospitals. One study found that only 4.9 percent of attending physicians and 2.0 percent of house staff felt that the emotional needs of the dying patient were being met by the hospital. And when asked if the hospital has an obligation to meet the emotional needs of dying people, 97.7 percent of doctors, nurses, and house staff replied that it did.[43] Commonly mentioned problems with regard to unmet emotional needs of the dying include staff avoidance of difficult communications, staff discomfort with their own feelings, and failure to refer the emotional needs of the patient for appropriate treatment.[44]

It is the growing societal belief, and the conviction among a small but vociferous minority of medical professionals, that bureaucratic technological death provides inadequate care for dying patients. The spread of disenchantment with the technological way of death spawned the hospice movement in America. In 1978, there were only fifty-nine hospice care programs. In 1981, that number increased to 440, and today there are approximately 1,300 organizations that either explicitly or loosely carry the label of hospice. This rather striking development of the hospice in America is an attempt to close the gap between the realities of hospital care of the dying and the realities of patients' needs. In this light, the growth of the hospice philosophy can be legitimately interpreted as a reaction to and against bureaucratic management of and excessive technological involvement in the dying process. The more people become unwilling to submit themselves and their loved ones to the dehumanizing requirements of medicalized death, the more they seek a spiritual and humanistic alternative. This approach emphasizes not machines and medicines, but rather the living out of one's days in a comfortable, dignified, peaceful environment.[45]

The Random House Dictionary defines "hospice" as a house of shelter or rest for pilgrims, strangers, and so on, especially one kept by a religious order. A hospice, then, going back to medieval Europe, was traditionally characterized

TECHNOCRATIC DEATH

Personal attention at the deathbed has been replaced by impersonal monitoring of machines in the intensive care unit.

by an image of warmth and generosity toward travelers, guests, and strangers. Thus, the etymology of the word itself suggests an image that is radically different from the definition of medicalized death. As we have seen, the core motivational force of the hospital is to cure—the giving of specific medical or surgical treatment for the purpose of repairing the bodily parts of an injured or diseased person. A hospice, notably similar to the idea of hospitality, is more concerned with care, namely, providing comfort and support to those who face the predicament of terminality. The hospice, in its present form of caring for dying people, seeks to create an environment where the emotional, social, and spiritual needs of a dying person and his or her family assume top priority. The fundamental differences between hospital control over dying and the hospice orientation of care for dying individuals is illustrated in Thanatology Snapshot 2.4.

The hospice's agenda in caring for the dying ideally aims to promote living and even to facilitate personal growth during the dying process. Too often, it is argued, dying and death are confused and the terminally ill are stigmatized as socially dead before being biologically dead. Thus, the hospice establishes

Thanatology Snapshot 2.4

Two Similar Words

It is a strange embrace, the one we now find welcoming us into the place called *hospital.* It is one which neutralizes instantly whatever life force it is that makes each of us into a unique individual. *Hospital* welcomes my body as so many pounds of meat, filled with potentially interesting mechanical parts and neurochemical combinations. *Hospital* strips me of all personal privacy, of all sensual pleasure, of every joy the soul finds delight in; and at the same time, seizes me in the intimacy of a total embrace. *Hospital* makes war, not love.

The modern hospice: a place of meeting, a way station, a place of transit, of arrival and departure. And yet, how different from the airport, the hotel lobby, the hospital. It is the difference in the quality of human life assumed and provided for, that makes the contrast. . . . People in hospices are not attached to machines, nor are they manipulated by drips or tubes, or by the administration of drugs that cloud the mind without relieving pain. Instead they are given comfort by methods sometimes rather sophisticated but often amazingly simple and obvious; and they are helped to live fully in an atmosphere of loving kindness and grace until the time has come for them to die a natural death.

It is a basic difference of attitudes about the meaning and value of human life, and about the significance of death itself, which we see at work in the place called *hospice.*

Sandol Stoddard,
The Hospice Movement
(New York: Vintage Books, 1978), pp. 3, 14.

---

for itself the tasks of maximizing the control of pain, providing fellowship for the dying person, promoting maximum independence for the dying person and his or her family, and providing support systems for the staff so that they can continue to offer compassionate care to the terminally ill. The components of this blueprint for a new approach to the modern organization of dying are summarized by DuBois:[46]

1. The aim is to manage physical symptoms and offer as much comfort as possible.
2. The unit of care is not a single individual but a community that includes the dying person, immediate relatives, and significant others.

3. After the death of a patient, support services are offered to bereaving families.
4. Institutionalized care is provided, with concern and support for the hospice staff.
5. Staff members are selected with close attention paid to the ability to provide strong support to dying patients and their loved ones.
6. The physical setting is designed to provide for privacy and offers the possibility of communal gatherings at the bedside.
7. The presence of children is encouraged.
8. Interaction among patients is encouraged.
9. An interdisciplinary care team includes doctors, nurses, clergy, volunteers, and others.

The major medical emphasis in the hospice is palliative, that is, to keep the dying patient as pain-free and comfortable as possible. The idea of pain control often gets lost in standardized hospital care, with its emphasis on disease control and death prevention. Pain is addressed in the hospice through the use of narcotics and analgesics. Medicines are customarily blended and/or adjusted to meet the needs of a patient's individual circumstance. The hospice's approach to pain control is aggressive. The staff does not wait for pain to start its ravages of the central nervous system and then try to catch up to it, but rather seeks to outmaneuver, anticipate, and stay ahead of pain from the moment of the patient's admission.

In addition to providing chemical treatment, the hospice addresses pain through psychosocial support systems. The hospice orientation to care promotes the idea that medicines given in an austere, cold, and alienating environment will have a less comforting impact than medicines given within a framework of support, warmth, and compassion. Cicely Saunders, the founder and director of St. Christopher's, the pioneering modern hospice in London, speaks of the importance of personalized attention to the needs of the dying:

> Personal, caring contact is the most important comfort we can give. . . . Often we are very busy, but there is always time for a brief word. Above all, we must never let the patients down, never just go by. The dying will lie with their eyes shut just out of tiredness when they are waiting for you. If you then fly past the end of the bed, rather pleased to find them asleep, they have lost that precious moment for which they have been waiting.[47]

The hospice, in many ways, seeks to reclaim the solace and comfort that were provided by the rituals of the Tame Death. Thus, a hospice is more than a place, more than a unit of beds where patients go to die. A hospice is a philosophy—a network of values and attitudes that attempt to minimize the indignities of dying so as to maximize the human life potential of the dying and their families. In this way, then, it is possible—albeit not consistent with the structure of the modern hospital—to bring hospice care to the traditional hospital setting by an interdisciplinary team of caring professionals, who may be properly termed a "hospice team." Perhaps more importantly and realistically, the hospice philosophy of care has been extended to bringing supportive services to those dying at home. As such, the home care program has become an essential ingredient of the hospice movement. Studies have indicated that if given a choice, four people to one prefer to die at home rather than in a hospital.[48] Nonetheless, about 80 percent of all deaths in America occur in an institution, namely, hospitals, nursing homes, and chronic care facilities. This statistic suggests that the resources of the American family are inadequate to cope with a relative's lingering death at home. The home care programs of hospices seek to provide assistance to those individuals and families who garner the resources needed and opt for home dying. This goal is expressed as follows:

> Maintenance of the family as a cohesive, supportive unit; provision for the relief of loneliness and separation anxiety; and symptom control for the maximum comfort and alertness of the dying patient are key objectives of the hospice staff in assuring accessibility and ancillary staff skills and in making arrangements for optimum care in a home environment.[49]

The hospice movement has gained many advocates over the past decade. However, before one concludes that the hospice is a dramatic and successful solution to many of the problems associated with dying in American society, several cautions must be noted. First, extensive and convincing scientific studies on the workability of the hospice as an institution and a philosophy of care are lacking. While hospice proponents assert a high success rate (in some cases 98 percent) in controlling pain, controlled scientific studies have yet to verify universally the validity of their claims. Studies are now being conducted to assess the program of care of the hospice, and more data should be forthcoming shortly. However, until such data become available, several important questions remain regarding hospice care of the dying. For example, does a special kind of person or family choose hospice care, and if so, are their special personal qualities mainly responsible for shaping the dying experience? Would these people have died in a significantly different fashion had they stayed in a nonhospice setting? Do hospices socialize their patients into behaving in accord with the ideals of the hospice? Do hospice patients feign dignity and report greater satisfaction with care because they feel that this response is expected from them? Are patients in a hospice program being "force-fed" a death-with-dignity philosophy that perhaps has become a new (bureaucratic) form of managing dying patients? These are just some of the important questions that future studies on the role of the hospice in society must address in depth.

In addition, before the hospice is extolled as the solution to the modern problem of dying, it must be recognized that the hospice is at odds with the concept of medicalization that is so comfortably at home in the technologically oriented 1990s. It is very difficult to envision how the hospice will gain support in a save-at-all-costs or avoidance-neglect framework. And the contexts of closed awareness, suspicion awareness, and mutual pretense preclude the possibility of applying the hospice ideal. Physicians who are trained in detached concern and who contribute to an atmosphere of denial will continue to have difficulty adapting their professional self-images to hospice care. It is perhaps too optimistic to believe that most physicians will be able to set aside technical intervention and cure for a redesigned goal of providing physical, emotional, and social comfort to their dying patients without sweeping changes taking place in their education, training, and work environment.

Finally, it is questionable whether American society is willing to support a widespread commitment to the notion of hospice programs. As indicated in Chapter 1, dying and death have increasingly become marginal in modern society and are often avoided, denied, and excluded from daily life. In a setting where the broader social structure and its citizens are uneasy with the presence of dying and death, is it realistic to expect the emergence of a broadly supported attitudinal and financial commitment to the hospice? Or, before this can take place, are changes in the structure of society and its commitment to a technological and medicalized way of life necessary?

In short, it is not difficult to see that the hospice represents an alternative philosophy of living and dying. It is, however, going to be difficult to reconcile its philosophy of care with some of the deeply grounded structural realities of the modern organization of dying discussed in this chapter. As a result, it may very well be that the hospice will coexist with, rather than replace, medicalized dying and will become an alternative that people can select in an era of increasingly pluralistic dying. To expect more would demand too many radical changes in the medical care system and the broader society in the foreseeable future.

A Concluding Statement

As the ritual and simplicity of traditional patterns of living have given way to modern complexities, the modern organization of dying, with its bureaucratic and technological foundations, has led to the dissipation of traditional patterns of human dying. Changes in the way dying is organized reflect changes in the society at large. Trends toward bureaucratization and technologization in the broader society have cast the dying process into a bureaucratic and technological mold. However, as we have seen, the movement toward hi-tech dying has elicited a public reaction against many of the indignities associated with medicalized death. Two outcomes of this reaction are the development of negotiated and pluralistic death patterns and the hospice movement, each of which has brought new sets of complexities to the death and dying scene.

The next chapter investigates how the trends and countertrends in the organization of modern dying affect the dying person and the dying process. Keeping in mind the framework developed so far, the nature of dying as it affects the lives of human beings will be discussed. How death and dying occur, what they mean to the people involved, and how patients and others respond to it will be focal issues of discussion.

3

The Dying Patient:
A Creation of the Modern
Organization of Death

So I'll continue to continue to pretend
my life will never end.

SIMON AND GARFUNKEL

The rise of the modern organization of death has had major implications for the dying individual. During the times of traditional Western death patterns, it was unheard of to define the dying person through his or her role in a system of medical care. During these periods, dying was a human and social phenomenon, not a medical one. Over time, a more scientific and technological definition of dying and death evolved, and the process of death became increasingly medicalized. This led to the redefinition of the dying person as a dying patient.

A dying person is defined and perceived through a framework in which personal and social relations, as well as social rituals, are dominant. A dying patient, by contrast, is defined primarily through a therapeutic and medical model in which his or her role is inherently shaped by the modern medical organization of death. This is not to say that social relations and rituals are nonexistent in the life of a dying patient, but rather that they are secondary to and subsumed under the framework of the medical, therapeutic model. In this way, then, the dominant forces that style and shape the life circumstances of the modern dying person are consistent with the broader social trends of technological progress and therapeutic individualism.

In this chapter, we will see how the transformation of the dying person into a dying patient has had sweeping consequences for the individual's experience of death. Consider, for example, the definition of dying. In traditional eras, the coming of death was easily recognizable. In illustrating the way in which people of traditional societies took charge of their own dying experience, Tolstoy

46

describes the death of an old coachman who himself understands (without the confirmation of a medical diagnosis) that he is dying and comments: "Death is here and that is how it is."[1] Unlike traditional definitions of dying, which relied on social convention and traditions, modern dying is subject to complex medical definition. A modern person is identified as dying only when the disease has been objectively and clinically diagnosed. In addition to the medical diagnosis, a subjective label must accompany the medical determination. This entails the recognition by physicians and other health care providers that, in all likelihood, the progress of the disease will lead to the death of the patient. Obviously, a person comes to recognize that he is dying when he accepts the subjective label that originates from the medical diagnosis. Therefore, a person may not legitimately declare herself to be dying solely on the basis of personal judgment and without the confirmation of medical doctors. Hence, legitimate identification of a person's dying necessarily entails evaluation, diagnosis, and definition of a medically validated terminal condition. The very definition of dying is therefore inseparably linked to the role of medical patients. A dying person is more correctly, in the modern context of death, a dying patient.

Elisabeth Kübler-Ross: Folk Heroine of *Death and Dying* and Bricklayer of Stages for the Dying Patient

Elisabeth Kübler-Ross, a psychiatrist by training, is the most prominent figure associated with the death and dying movement. Not only has the work of Kübler-Ross received widespread attention among scholarly and academic audiences, her name and her formulation of the five stages of dying have significantly penetrated the everyday workings of American culture. She has been accorded widespread media attention, and the idea of her five-stage process of dying has become a standard image of death for the general American public. Much of the recent popular attention given to the issues of death and dying has been spawned by the appeal and acceptance of Kübler-Ross' message to the American people. In important ways, Dr. Kübler-Ross has become a popular professional figure who is not only exalted by thousands of faithful followers, but who has also received unparalleled public recognition as the prevailing "expert" and spokesperson for the death and dying movement. Perhaps even more significantly, her five-stage theory has become established and institutionalized as the unopposed American credo of human dying. Simply put, Kübler-Ross has become the most significant personality among those shaping the thanatology revolution and advocating the rights and needs of dying patients. Any serious discussion of the plight of the dying must begin with a thorough evaluation of her contributions to terminal patient care.

In her most famous book, *On Death and Dying*, Kübler-Ross begins by lamenting the lack of sensitivity and compassion in present-day medical treatment of the dying patient. Her message is very simple, namely, that increased reliance on scientific and technological factors in the treatment of dying patients has led to a standard of care that is increasingly mechanized, callous, and dehumanizing. She contrasts the sterility of death in the intensive care unit with a personal experience from her youth, the death of a farmer who had fallen from a tree and was not expected to live:

> He asked simply to die at home, a wish that was granted without question. He called his daughters into the bedroom and spoke with each of them alone for a few minutes. He arranged his affairs quietly, though he was in great pain. . . . He asked his friends to visit him once more, to bid goodbye to them. Although I was a small child at the time, he did not exclude me or my siblings. We were allowed to share in the preparations of the family just as we were permitted to grieve with them until he died. When he did die, he was left at home, in his beloved home which he had built, and amidst his friends and neighbors who went to take a last look at him where he lay in the midst of flowers in the place he had lived in and loved so much.[2]

In one brief paragraph, Kübler-Ross not only describes an experience that is of obvious significance to her personally, but sets an agenda for the remainder of the book and for her professional work in the area of death and dying. The death of her farmer friend becomes her ideal standard of dying. Much of her work is committed to making this ideal death of hers possible for every American. Her explicit purpose, from this point on, is twofold: first, to promote an acceptance of dying and death so that dying patients can readily become sources of inspiration and learning for others; and second, to find ways for dying patients to find fulfillment in and to grow from their experience of living with dying.

In an age of hi-tech dying there is tremendous appeal and comfort in this message. However, two salient problems with her agenda are that she fails to pay adequate attention to the social forces that define the prevailing images of death in a given time and place, and she fails to specify the relationship that exists between human beings and those forces. As already discussed, the social fabric of community and ritual during the times of traditional death was clearly conducive to widespread occurrence and acceptance of the type of death Kübler-Ross idealizes. On the other hand, the medicalization of death, the individualist ethos, the general societal unfamiliarity with death, and the absence of death rituals that typify the modern social setting are not conducive to widespread existence of Kübler-Ross' ideal death scenario. Her attempt to establish a model for peaceful and tranquil death overlooks the strong influence which social structure and prevailing styles of living have on establishing the styles of death.

Thanatology Snapshot 3.1

On Rejecting Technological Death

One of the most important facts is that dying nowadays is more gruesome in many ways, namely, lonely, mechanical, and dehumanized; at times it is even difficult to determine technically when the time of death has occurred.

Dying becomes lonely and impersonal because the patient is often taken out of his familiar environment and rushed to an emergency room. . . .

He may cry for rest, peace and dignity, but he will get infusions, transfusions, a heart machine, or tracheotomy if necessary. He may want one single person to stop for one single minute so he can ask a single question—but he will get a dozen people around the clock, all busily preoccupied with heart rate, pulse, electrocardiogram or pulmonary functions, his secretions or excretions, but not with him as a human being. . . . Those who consider the person first may lose precious time to save his life! At least this seems to be the rationale or justification behind all this—or is it? Is the reason for this increasingly mechanical, depersonalized approach our own defensiveness? Is this approach our way to cope with and repress the anxieties that a terminally or critically ill patient evokes in us? Is our concentration on equipment, on blood pressure, our desperate attempt to deny the impending death which is so frightening and discomforting to us that we displace all our knowledge onto machines, since they are less close to us than the suffering face of another human being which would remind us once more of our lack of omnipotence, our own limits and failures, and last but not least, perhaps our own mortality?

Elisabeth Kübler-Ross,
On Death and Dying
(New York: Macmillan Publishing Co., 1969), pp. 8–9

The first phase of Kübler-Ross' journey toward ideal death is denial. Based on her interviews with 200 dying patients, she notes that almost all patients use denial as a buffer against the devastating news of their terminal condition. Most patients, according to Kübler-Ross' framework, will only temporarily employ the defense mechanism of denial. Generally, she argues, patients will drop their denial and move on, using less radical defense mechanisms to protect themselves during the course of terminal illness.

During this discussion, Kübler-Ross, perhaps rather naively, suggests that health care personnel should use their contact with terminal patients as an opportunity to reflect honestly on their own patterns of denial in relation to the denying dying patient. Not only does she see greater awareness of his relationship to denial and death as an important factor contributing to the well-being of the dying patient, she argues that candid self-reflection on death and denial can only serve to improve the development and maturity of the health care provider as a person and as a professional. Kübler-Ross, however, never mentions how such awareness can develop, nor does she address the structural realities inherent in the training of physicians and the everyday performance of work tasks, which strongly inhibit open and honest self-contemplation about human mortality. She leaves the reader with a hopeful and unrealistic view that fails to address the myriad complex factors that influence death denial for patients, families, and care providers. Everything that Kübler-Ross feels is necessary regarding the denial stage is stated in eleven and one-quarter pages. This is not to say that brevity in itself is a failure, but rather that her sweeping, generalized use of the concept of denial provides the reader with an oversimplified treatment of the relationship between denial, dying, and death.

Kübler-Ross then moves on to a description of the second phase of their journey: anger. "When the first stage of denial cannot be maintained any longer, it is replaced by feelings of anger, rage, envy and resentment."[3] Recognizing that the angry dying patient is difficult to deal with for both the family and the medical staff, she proceeds to offer helpful and therapeutic suggestions on how to assist the patient in transcending her anger. Primarily, she emphasizes the importance of respecting and understanding, not judging, the patient, and argues that this will enable the patient to begin to move down the path of dying toward the final destination of peace and acceptance that typifies her ideal death scenario. Kübler-Ross also uses this opportunity to offer a warning and a directive to medical personnel:

> Needless to say, we can only do this if we are not afraid and therefore not so defensive. We have to learn to listen to our patients and at times even to accept some irrational anger, knowing that the relief in expressing it will help them toward a better acceptance of the final hours. We can do this only when we have faced our own fears of death, our own destructive wishes, and have become aware of our own defenses which may interfere with our patient care.[4]

Unfortunately, Kübler-Ross does not consider the circumstances of professional training that prohibit, facilitate, or hinder the development of such a response on the part of medical professionals. Instead of developing her theory and connecting it to established sociomedical research, Kübler-Ross supplements her nine-and-one-half-page discussion of the stage of anger with a twenty-two-page transcript of an interview with a young Catholic nun hospitalized with terminal Hodgkin's disease. Her discussion with the nun is offered as evidence of the therapeutic value of permitting a dying patient to express her anger. Kübler-

Ross speaks about how her interview met several of the nun's important personal needs: "She was understood rather than judged. She was also allowed to ventilate some of her rage. Once she was able to relieve this burden she was able to show another side of her, namely of a warm woman capable of love, insight and affection."[5] It was not long after the interview that the beneficial effects of Kübler-Ross' approach became apparent:

> A few days later, she visited me, fully dressed, in my office to bid farewell. She looked cheerful, almost happy. She was no longer the angry nun who alienated everybody, but a woman who had found some peace if not acceptance and who was on her way home, where she died soon thereafter.[6]

Kübler-Ross feels it was her therapeutic intervention that enabled the nun to let go of her anger and embrace the prospect of her upcoming death with love and acceptance. She fondly speaks of the lessons the nun had taught many of them through her love and acceptance, but one is left with the unmistakable impression that the nun's special value to Kübler-Ross' book was based on the nun's ability to transcend her anger and bitterness and *become a more cooperative patient.*

In this vein, and in reading Kübler-Ross carefully, it becomes evident that she has established in *On Death and Dying* an Imperial Journey of Dying that monolithically defines a good death as one grounded in peaceful, tranquil acceptance. This framework identifies for the therapist or health care professional the task of assisting the patient in his journey toward dying in the same framework of peace and tranquility as that which surrounded Kübler-Ross' farmer friend. In this way, Kübler-Ross, and those practitioners who accept her ideal of the "good death," have become travel agents for the dying, offering therapeutic intervention to a singular destination: tranquil, peaceful death.

The third stage, bargaining, seems so unimportant to the Imperial Journey that one wonders why it is mentioned at all. Bargaining is the attempt to negotiate with God for more time. It is the attempt to trade the promise of "good behavior" for the opportunity to live long enough to accomplish some goal (e.g., to see one's grandson graduate). While Kübler-Ross recognizes that bargaining may be a psychologically useful defense mechanism for the patient, she also indicts the process as being infantile, noting that similar attempts to manipulate and bargain with parents are regularly used by children. She contends that children can never be expected to maintain their end of the bargain once they have achieved their particular goal. She sees bargaining with dying patients in a similar light and notes that if they live to the point they bargained for they are still not ready to accept death, as they indicated they would be when arranging the deal, and always seek to negotiate for more time. In this way their bargain is dishonest and manipulative–just like a child's.

It is interesting to note that Kübler-Ross covers the stage of bargaining in a mere three pages. But of even greater significance is that she fails to show, through use of systematic interviews and observations of terminal patients, that bargaining is a typical part of the dying process. In addition, as in her discussion of the other stages, she fails to examine variations that may exist in this so-called stage of bargaining. For example, it would be essential for a comprehensive and viable study to see if religious and nonreligious patients bargain in ways that are different, or perhaps do not bargain at all. It is equally important to see how the age, sex, and ethnic background of the patients, as well as their particular family structures, affect their experience in the stage of bargaining and in the other stages as well. Kübler-Ross also fails to tell us how long a patient can typically be expected to remain in the stage of bargaining. In this way, if one reads Kübler-Ross seriously, it is evident that her work lacks attention to scholarly detail. The task in her entire body of work on death and dying is not to study, explore, and analyze but rather to inspire, convince, and promote an acceptance of her vision as being the definitive word on the nature of the terminal process. Thus, her work assumes the character of a manifesto.

The fourth stage of the dying process is depression. In identifying the causes of depression for the dying patient, Kübler-Ross defines two types of depression that are relevant to the dying process: reactive and preparatory. Reactive depression is a response to those things already lost during the process of dying (self-esteem, financial stability, physical self-image, etc.). Preparatory depression does not result from a loss that has already occurred, but results from the anticipation of impending loss. Unfortunately, Kübler-Ross never probes, in depth or detail, the nature of depression and the psychosocial variables that may affect a patient's experience. She never grounds her discussion in serious, clinically verified knowledge. Instead, in her quest for ideal and simple death for everyone, she offers a simple solution to the problem of depression:

> The patient is in the process of losing everything and everybody he loves. If he is allowed to express his sorrow he will find a final acceptance much easier. . . .[7]

Anyone who has worked with dying patients or studied their lives knows that the ensemble of psychological, social, and physical variables that affect the dying process preclude any easy transcendence of the problem of depression—or virtually any other problem, for that matter. Yet again, Kübler-Ross describes, defines, and analyzes the stage of depression in simply three pages, supplemented by a twenty-page interview with a depressed patient and three pages of commentary on that case. A reader must accept Kübler-Ross' message, not on the basis of scholarly evaluation and documentation, but on her word that the things she says are so. Ironically, it is precisely the superficiality of her observations which make them so popular as it is much easier for the reader and the public to relate to a message unencumbered by complexity and ambiguity. Thus, while Kübler-Ross' simplifications fail to consider the complex realities of the dying process, they have been extraordinarily successful in convincing large and diverse numbers of people of their validity.

The fifth stage of dying, the final destination in the Imperial Journey of Terminal Illness, is acceptance. If the dying patient has been the recipient of the therapeutic ministrations of a competent travel agent, or has been able to heroically gather the resources by personal ability and effort, he will be able to transcend the earlier stages of denial and turmoil and arrive at a quiet acceptance of his fate. Emphasizing that the patient's needs at this time are for rest and dignity, Kübler-Ross comments:

> [T]here comes a time for "the final rest before the long journey" as one patient phrased it. . . . This is the time when the television is off. Our communications then become more nonverbal than verbal. The patient may just make a gesture of the hand to invite us to sit down for a while. He may just hold our hand and ask us to sit in silence. Such moments of silence may be the most meaningful communications for people who are not uncomfortable in the presence of a dying person. We may together listen to the song of a bird from the outside. Our presence may just confirm that we are going to be around until the end. We may just let him know that it is all right to say nothing. . . . It may reassure him that he is not left alone when he is no longer talking and a pressure of the hand, a look, a leaning back in the pillows may say more than many "noisy" words.[8]

The hallmark of the stage of acceptance, according to Kübler-Ross, is dignity. Although she never offers a precise definition of what dignity entails or how exactly it may be achieved, she implies that a dignified death is a death that is not emotionally turbulent and is typified by an attitude of courage, joy and love. A description of a model patient who had been able to face her death with such dignity gives a better idea of what Kübler-Ross means when she speaks of dying-with-dignity:

> She complained very rarely and attempted to do as many things as possible by herself. She rejected any offer of help as long as she was able to do it herself and impressed the staff and her family by her cheerfulness and ability to face her impending death with equanimity.[9]

There are many problems with the journey toward acceptance delineated in *On Death and Dying.* First of all, the expectation that all patients should be able and willing to accept death with courage and dignity overlooks the fact that there are several patterns of evasive interaction that shape the context of dying for many individuals. As discussed in the preceding chapter, the acceptance of death is at odds with the realities of the contexts of closed awareness, suspicion awareness, and the drama of mutual pretense. Kübler-Ross assumes

Thanatology Snapshot 3.2

Going Out and Beyond

I was out of my body and I was gone. I had an incredible experience as if there were a lot of beings who took all the tired parts out of me and replaced them with new parts. I'm not terribly car-oriented, but I felt as if somebody brought the car to the shop, and ten people, a dozen people worked on it, and everybody worked on one part so that within a very short time every rusty part was replaced with a new one.

It was an incredible experience—everybody worked on the machine, on me, and I was just floating and taken care of. And half an hour later I woke up and felt as if I were twenty years old—no tiredness, no fatigue. I felt super healthy, and I had no idea what happened to me—it was the first experience of my life of this kind. And this woman sat there with her mouth open. She said, "You know, anybody who walked into this room would have been absolutely convinced that you were dead. There was no respiration, no breathing, no pulse. You were warm but there was nothing. . . ."

Then our patients who had come back to life started telling us what they experienced when they were declared dead. This was often after accidental death: sudden deaths where they had no preparation, they couldn't go through what we call the stages of dying. They told us at the moment of death they had an incredible experience where they just left their bodies.

In a car accident, say, they can see the scene of the accident: they see the rescue team, they see the people who try to get them out of the car with blow torches, and they look as if they're in terrible pain and agony and going through a dreadful nightmare. Meanwhile, though, they are floating above the wreck, having a beautiful sense of peace and equanimity. They watch what's going on with a detached kind of observer's point of view, but without any of the effect, except peace and sudden surprise when they realize that that person is really them, and sometimes there's some confusion about that.

I was dying to see if it was really possible to induce OOB [out of body] experiences. . . .

The second I was hooked up [to a polygraph machine], I gave myself the command: I'm going to go faster than the speed of light and I'm going to go further than any human being has ever been. That was my command to myself and no sooner did he get going, I took off with the speed of light until it dawned on me that I'm going this way—I'm going horizontally! My God, by then I had gone a few hundred thousand miles and the second I was going horizontally and that's wrong—just the thought—I turned at a right angle and went vertically. Stuff like this excites me! And I went so far and so fast that nobody could ever catch up with me. I felt very safe: nobody could find me. I was really where nobody had ever been and from then on I have no recollection. I knew I had probably gone where nobody had ever been. I felt super. . . .

The second night after . . . I tried to sleep but I couldn't—I remember tossing back and forth. It was like in a fever. I was delirious, fighting to sleep, wanting to sleep, but knowing that I couldn't fight it much longer—that kind of a turmoil. And then it hit me like lightning—the whole experience. To describe it in words is . . . What happened is that I went through every single death of every single one of my thousands of patients that I had seen by then. And when I say I went through their death processes, I mean this literally—I had every experience every patient ever had and the bleeding and the pain and the agony and the cramps and incredible pains and tears and loneliness and isolation—every negative aspect of every patient's death. And this repeated itself a thousand times—always in a different version but with the same agony. And during this endless incredible agony (which was physically very real: I couldn't breathe, I couldn't catch my breath, I couldn't even finish a thought because it would hit me again—a cramp, bleeding, or shortness of breath), somewhere in the middle I was able to say one sentence, and I asked for a shoulder to lean on. . . . And the minute I finished this thought—this prayer perhaps, to give me a shoulder to lean on—this incredible voice came from everywhere: "You shall not be given. . . ."

What followed cannot be put into words. I was lying on my back. The room was illuminated by the night light. My abdominal walls started to vibrate at a very high speed—it was going super, super fast—and I looked at my belly and what I saw was anatomically impossible. (I felt this scientifically even while I was going through it. It was as if I had an observing ego watching the whole thing.) And every time I watched a part of my body, it started to follow, vibrating. And then I looked at the closet, and it started to vibrate. And the walls started to vibrate, and the whole world . . . I had a vision of the whole universe, everything vibrating. And in front of me something opened. It was a visual image. At first I thought it was a vagina. The moment that I focused on it, it turned into a lotus flower bud. It had the most incredible colors—beauty that I cannot put into words. I watched this and the vibrations going on in the whole room. Behind this flower bud came something like a sunrise—an incredible light. And this bud opened up into the most fantastic lotus flower. I looked at all this in utter awe. There were sounds and colors and visions beyond description.

The moment the light was at the peak and the flower was wide open, all the vibrations stopped and all the million molecules (it's as if somebody put a million pieces of a puzzle into one—that's the best way I can describe it) became one and I was part of that one. I really cannot describe this experience.

Elisabeth Kübler-Ross,
in Peggy Taylor and Rick Ingrasci, "Out of the Body:
An Interview with Elisabeth Kübler-Ross,"
excerpted from *Out of the Body*,
New Age, November 1977.

that the only acceptable and dignified form of death is that which takes place within open awareness. This not only restricts the patient's right to carve out a different way of dying, but also ignores the realities of the closed patterns of interaction that shape the dying experience of modern patients. As Glaser and Strauss carefully show through their observations, the contexts of action and interaction for dying patients are characterized by personal, social, and medical complexities. A simple and singular destination for all dying patients is clearly unfeasible and may not necessarily be desirable.

A second major problem with Kübler-Ross' perspective is that it is internally inconsistent. She begins her book by claiming that the medicalization of death establishes a style of dying that is insensitive and dehumanized (see Thanatology Snapshot 3.1). Thus, Kübler-Ross accepts the argument that Ariès, Gorer, Illich, and myself have made regarding the technologization of death and its consequences for death denial and patient care. Yet, just seventy-five pages later, she speaks in highly descriptive and romantic terms of the ways in which patients courageously accept their fate. In describing the peace, tranquility, and dignity that she says the patients she interviewed had experienced, she appears to be contradicting her earlier thesis that medicalized death has made dying lonely, mechanical, and inhuman. Indeed, it is especially difficult to reconcile Kübler-Ross' reports of unanimous acceptance and dignity among her patients with the sociomedical realities of pluralistic death patterns and closed-awareness contexts discussed in Chapter 2. If Kübler-Ross' accounts of patients' dignity and acceptance are true and accurate, then the modern organization of death is neither as insensitive nor as mechanical as she claims. Or, is it possible that she is implying that by her involvement in the lives of dying patients during the interview process, she was miraculously able to provide the means for alienated and dehumanized patients to travel the road to dignity and acceptance? And, even if there were substantial truth to such an assertion, it would surely reflect the argument that the system of care is not as inflexible and inhuman as claimed. For after all, it was within this system of hospital-based care of the dying that all of the alleged dignity in dying took place.

Nevertheless, the appeal of Kübler-Ross, in light of and despite the deficiencies in her work, is a phenomenon in itself that is worthy of our attention. There are three salient explanations for her widespread appeal: her adoption of the clinical/therapeutic model in her work with dying patients, her ability to assimilate her perspectives on death and dying into symbols and ideals already valued in the broader American culture, and her ability to motivate others through her personal charisma.

Although Kübler-Ross offers a description of the process of dying, much of her energy and attention is directed toward the goal of patient management.[10] She speaks directly and indirectly to health care providers in her descriptions and accounts. Her goal is to sensitize medical personnel to the requirements of effective management of terminal patients. The idea of patient management is a central focus of all medical activity, and by injecting her work with dying patients into the mainstream of medical activity, Kübler-Ross is taking a giant step toward legitimation of her goals and perspectives. It is no secret that life on the units, floors, and wards of hospitals is organized for the convenience of staff and for the purpose of facilitating the flow of work. It is also no secret that dying patients can be a source of distress and disturbance for medical staff. The way Kübler-Ross has identified the stages of dying has become an appropriate and useful method to protect the vested interest of the staff in maintaining a smooth, operational harmony among the patient population. Her script for the dying can also prove useful to the medical management of dying patients in at least two major ways. First, during the early stages of the Imperial Journey, when the dying person may be difficult to handle due to expressions of denial, anger, and depression, medical management can legitimately introduce psychiatric, therapeutic intervention. This therapeutic intervention into the stressful life circumstances of the dying patient is not only useful for managing patient discontent and turbulence, but is also consistent with society's tendency to identify and treat serious problems and sources of distress as the private troubles of individuals. The corresponding response, then, is to address the private troubles of individuals through personalized schemes of medical and therapeutic management and to avoid addressing structural or institutional causes of the problem.

In addition to responding to problematic dying patients in a therapeutic manner, the medical staff's management difficulties could be resolved if the dying patient, through her own resources or through the guidance of the therapeutic travel agent, could be transported to the stage of acceptance. In this way, the idea of acceptance as a universal, ideal form of death becomes an attractive vehicle of patient management to the profession of medicine. An opportunity is then offered to circumscribe and contain inappropriate and disturbing behaviors among dying patients.[11] Not only does the Kübler-Ross approach succeed by invoking the medically and culturally valued therapeutic model, but it also serves to protect that model in the daily routine of hospital work from inappropriate dying-patient behavior. In these ways, Kübler-Ross' work reflects and supports the sociohistorical forces in contemporary society, in which the therapeutic and psychiatric paradigm has become a dominant perspective in medical care and in the wider society.[12] The natural interconnection between already established social and medical mores and Kübler-Ross' view of the ideal death greatly enhances the appeal of her message.

Kübler-Ross' favorable public standing also reflects the American value of individualism. As de Tocqueville, in his classic discussion of *Democracy in*

Thanatology Snapshot 3.3

Going . . . Going . . . Gone

. . . Things have gotten out of hand: Kübler-Ross herself has become the guru to a nationwide network of death 'n' dying centers called "Shanti Nilaya"; the "Conscious Dying Movement" urges us to devote our life to death awareness and also opens up a "Dying Center"; a video artist kills herself on public television and calls it "artistic suicide"; the EXIT society publishes a handy, do-it-yourself Home Suicide guide that can take its place next to other recent Home Dying and Home Burial Guides; a pop science cult emerges around the "near death experience," which makes dying sound like a lovely acid trip (turn on, tune in, drop dead); attempts at two-way traffic with the afterlife abound, including a courier service to the dead using dying patients and even phone calls *from* the dead; belief in reincarnation resurfaces as "past lives therapy. . . ."

Dividing dying into stages was a stroke of genius. Kübler-Ross brought forth her five stages at just about the time when people were dividing life into "passages," stages, predictable crises. Getting dying properly staged would bring every last second of existence under the reign of reason. . . .

What's been lost in the general approbation of Kübler-Ross's five stages is the way her ordering of those stages implicitly serves a *behavior control* function for the busy American death professional. The movement from denial and anger to depression and acceptance is seen as a kind of spiritual *progress*, as if quiet acceptance is the most mature, the highest stage to strive for.

What Kübler-Ross calls bargaining, others might call a genuine search for reasons to live, to fight for life. But she has no patience with dilly-dallying by the dying. She disparages "bargaining" that goes on too long, describes patients who don't resign themselves to death after they've gotten the extra time they bargained for as "children" who don't "keep their promises" to die.

America wrote, individualism and personal choice have deeply penetrated, defined, and shaped the character of American citizens. In their recent evaluation of individualism in American society, Robert Bellah and his colleagues note that Americans tend to define the ultimate goal of a good life through a framework of personal choices.[13] And, as discussed earlier, individual fulfillment has become a moral imperative of modern social living. By establishing a psychosocial framework for death with dignity, the horror of death can be transformed into a final frontier for self-expression and self-exposition.[14] If one is able to die in the manner prescribed by Kübler-Ross, one is theoretically able

Thanatology Snapshot 3.3 (*continued*)

Yet by "acceptance," Kübler-Ross means the infantilization of the dying: "It is perhaps best compared with what Bettelheim describes about early infancy," she says. "A time of passivity, an age of primary narcissism in which . . . we are going back to the stage we started out with and the circle of life is closed."

Certainly this passivity makes for a quieter, more manageable hospice. Crotchety hospice guests who quixotically refuse to accept, who persist in anger or hope, will be looked on as recalcitrant, treated as retarded in their dying process, stuck in an "immature" early stage, and made to feel that it's high time they moved on to the less troublesome stages of depression and acceptance. . . .

Kübler-Ross . . . boasts of some new benefits troubled people can look forward to as soon as they die. Death is a cure-all, the ultimate panacea: "People after death become complete again. The blind can see, the deaf can hear, cripples are no longer crippled after all their vital signs have ceased to exist."

If this encouragement to euthanasia as a quick solution to all physical imperfections were not so stupid and dangerous, it would be an occasion for regret: death has claimed another victim, the mind of Kübler-Ross. Another sad but predictable triumph of death over reason, another case of an interesting mind committing suicide. It begins to seem that thinking about death is, like heroin, not something human beings are capable of doing in small doses and then going about the business of life. It tends to take over all thought, and for death 'n' dying junkies, the line between a maintenance dose and an O.D. becomes increasingly fine. When Kübler-Ross finally makes her "transition," I'm certain all her nagging physical ailments will clear up, just as she says, but they'll have to mark her mind D.O.A.

Ron Rosenbaum,
"Turn On, Tune In, Drop Dead,"
Harper's, July 1982, pp. 32–42.

to grow from the experience. By facing up to death with dignity, one is able to show others the strength and courage of one's personality. Kübler-Ross' prescriptions have thus been popularly embraced as a blueprint providing dying patients with a final opportunity for self-enhancement.

Dennis Klass, in an important article on the appeal of the five-stage Imperial Journey of Death, takes the argument even further. He suggests that Kübler-Ross' indictment of the cold, oppressive system of technological care of the dying came at a time when American society was ready to challenge and critique the rational world of technological progress. He adds that her schema for dying a good death arrived at a time when the public was becoming increasingly receptive to the ideas of emotional expression and harmony with nature. Kübler-Ross was thus able to associate with a symbol system in American culture that extolled the value of natural living and self-actualization and emphasized the potential of the individual human being. Klass argues that at the time of publication of *On Death and Dying* in 1969, a significant part of the American population had begun to understand themselves and the world through the symbols of self-liberation. Freedom from a technologically repressive system and movement toward self-actualization for the dying patient were legitimate ideas already nurtured in the wider society. The words of Klass are effectively to the point:

> She gave a symbol by which the depersonalizing, masculine, rational world of technology could be fought with maternal nurturance. As the home had become a haven in a heartless world, for Kübler-Ross, the acceptance of death could become a haven in the manipulated technological hospital world. . . .
> Without knowing it, then, Kübler-Ross spoke to a people prepared by a tradition of symbols. . . . The five stages organized psychological processes so death could be a kind of trip (albeit one way), another experience to be experienced. The trip was predictable because the subjective states of anger, depression and bargaining could be checked against the normal response. At the end of the trip was a blissful primary narcissism where "we experience the self as all." Such bliss is at harmony with a romanticized nature.[15]

Thus, the spirit of the times, as well as the cultural diffusion and acceptance of the idea and symbols of self-actualization, had carved out a foundation of acceptance for her work even before it was published.

A third factor that accounts for Kübler-Ross' popularity is her personal charisma, which benefits from a favorable social climate that is eager to verify the legitimacy of her guidelines for the dying. In many ways, Kübler-Ross herself has become synonymous with death and dying. Her counseling work and her workshops for the dying have been complemented by the establishment of a growth and healing center, Shanti Nilaya (the "Final Home of Peace"). This center, now located in Virginia, is a spiritually oriented retreat that has special facilities for quadriplegics and terminally ill children, and that serves as a place where "youngsters and adults of all ages and backgrounds come to learn how to live until they die."[16] In addition, Kübler-Ross has been able to legitimize, at least partially and to a selective audience, such transcendental ideas as out-of-body experiences, the value of near-death experiences, and the possibility of communicating with the spirits of the dead. While they might seem bizarre to many of us, such ventures into the transcendental and spiritual realm became commonplace for Kübler-Ross and her entourage of death enthusiasts. In fact, Kübler-Ross has claimed to have four spiritual entities attending her, whom she called "Mario," "Anka," "Salem," and "Willie," and is a fervent believer in faith healing. She has also plunged dead center into a sex scandal involving a man whom she dubbed as "the greatest faith healer in the world." Of special note is that these "oddities" have done little damage to her popularity and popular credibility.[17] In this way, Kübler-Ross has become a charismatic spiritual leader who has been successful in identifying the legitimacy of her message with her personal self and its aspirations. It is no wonder that her coterie, motivated by her presence and charismatic authority, enthusiastically and unquestioningly follow "Elisabeth" (as she is affectionately called) into controversial realms of spiritual and religious experience. As Klass and Hutch put it:

> . . . [R]eligious leadership . . . is Kübler-Ross' work. . . . There can be little doubt that Kübler-Ross' influence on the death awareness movement has been very strong. The early hospices, several of the self help groups, and many of the professionals working in the field exist and remain motivated because of her. Basically, she is a lecturer, a leader of workshops, and a personal counselor. Her publications actually amount to a very few materials. In this sense, the general death awareness movement which she initiated in western culture is itself the institutionalized form of Kübler-Ross' charisma. Apart from the "ethos" of the death and dying movement, the only thing holding together her religious movement, especially the organization of Shanti Nilaya, is her person, Kübler-Ross herself. The rituals by which her truth is carried to the general public remain fundamentally tied to her personal presence, and even mentioning her name at this time seems to evoke a sense of the truths associated with part of the culturally-based death and dying movement.[18]

In short, the attraction of the Kübler-Ross school is rooted in the prevalence of the therapeutic model and the individualist ethos that are dominant in American society. Her personal presence and charisma enhance the popular acceptance of her work. Yet, as we have seen, there is a danger and a potential for distortion inherent in her perspective. The nature of the modern dying experience is so vast, and shaped by so many combinations of variables, that it is inappropriate to propose that one model of death can comprehensively explain it. Moreover, in the modern context of pluralistic-negotiated death patterns, it is absurd to set forth a uniform, standardized definition of a good and appropriate death for all. Nevertheless, Kübler-Ross remains an exalted leader of the death-and-dying movement. Her Imperial Journey, and the precepts contained in her writings and lectures, have assumed the status of modern thanatalogical commandments showing all dying patients the way to a good and appropriate death.

Trajectories of Dying

The perspective emphasized in this book argues that in order to understand the life events and circumstances of the dying patient, one must consider the process of dying in the broader context of American society. It is important to understand that an individual patient's dying embodies and reflects values and conditions of society at large. Specifically, two salient factors that affect the dying patient's life course are (1) the growing incidence of chronic illness and (2) the institutionalization of the dying person and the role of medical technology. It is the complex interplay of the types of illness and the technologies that are used to treat them that significantly shape the course of death. The potential for complexity or simplicity in dying is related to such factors as multiplicity of diseases, pain levels and manageability thereof, the number of medical specialists involved in the care of the patient (the interests of each and the conflict of one interest with another), the treatments being used, the available options, the possibility of side effects and complications of the treatment, the amount of machinery-technology being used, the response of the disease state to medical intervention, the aggressiveness of the medical staff in managing the illness(es), and the role of the patient and his family in shaping the form of dying. It is not difficult to see that many intricately interwoven factors distinctly influence the process of dying. While some patients may die smooth, routine deaths, many will experience lengthy, complex, and turbulent ones in this age of chronic-illness dying and hi-tech medical response.

An individual's course of dying consists of much more than the characteristic phases of a particular disease. It involves disease processes in dynamic relation to medical response. The blending together of the course of an illness with the response of medical professionals, added to the impact that this has on the dying patient, gives rise to what has been termed the trajectory of dying.

The idea of dying trajectories has been introduced into the thanatology literature by the pioneering work of Glaser and Strauss. In their important study *Time for Dying*, they discuss the variable trajectories that the course of dying may assume. They begin by noting that the dying trajectory of each patient has duration, that is to say, it takes place over a period of time. Second, the dying trajectory has a particular shape for each patient: plunging straight down; moving slowly but steadily downward; vacillating slowly, moving slightly up and down before diving radically; moving slowly down at first, then hitting a long plateau, then plunging abruptly to death and so on. As Glaser and Strauss describe:

> Since dying patients enter hospitals at varying distances from death, and are
> defined in terms of when and how they will die, various types of trajectories are

commonly recognized by hospital personnel. For instance, there is the abrupt surprise trajectory: a patient who is expected to recover suddenly dies. A trajectory frequently found in emergency wards is the expected swift death: many patients are brought in because of fatal accidents and nothing can be done to prevent their deaths. Expected lingering while dying is another type of trajectory; it is characteristic, for example, of cancer. Besides the short term reprieve there may also be the suspended-sentence trajectory. . . . Another commonly recognized pattern is entry-reentry: the patient slowly going downhill returns home several times between stays at the hospital.[19]

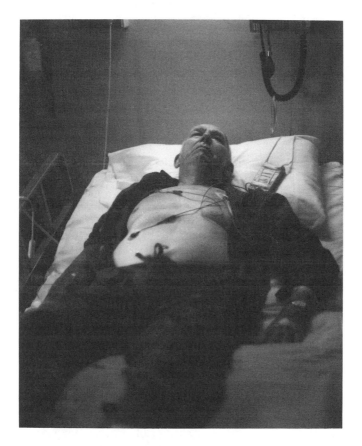

THE LONELINESS OF THE DYING

Isolation, emptiness, and loneliness are dominant experiences of the dying patient. This patient, hospitalized in his battle with cancer and heart disease, received absolutely no visitors during his ten-day hospital stay.

The combination of a wide range of possible shape and duration factors makes the dying trajectories of modern patients highly variable and individualized. Yet, certain types of patterns emerge, and these are useful because they identify trajectories that typify the modern dying experience. The three categories of dying trajectories delineated by Glaser and Strauss are lingering trajectories, the expected quick death trajectory, and the unexpected quick death trajectory.

The idea of lingering trajectory is in itself multifaceted and variable. There are many ways in which lingering trajectories can take place. Patients may move toward death slowly in a hospital setting, nursing home, state institution, or even in their own homes, or some changing and variable combination of locations. Thus, slow dying is not one singular way of death but takes its own unique form from a combination of possibilities. It is important to recognize that each particular circumstance of the slow-dying trajectory has its own special requirements, responsibilities, and consequences for the dying patient.

The burden on the family may be heavier for those slowly dying-at home, which would, in all probability, require special arrangements for home-care nursing. If a patient lingers at home and requires unpredictable but regular periods of hospitalization, breaks in routine and the anxiety of being transferred from one setting to another have to be dealt with. In addition, upon reentry into the hospital setting, the locus of control of the dying person shifts from herself and her family to professional caretakers. It is also possible, in large part due to medical technology, for patients to linger beyond expectable limits and for family members to become stressed and distressed by the unrelenting responsibility and drain of energy that come from caring for a dying loved one at home or in an institution. Thus, while it is beyond the scope of this chapter to discuss all of the possible trajectories of lingering death, it is clear that slow death can assume many different forms, requiring different types of care and work responses, having different kinds of implications for the dying patient and her family.

The second type of dying trajectory analyzed by Glaser and Strauss is the expected quick death. While the idea of a quick death may, at first glance, seem more stable than the lingering trajectory, it also has its own fluctuations and variations.[20] For example, the "danger period quick trajectory" involves watching the patient through a determined period of time and, if he makes it, trying to intervene at that point to prolong the dying.[21] In the "crisis trajectory," a patient is so critical that she must undergo a medical procedure that may result in death. All of these trajectories are changeable into a "save-loss" quick trajectory, whereby

[in] the course of trying to save the patient the staff succeeds enough so that they do not expect a quick death, they feel he has been saved. Then, suddenly

he starts to die quickly—and perhaps they save him again. This up and down shaping of his trajectory can go on for days, resulting in indeterminate expectations as to whether he will live or die.[22]

Again, as is evident, the response of staff and the duration and the consequences of the quick trajectory vary from situation to situation and from patient to patient.

The third type of dying trajectory is the unexpected quick death. A variety of factors are also involved in shaping the contours of this trajectory. Foremost among them is the perceived status and medical condition of the patient. The definition of the situation of unexpected death, and its responses and implications, is determined by whether the person was thought to be dying, or whether it was a totally unexpected crisis that emerged and resulted in death on the operating table during routine surgery. The impact of the unexpected death on the medical staff may also vary according to where the death takes place. For example, emergency rooms and intensive care units are geared to respond quickly to trauma-type situations and are therefore more prepared for the possibility of sudden and unexpected death than other medical wards, where the occurrence of this type of death may precipitate a crisis for the staff. In any event, the unexpected quick death trajectory is made up of a variety of possible situations, responses, and implications.

The process of dying is clearly multifaceted and intricate. Not only does the dying patient face an anxiety-producing labyrinth of possibilities and situations, but the emotional and social needs of dying patients and their families are similarly complex and variable. The emergence of pluralistic-negotiated death patterns in modern society has already been discussed. Dying trajectories, and their complexity and variability, are very much consistent with the idea of pluralistic dying and death. And it would be remiss at this point not to mention that the myriad combinations and complexities of modern dying trajectories illuminate how oversimplified Elisabeth Kübler-Ross' standardized equation for dying a good death actually is.

However, despite its monolithic nature, Kübler-Ross' model does have some relevance for understanding the illness trajectories of dying patients. Part of what forms the dying trajectory is the patient's awareness of, perceptions of, and response to his own terminality. The view that patients should muster the courage and composure to sustain and drive themselves toward dignity helps to define and shape patients' own responses to their dying process. The expectation that patients should be compliant and cooperative is established so as not to disrupt the work routine of the medical staff. This attitude is often internalized by patients as the way they feel they should respond to their illness. The degree to which patients are defined by the medical staff as complying with the requirement of cooperation will, in turn, influence the staff's responses to patients. It is precisely this interaction of patient and staff attitudes that will influence and shape the course of the dying trajectory. In the *Social Organization of Medical Work*, Strauss and his colleagues comment:

> There are more subtle kinds of patient work, which if not done . . . properly, get patients into trouble with the staff. We have in mind, for instance, the complex situation which can arise when the staff knows that a terribly ill person now knows he or she is dying. The patient is, in our terminology, expected to do certain (unrecognized) work involving maintaining reasonable control over reactions which might be excessively disturbing for the staff's medical work, for its composure, too, and, perhaps, even disruptive of other patients' poise.[23]

In an important way, one of the unspoken rules of dying is that the dying patient should exude calm or at least stoicism in the face of her own death. It is at this point that Kübler-Ross' model is applicable. The standard of a good death and the requirement of death with dignity, which are central to the perspective of Kübler-Ross, become useful in establishing rules and expectations for the dying patient. The adherence to or violation of the rules has significance for the contour of a patient's dying trajectory. It is worth noting that not only are patterns of dying highly individualistic, pluralistic, and personally carved out, at least in part, by the dynamic interrelation among family, staff, and the dying patient, but expectations of composure and dignity for the dying patient are a useful sociomedical strategy for organizing, containing, and managing death.

Thanatology Snapshot 3.4

Enforced Expectations for the Dying Patient

. . . Patients who do not die properly . . . create a major interactional problem for the staff. The problem of inducing them to die properly gives rise, inevitably, to a series of staff tactics. Some are based on the patient's understanding of the situation: staff members therefore command, reprimand, admonish, and scold. . . . These negatively toned tactics . . . are supplemented, and often overshadowed, by others through which personnel attempt to teach patients how to die properly.

Barney Glaser and Anselm Strauss,
Awareness of Dying
(New York: Aldine Publishing Co., 1965), pp. 90–91.

In this way, then, one begins to see how the organization of death in the broader society is brought to bear explicitly on the life circumstances of individual dying patients.

As noted above, one of the salient factors relevant to modern dying trajectories stems from the prevalence of chronic illness. It is now important to look at the way in which chronic illness trajectories are shaped by the medical profession's technological orientation and its related values of activism, cure, and treatment. A study by Mumma and Benoliel found that, despite a terminal diagnosis and the fact that many dying people are very sick, there is a clear and definite tendency among medical staff to intervene actively in the dying process for the purposes of treatment or cure:

> The findings suggest that the life-saving/prolonging ethic was quite powerful at this teaching hospital. Despite the fact that the majority of patients had been designated no code and had conditions labelled by their physicians as either grim or terminal, the treatment orientation was overwhelmingly toward the cure end of the comfort–cure continuum. . . . These findings are consistent with the literature related to the impact of technology on the care of hospitalized persons with life-threatening illnesses. More than fifty percent of the patients in this study died in critical care units, and to be hospitalized in a critical care unit in a teaching hospital all but guarantees the use of life-saving/prolonging technology. This push toward cure is likely despite a grim or even terminal diagnosis.[24]

Mumma and Benoliel correctly point out that one important consequence of the medical profession's active-intervention orientation is the prolongation of the dying process, which therefore naturally increases the incidence of lingering dying trajectories in the hospital setting.

Another important consequence of active medical intervention has to do with the way this technological mediation of death manages and arranges the human side of dying in a restricted, technical framework. In an age where the incidence of chronic terminal illness is high, there are many opportunities for medical intercession in the disease process. The complexity of chronic illness and the multiple symptoms associated with it enable physicians to select and target specific symptoms for treatment. In this way, despite the fact that the ultimate fate of death cannot be avoided, specific symptoms (such as body swelling, size of nodes, red blood count, etc.) can be aggressively treated by medical intervention, the medical result of which extends the length of the dying trajectory. The value of active intervention for the medical profession is that it brings the standard medical response and technological activity to the forefront of a physician's relationship with a dying patient. In this way, the personal and human issue of dying is transcended by the medical concern of symptom management of chronic illness.

The technically defined prolonged dying trajectory of chronic terminal illness will be characterized by peaks, valleys, twists, turns of medical and emotional turbulence. As I show in *On Death Without Dignity*,[25] the technical care of dying patients enables physicians and other medical staff to treat the dying patient with the same set of expectations and responses applied to other medical patients. It is for this reason that dying patients with a lingering trajectory are often cast in the sick role, and attention is then focused primarily on the management of physical symptoms. By focusing on the manageability of specific symptoms, the unmanageable issue of dying is superseded and deferred. As a result, the process of dying is prolonged in such a way that it is often filled with uncertainty and ambivalence. As the patient embarks on this roller-coaster journey of death, symptoms may temporarily improve and hope of recovery, although truly not warranted by the patient's overall condition, is kindled. As symptoms progress and worsen, despair emerges. An intensified medical response to worsening symptoms may be successful in ameliorating the symptoms and rekindling hope. The course of lingering dying, with its orientation to symptom intervention and management, sweeps the dying patient onto a roller-coaster journey of rising hope, plummeting despair, and overriding ambivalence. The illustration of this roller-coaster journey of death in Thanatology Snapshot 3.5 demonstrates how anxiety, confusion, uncertainty, unrealistic expectations, and socioemotional instability are typically found in lingering death trajectories.

Suffering in the Face of Dying

This chapter has been dealing with the ramifications of individualized, technocratic society for the role of the dying patient. Closely bound up with the chronicity of the modern dying experience, and its roller-coaster turbulence and medical prolongation, are the problems of pain, debilitation, and suffering.

The fear of physical pain is one of the most deeply felt anticipations associated with the modern dying experience. All of us have experienced "normal pain" during our lives. When a hot stove is touched, pain is experienced but then subsides and goes away. The modern dying experience brings a different type of suffering to the life situation of the dying patient, namely, chronic pain. Unlike normal pain, chronic pain stubbornly persists; its presence is regular, constant, and often inescapable.[26]

In a society where unwanted pain is defined as an unacceptable and purposeless disruption of a person's internal and external equilibrium, the dominant response is to attempt to manage and control pain. Thus, modern society's response to pain is defined by the therapeutic model, which seeks alleviation through drugs, surgical intervention, or other kinds of medical treatment. This

Thanatology Snapshot 3.5

A Roller-Coaster Journey for the Dying Patient

Let us consider the emotional roller-coaster ride of heightening hope and plummeting despair as illustrated by the following case study. A patient with a relatively unsuccessful history of lymphosarcoma was admitted into the hospital with a severe swelling of and pain in her abdomen. Her physician discusses some test results with her:

DR.: The preliminary results of your CT scan show a growth of nodes, which are causing an obstruction. You have been urinating today— losing some weight—which is the first time you've done that since coming in. If that continues, we'll continue you on lasix to get the swelling down. If not, I'll ask Dr. _____ to radiate your belly (he points to the spot where radiation would occur) to see if we can't get the nodes to reduce. Then we can get you started on chemotherapy.

PATIENT: What will the medicine be?

DR.: Some VMF, methotrexate, some . . .

PATIENT: Could you write that down for me?

DR.: When it comes time to start treatment, I'll go over the whole thing with you. Don't worry, you'll have plenty of advance warning.

PATIENT: For how long will I be getting the chemo?

DR.: I'm not exactly sure what the protocol* calls for. I have to check it out. It'll be something I give you every three or four weeks.

PATIENT: Oh, so then we are talking about something overnight?

DR.: Yes, . . . or I can give it to you in the office.

PATIENT: Thank you, Doctor. Oh, thank you so much, Dr. _____ .

It is important to note how the patient selectively overlooked the initial concerns discussed by her physician and focused her attention on the specifics of her chemotherapy regimen, placing hope in the curative potential of the protocol drugs. The scenario resumes the following day after a physical examination and a review of specific symptoms and test results. The patient begins:

---

*"Protocol" refers to a combination of drugs and/or procedures that have not yet been approved by the Food and Drug Administration for regular use but have been approved for use in experimental studies. The patient was being started on a protocol because she had been placed on all the regularly prescribed regimens for her type of cancer, and each of those had ultimately proved ineffective.

(continued)

Thanatology Snapshot 3.5 (*continued*)

PATIENT: Doctor, so this isn't serious? (pointing to her severely swollen stomach).

DR.: Serious? (he hesitates). Well, your disease is serious. You do have extensive nodes (he points to her shoulder) and your swelling is probably due to node growth . . .

PATIENT: (interrupting) But, it won't require surgery . . . ?

DR.: No, that I can tell you for sure.

PAUSE

DR.: The critical thing is your swelling. We want to get that down with medication and/or radiation. That is the really important thing for now. Then we can begin to start you on some chemotherapy which will have a chance to work.

PATIENT: (focusing on the last phrase and showing visible relief) Okay, Doctor, good. That'll be fine. Thank you.

The patient is locked into focusing on her hopes for the future. She is placing her faith in and relying on her expectations of the benefits to be derived from her anticipated chemotherapy treatments. She is, however, "naively unaware of how serious her condition is," as one doctor put it. Clearly, her doctors have been unsuccessful in communicating the nature of her diagnosis to her, and/or she has been selectively screening out indicators of bad news. In any event, this naiveté on the part of the patient was not very well understood by her attending doctors. Her primary oncologist, shaking his head in disbelief, commented to her radiologist in a hallway conversation: "She really thinks she is going to get well." And the radiologist responded: "Do you mean she hasn't caught on?"

The following conversation initiated at the request of the patient's parents, between the patient and a staff associate in the cancer center, where she was being treated, illustrates that whatever its etiology, the patient had not fully embraced the realities of her condition:

PATIENT: I'm just so relieved that they are going to start chemotherapy tonight.*

*Chemotherapy had been discussed as a possibility with the patient, and as a result of miscommunication between her attending physician, the house staff doctors, and herself, the patient had come to believe that the start of treatment was imminent. As it turned out, the intern and resident in charge of her case had wrongly informed the patient that she was to start on chemotherapy. The attending physician, upon hearing this, reversed the order, citing two reasons. First of these was that his concern, at this stage of the patient's dying, was to make the patient as pain free and comfortable as possible. The interns aggressively argued with the attending physician on this point, emphasizing that they had an obligation to "try everything to save her." The physician was getting nowhere with the argument that treating her pain and other symptoms of discomfort was

ASSOCIATE:	You seem very enthusiastic about starting treatment.
PATIENT:	Yes, (with a deep sigh of relief). I just want to get better.
ASSOCIATE:	That is the hope of many people, but you should realize that being placed on the protocol does not mean an automatic, miracle cure.
PATIENT:	What do you mean?
ASSOCIATE:	Dr. _____ told you of the seriousness of your condition two nights ago. Your disease is extensive, and he said it is spreading. It's important for you to recognize that the treatment will not make you well again, overnight.
PATIENT:	It won't?!? (disbelief)
ASSOCIATE:	You've been on other combinations of medicine before, and you know that they work slowly, not rapidly. I hesitate even to discuss this with you, except that your doctors and I do not want you to be hooked onto expectations that are unrealistic and be shattered because of it.
PATIENT:	Yeah.
ASSOCIATE:	If the treatment will have any positive effect, it will be only with you fighting along with it. The treatment will not instantaneously make you well.
PATIENT:	Oh.
ASSOCIATE:	Have you thought about the seriousness of your condition, at all?
PATIENT:	No. I thought he said that the chemotherapy would work and that I'd get well . . .

The swelling in the patient's stomach was increasing, and the attending physician decided to begin radiation treatment in order to reduce the nodes that were

---

all that should be done. He then raised a second point, namely, that one of the requirements of the protocol was that the patient not be edematous. Since the patient had severe abdominal swelling, she was not, at this point, eligible for the protocol. It was this technical requirement of the protocol that prevailed upon the consciences of the house staff doctors. At this point, they defined their professional obligations in terms of reducing the patient's edema so that active treatment could begin. Again, from the conflicting perspectives of the doctors involved in the care of this patient, we see how normlessness prevails regarding the ways in which patients should be managed technologically and in the definition of the ultimate goal of the medical management of the terminally ill patient. Yet, we also see the centrality of technology to the divergent and conflicting forms of patient management.

(continued)

Thanatology Snapshot 3.5 (*continued*)

obstructing her kidneys, inhibiting the passing of fluids and causing her swelling. The early stages of radiation treatment brought some modest success. She was losing some fluid weight, but she was feeling increasingly weak and was in intense pain. At this point, as her swelling was being diminished, her doctors looked again toward the protocol. It was discovered, however, that an additional requirement of the protocol was that the drugs could not be administered if radiation treatment had been received within the last thirty days. Her attending physician, recognizing that chemotherapy could not begin for at least a month, told the patient that he planned to finish up her radiation treatment and discharge her at the end of the week.

In recognizing that the hoped for and "promised" beginning of chemotherapy was not going to take place, the patient became increasingly depressed over the seriousness of her condition. A brief encounter with her mother expresses the re-emergence of a sense of despair and its superseding of the hope which had been previously alive and well:

PATIENT: I'm just so tired and feel so terrible.

FAMILY MEMBER: But, Dr. _____ said this wasn't serious.

PATIENT: I know I'm going to get better, but I just feel so lousy. Feeling this bad, I just don't know. I'm scared about this new protocol. That it may not work. I'm just so frightened.

The faith which previously had been placed in the new chemotherapy regimen has now been shaken, and the fear of not getting well became the dominant expressive and behavioral concern of the patient.

The patient was released from the hospital at the week's end, as there was nothing more medically that could be done for her. She was, however, readmitted several days later, having difficulty breathing, visibly weak and in extensive discomfort and pain, and her abdominal swelling had increased again.

She was perceptibly nervous and disturbed by her worsened physical condition. This increased worry over her illness became explicitly expressed as an acute sense of fear:

. . . When will I be going home again? It's just taking so much longer than I thought. I just want to go home so bad. Do you think I'm ever going to feel well again?

I can't even describe the fear. There are no words. . . . It's too frightening for me to be able to describe. I'm just so afraid.

During the first two days of this readmission, it seemed that death was a likely possibility. Specific symptoms and key indicators were not encouraging. Her physician, at this point, was resigned to trying to make her as comfortable as possible and hoping for the best. Her fears during this period grew and her sense of despair heightened:

> I'm afraid of dying; it really scares me. When I told you a month ago that I wasn't scared to die, I guess I was just denying it. Now, I'm so anxious, nervous, depressed. I don't know how much more of this I can take.

On the third day of this admission, her symptoms showed improvement. She had lost some fluid, her white blood count was up, creatinine clearance was improving and her pain was easing a bit. She was generally feeling better, but it needs to be stressed that her cancer remained very widespread* and she was very, very sick. . . .

Her physical symptoms continued to improve over the course of the next several days, and she and her physician hooked onto these particular improvements. Her physician, who never even expected her to live this long, expresses his optimism over these improvements:

DR.: I think you are doing incredibly well. Your lungs and your belly have responded tremendously to the radiation we gave you. Certainly better than anyone ever expected you to.

PATIENT: When am I ever going to get on chemotherapy and go home?

DR.: We can't start chemotherapy until your creatinine clearance improves.

PATIENT: Cre . . . ? Creatinine?

DR.: Creatinine is a waste product that is cleared by the kidneys into the blood. Right now your kidneys are not operating sufficiently to clear creatinine as rapidly as should be. As soon as we correct that, we'll get you on chemotherapy.

PATIENT: And that will help?

DR.: Oh, yes. It'll help you get rid of your lumps here and there (he gently points to the patient's kidneys and shoulder).

PATIENT: These medicines work? You've used them before?

DR.: Yes, they are excellent medicines. No magic. They are just good medicines, and when used in combination, we hope to get beneficial results from them.

PATIENT: When can I get started?

---

*She had obvious and visible growths of tumor on her forehead, around her armpit area and on her vagina, in addition to the internal growth of nodes.

(*continued*)

Thanatology Snapshot 3.5 (*continued*)

DR.: We want to get your belly flat first.

PATIENT: I think it's flat now.

DR.: No, you've got another five pounds of fluid left, plus something over here (he probes her stomach). We'll get that out, monitor the creatinine carefully and then get you started.

PATIENT: So, by next Monday you'll be treating me?

DR.: I hope so.*

PATIENT: And then you can figure I can go home by next Wednesday or so.

DR.: Yes, I hope so.

PATIENT: Okay, Doctor. Thank you. Thank you very much.

DR.: (preparing to leave) You're doing incredibly well. Your lumps in your belly are rapidly diminishing. Your legs are doing well and your belly has been really flattened by the radiotherapy. You are doing spectacularly, again, far better than anyone thought you would.†

At this point, the despair which the patient brought to the hospital and deeply felt early on during her readmission has been replaced by a rekindling and stirring of the fires of hope. The patient, however, died three weeks later, never having been placed on the much-hoped-for chemotherapy regimen. Her attending physician was on vacation at the time. . . .‡

David Wendell Moller,
On Death Without Dignity
(New York: Baywood Press, 1990), pp. 88–93.

*The patient had become so sick that the attending physician, with full support of the house staff physicians, was considering giving her a chemotherapy treatment despite her having been radiated within the past thirty days.

† In terms of the management of specific symptoms, the use of the words "spectacular" and "incredible" may have been technically accurate, but in terms of the more penetrating issue of life and death, they are clearly inappropriate.

‡ It's amazing how abruptly the lingering and the roller-coaster journey of dying can come to a halt.

---

medicalization of the pain experience is useful and effective when dealing with normal types of pain that are responsive to medical management. There is, however, a large incidence of stubbornly resistant, chronic pain that is often unresponsive to medical management and control. It may very well be that successful management of the chronic pain experience requires understanding of and control over many aspects of the dying person's life, such as family dynamics, fear of dying, potential for physical deterioration and debility, level of productivity and activity, support systems, and religious beliefs.[27] The intensity, duration, and depth of a terminal patient's chronic pain are far removed from the range of ordinary-normal experiences of many Americans.[28] As this experience with pain is foreign to most people, and as this type of pain is treated predominantly within a medical framework, the dying person suffering with pain is often left with feelings of social isolation and anxiety. The absence of cultural guidelines to support and direct family members and loved ones through the duration of chronic-dying pain can also result in feelings of helplessness and confusion. Such feelings can contribute to the abandonment of the dying person by friends, acquaintances, and loved ones. Sometimes it is beyond human comprehension and endurance, especially in modern society, for a loved one to watch the dying one's pain—to observe it occurring without mitigation even by sleep, to not know what to do or how to act, and to go on without understanding the purpose of it all. The suffering involved for friends, acquaintances, and family members may prompt a decision to minimize their presence at the bedside of the dying person, thereby increasing his isolation.[29]

In addition to dealing with the possibility of intense pain, the dying patient is confronted with the prospects of loneliness, helplessness, physical deterioration, and associated feelings of blemishment, loss of social identity, and loss of social utility. As Thanatology Snapshot 3.6 illustrates, when we speak of the life situation of the dying patient, we must speak in terms of multiple combinations of predicaments and factors that threaten to dishevel, fragment, splinter, depress, break down, and decimate his psychosocial identity. It is this process of personal, social, and physical devolution that defines the dying patient as a second-class citizen and significantly shapes his suffering.

American culture is not designed to respond openly to the human experience of pain and suffering. It is not difficult to see how alcohol, aspirin, drugs, or even shopping and entertainment are readily used to disengage pain and suffering from everyday social living. On a sociomedical level, pain and suffering are typically transformed into technical matters requiring technical intervention. In this way, the standard medical response to pain and suffering is to treat these human experiences as objective-therapeutic entities and to prescribe drugs, psychotherapy, or other medical-therapeutic maneuvers to nullify human suffering and pain.[30] The expectation of death with dignity becomes a culturally prescribed method of encouraging the dying patient to transcend her pain and suffering in the pursuit of a good death. Suffering is also typically avoided in the roller-coaster journey of the lingering-dying patient by focusing concern on the technical manipulation of specific physical symptoms. The psychosocial sufferings of the dying patient are hence defined either as being

Thanatology Snapshot 3.6

Voices of the Dying Patient

Pain, pain, pain. . . . Sometimes it's so hard to describe the pain. It's a pain like—well, the way I feel when pain gets over me. . . . It's like . . . EATING AWAY AT MY BONES.

[The pain] makes me shaky and nervous. It's an incredible kind of tension. When I get that pain, I sometimes become angry and nasty. I get so desperate that I act in ways I shouldn't act. When I'm going through it, I just get so desperate.

Going on and on for months and months in pain all of the time—that scares me. Deteriorating! Sometimes it gets so bad that I wish I could go just like that [snap of fingers] instead of suffering, and that's what you do with cancer. You just go down and down. Deteriorating, being in pain. Yes, that worries me very much.

The pain is evil. It's destructive, bad and even demonic. The cancer makes me nervous, anxious, obsessive. The pain and suffering is so bad that it must be evil.

Really, there is no meaning to the pain. Pain like this has to come from the devil. At certain times I feel the pain is punishing me. But for what? That I just can't see.

I've had a good life. Now I'm shot though. This tumor—or cancer—or whatever. Well, let's just say I'm three-quarters shot. . . . Oh, to be able to do the things I used to do.

I'm no good to anybody. Why am I living? Why doesn't God just let me die? I feel so useless, and I'm a burden to everyone. This is no way to live. The pain, oh, why? I'm just no good.

I'm useless! I'm of absolutely no use to anyone.

I'm not myself anymore. Oh, the way I used to be. I can't even stand to look in the mirror anymore. [tears begin to stream from her eyes]

I used to feel pretty confident—have a lot of confidence and be pretty outgoing. Now I feel shy and ugly—trying not to be noticed too much. . . . It seems that everything keeps going back to this cancer. It makes me feel so ugly, and it's just so depressing.

If I could just go back to the way I was—used to be—then I'd be ready to die.

Words of terminally ill cancer patients,
as reported by David Wendell Moller,
On Death Without Dignity
(New York: Baywood Press, 1990), pp. 70–81.

irrelevant to medical treatment or as being amenable to therapeutic ameliora-tion. The consequence for the dying patient is that the experience of suffering becomes increasingly privatized as he seeks the internal resources to endure grief, loss, and stigma without the strengthening support of traditions and rituals. In an important way, the individualization and privatization of suffering, along with technical management and control, is a way of containing the expression of human suffering and ensuring that the sufferings of the dying patient, like those expressed in Thanatology Snapshot 3.6, do not unduly blemish the professional domain of medical work or enter into the mainstream of social activity.

The Dying Patient: A Conclusion

As the process of dying has evolved from the social-moral order of ritual, tra-dition, and community to the technical order of medical and therapeutic man-agement, the life circumstances of the dying patient have become complex, variable, often terrifying, and degrading. Leslie Thompson summarizes this situation well:

> The startling medical advances of this century have created unparalleled oppor-tunities for ill and suffering people, but these same forces have combined to make the experience of dying a terrifying, fearful, lonely vigil for many. Devoid of traditional myths, rituals, and family support, many patients now die in sterile institutional settings often appearing as mere appendages to life-supporting machines. This shift from the moral to the technical order manifests itself in doctors' fascination with gadgets, the emphasis upon parts of the body, and a concomitant blurring of distinctions concerning death, personhood and individual rights. The patient is reduced to a secondary role in his or her own death, thus engendering a widespread desire for a sudden death. In light of these circum-stances, health care personnel need to create circumstances that would give people more opportunities to die in styles commensurate with their life styles. Society must seriously study the implications of the unthinking treatment of people; and routines, policies, and procedures based on matters of efficiency and techno-logical convenience must be replaced by those human and humane ceremonies, attitudes and policies that must assure that technology's magnificent achieve-ments do not obscure the human need for individuality and spiritual growth even during the dying process.[31]

As we saw in the preceding chapter, the technical-bureaucratic style of death is the dominant model of death in contemporary society. However, as pointed out earlier in this chapter, a growing cultural resistance to technologized death has emerged during the past two decades and has promoted alternative styles of dying. Largely through the leadership of Kübler-Ross, the death-with-dignity movement has continually voiced the idea that death and dying offers a final and ultimate opportunity for self-growth. Indeed, the dominant alternative model of death that has been institutionalized by the death-with-dignity movement is based on Kübler-Ross' idea that a good death is that of a person who courageously accepts her situation. In this way, the exclusively technologically oriented management of the dying patient in the model of medicalized death is both opposed and complemented by an alternative model of courageous and dignified dying.

The experience of the dying patient is thus framed by the ambivalence and normlessness that flow from the simultaneous existence of the technocratic and dignity models of dying. In this way, dying has become largely pluralistic, characterized by both variable trajectories and differing meanings. There is ambivalence resulting from being torn between the forces of technical death and dignified death, as well as from the turbulence that results from the roller-coaster journey of the lingering dying experience.

In addition to, and partly due to, the pluralism and ambivalence of dying, many of the human and social needs of the modern dying patient are largely unmet by either the medical-technical or therapeutic-dignity models of death management. The voices of dying patients in Thanatology Snapshot 3.6 clearly show how the technological management of the dying patient leaves a whole spectrum of needs unaddressed. It also illustrates that the needs, sufferings, and concerns of dying patients are so deep and intricate that it is unrealistic to expect that they can be met by a quick, simple journey to the stage of dignity and acceptance. Indeed, any serious scholarly evaluation of the human situation of dying patients will undoubtedly point to the simplicity of Kübler-Ross' prescription for dying a good death and demonstrate how the psychosocial needs of dying patients are neglected by their technological management.

It is this situation that leaves the dying patient trapped in a fluctuating web of neglect, unrealistic expectations, and, above all, uncertainty and ambivalence. If a dying patient is to succeed in adjusting to and enduring her course of dying, she will do so largely on the basis of her own personal resources and social circumstances. In this way, "successful dying" becomes the result of individual achievements of the dying patient. Not only, then, is dying variable and individualized, but it also becomes a personal challenge to every dying patient—a challenge, however, that the modern individual is neither adequately trained nor equipped to accept.

4

Funerals as Social Facts

Ceremony is the link between the sacred, society, and the individual. For whatever else it does, ceremony permits man not only to reestablish the purity of the relationship between himself and the sacred but also between himself and his fellow man.

ROBERT FULTON

The American funeral—with its vulgarity, sacrifice of spiritual values to materialistic trappings, immature indulgence in primitive spectacles, unethical business practices, and overwhelming abnegation of rational attitudes—has become for many students of the national scene a symbol of cultural sickness.

RUTH MULVEY HARMER

Obviously, funerals are a fact of daily social life. Less obviously, they are ceremonial emblems of humanity's attempt, on both an individual and a collective basis, to respond to the turmoil generated by the deaths of individuals. Obviously, they provide legally and culturally sanctioned ways of disposing of dead human bodies while reinforcing systems of support for grieving survivors. Less obviously, funerals are an embodiment and a reflection of social life in a given time, place, and culture. In this way, the patterns of living in a given societal context give rise to particular patterns of funeral customs, and the funeral customs of a given era reflect and uphold the styles of life. The purpose of this chapter is to explore the funeral as a social fact, paying special attention to the social uses of funerals and to the emergence of controversy over contemporary funeral practices.

A Historical Overview

In order to understand how the contemporary American funeral reflects the patterns of modern living, it is essential to understand from a comparative historical perspective how present-day funeral practices evolved. Without such an understanding, we cannot fully grasp the idea that the funeral is a social phenomenon that is intimately linked to the values and institutions of the broader social order. It is important to illuminate in detail how changes in funeral customs are precipitated by historical and social changes. Through understanding the traditional structure and meanings of American funerals, we can perceive how these customs have changed over time, and how the practices in contemporary society are logically consistent with modern lifestyles and folkways.

Funerals in colonial America were models of simplicity. The earliest funerals in New England entailed a procession to the gravesite, where mourners stood silently by as the simple pine box was lowered into the grave. For example, during the seventeenth century:

> At burials nothing is read nor any funeral sermon made, but all the neighborhood or a goodly company of them come together by tolling the bell, and carry the dead solemnly to his grave and then stand by him while he is buried. The ministers are most commonly present.[1]

As the seventeenth century came to a close, more extensive social and religious ceremonies became associated with funerals. As Haberstein and Lamers point out, mourning began to take on an extensive social character.[2] Rings, scarves, gloves, books, verses, and needlecraft were all commonly used to pay tribute to the dead. For example, it was not unusual for mourners in attendance to be given mementos of the funeral, such as badges, scarves, and gloves. Indeed, at one funeral of the spouse of a political figure, 1,000 mourning gloves were given away. (A favorite avocation of some people in colonial America was to accumulate the maximum number of mourning gloves for their private collections; the larger the collection, the greater the social prestige.) Thus, despite formal efforts to restrict lavish displays of mourning, as the 1700s progressed, colonial Americans increasingly celebrated death through social rituals. Let us trace this historical evolution from the earliest patterns of simplicity to the proclivity toward greater complexity and lavishness.

The original American funeral was a communal affair. Historically in towns, villages, and communities of agrarian America, the response to death was both direct and personal. It was the custom in colonial America for neighbors to wash the body of the deceased, dress the corpse, and prepare the body for burial. The local cabinet maker built the coffin, and the body might or might not be laid out in state for several days.* Funeral services typically began in the church, followed by a procession on foot to the gravesite. After the burial, neighbors and relatives returned to the house of the deceased, where food, drink, and

---

*If the body was not buried shortly after death and was laid out in state, this was done in the home. It was customary for survivors to keep a constant watch over the body; in shifts, they watched it all day and stayed awake all night with the deceased until the time of burial.

tobacco were consumed with an air of festivity that tempered the pall of gloom hovering over the mourners.[3] Margaret Coffin describes the festivities that took place during the funeral feast for a deceased man in Albany, New York:

> [I]n 1756 Lucas Wyngard died in Albany, was buried, and his funeral feast began. This became an all night celebration and free-for-all. A pipe of wine was drunk, dozens of pounds of tobacco were smoked, grosses of clay pipes were used and smashed, and not a whole glass or decanter remained in the house by morning. The pall bearers finally kindled a fire in the fireplace with the scarves they had been given as tokens of their office.[4]

Joviality, festivity, and plain drunkenness were typical at funerals and postfuneral parties. Indeed, in planning for one's own funeral expenses during the 1700s and 1800s, it was not at all uncommon to provide for generous amounts of liquor, wine, and food for the attendants and guests. The "Irish wake," with its excesses of food, drink, and jollity, is a classic example of the party atmosphere that surrounded death and burial in traditional American society.

Nevertheless, the American funeral was far from frivolous. To the contrary, the death of a member of the community was a serious and solemn occasion. The vigil over the body in the home, the arrival of guests from out of town, the preparation of the body for burial, and the procession to the gravesite were activities carried out with a depth and degree of sadness and solemnity virtually unrecognizable in modern funeral proceedings. The ubiquitous wearing of mourning clothes, both during the funeral and for extended periods of time after the burial of a loved one, was also an important expression of the solemn and ritualistic response that death traditionally evoked in American society.

As early American funerals were moored in simplicity, so was early American undertaking. Funeral undertaking in the sense that we know it today did not exist until the nineteenth century. Prior to then, various tradesmen were involved, first in providing material goods for funerals and later in offering related services as well. The cabinet maker or village carpenter traditionally made coffins individually upon demand. Additionally, general stores sold coffin hardware: heavy screws with heads fashioned like weeping cypress trees; nails shaped like crosses; long steel tacks; and copper, brass, tin, or silver plates that would be inscribed with a person's name or title, such as "Mother" or "Our Daughter."[5]

Throughout the eighteenth century, it was not uncommon to find carpenters/coffin makers expanding their business to include other services related to funerals, such as renting out funeral hearses and carriages or providing personal services at the gravesite. Slowly, almost haphazardly, the undertaker began to emerge as a professional involved solely in the disposal of the dead.

During the nineteenth century, the time was ripe for the professional coordination of death. As people increasingly took up residence in American cities, it became structurally and physically inconvenient for them to prepare a body for burial and to lay out the dead in their own homes or apartments. A second factor encouraging the professionalization of death and the creation of the modern undertaker was the change in embalming techniques. Embalming, as a technical procedure to delay the onset of visible decomposition, originated during the Civil War. Thomas Holmes, the father of modern American embalming, successfully embalmed thousands of Civil War soldiers whose families wanted their remains preserved during the journey back to their home towns for burial. Although embalming was not a widespread practice during the late 1800s, the successful embalming of President Lincoln and the preservation of his body throughout the funeral procession that journeyed by train through several states brought greater public acceptance of the idea of embalming and bodily preservation. Even as recently as the early 1900s, the dead were generally embalmed at home, with the family participating in the process. While the idea of embalming a loved one personally in the home undoubtedly sounds bizarre today, this practice must be considered in light of everyday life in America during the nineteenth and early twentieth centuries. As Pine points out, many families had been regularly immersed in "the blood and guts" of everyday life through regularly slaughtering cows, pigs, and chickens. Additionally, children were raised in a cultural context that included them in death watches and mourning rituals. Thus, as individuals grew into adulthood, they did so with exposure to and familiarity with death. Therefore, the idea of dealing with the blood of a deceased family member, especially someone who had been cared for at home during the duration of her illness and dying, was not likely to be as abhorrent as it would be to the sensibilities of modern persons, who are largely isolated from blood, disease, dying, and death.[6]

However, as people moved to the cities during the nineteenth century, two major social changes affected American embalming practices. First, urban life, with its distance from the folkways of rural, agrarian America, diminished the willingness of family members and neighbors to be directly involved in embalming the body of a deceased loved one. Initially, the result was that an undertaker was called to the family's home to perform the embalming. In the late 1800s and early 1900s, however, technological developments made the process of embalming more complex and sophisticated. These changes also entailed the emergence of specialized, technologically up-to-date facilities to prepare the dead for viewing and burial. Hence, the widespread and growing acceptance of embalming, increasingly sophisticated embalming techniques, and the development of political legislation to regulate embalming practices contributed to the emergence of the American undertaker as a professional specialist in the care and handling of the dead. These changes in the broader society—namely, urbanization and technological progress—ultimately gave direction and shape to the modern funeral.

Although this brief presentation of the traditional American funeral does not even begin to touch on all of its nuances and variations, it does identify several key characteristics of traditional folkways of both life and death: extensive community and family participation, an overall stress on simplicity despite some yearning for extravagance, the recognition and celebration of death as an important social and religious event, and a core of solemnity and seriousness.

A Portrait of the Modern Funeral

In its style, substance, and social significance, the modern funeral is a radical departure from the traditional American funeral. Upon the cessation of life, whether one dies at home or in an institution, death and its causes must be certified by a physician. In the bureaucratic scheme of things, death does not exist unless the requirements of certification have been met. In other words, a family's private physician must sign the death certificate, or death will be certified by the county medical examiner, or some other physician.

After this process has occurred, the role of the medical profession ends (unless, of course, an autopsy is to be performed) and that of the undertaker begins. A call is made by the family or the hospital to a funeral home of the family's choice. Regulations and prevailing folkways generally prohibit the release of a body from a nursing home or hospital to anyone except a mortician or mortuary. From the very onset, we can see that the contemporary funeral is radically different from the traditional funeral. Strangers who have never known the deceased or his family will enter the hospital room, morgue, or family's home and will gather the dead body in their arms, put it in a body bag, place it on a cot, and remove the body to the funeral home. The body will then wait in a back room, basement, closet, garage, or other storage area until the family has made the necessary arrangements for the funeral with the funeral director.

Making arrangements for the funeral includes making a decision as to whether embalming is desired (although many funeral directors just assume that the family desires it) and about the number of days and the specific visiting hours that the body will lie in the "reposing room." Some funeral directors offer a package of services with standardized hours and days for visitation. Others generally do not set limits but charge their customers by the day. The family will be asked to decide on a place for the funeral service, which can be held in the funeral home or in a local church, and they will be asked to decide pallbearers, who can be friends, relatives, or strangers hired by the funeral home.

A hearse is usually rented to transport the body to the church for the funeral service and to the cemetery, and flower cars and limousines are available. The family will also be asked about the clothes that they wish their loved one to be buried in, notices for newspaper obituaries, choice of cemetery, and ownership of a plot.

A major part of the process entails choosing a casket and vault for burial. The "selection room" (as it is often fondly referred to) is generally separated from common access by closed doors or stairs. In this room, a display of caskets —ranging from cloth-covered pressed wood (the cost of which is in the low hundreds of dollars), to various types of colored metal caskets (from the high hundreds to low thousands of dollars), to elegantly finished hardwood caskets (from the middle to high thousands on a $10,000 scale), to perhaps a solid cast bronze sarcophagus—offers a variety of choice. At this point, the funeral director clearly adopts the role of a salesperson. The director's goal is to sell the customer an above-average-priced casket. To assist the funeral director in this enterprise, some features of the better caskets may be tastefully outlined by advertising literature placed inside the casket: "Guaranteed airtight for fifty years" or "Buy this casket and a tree will be planted in a national forest in memory of your loved one." In addition, caskets in the selection room are typically arranged according to a merchandising strategy designed to accentuate and promote the sale of more expensive caskets.

The role of the undertaker began in American society as a provider of merchandise for death, and in this regard the contemporary image of the undertaker as a salesperson and merchant of the artifacts of death is consistent with the historical viewpoint. Except for changes in styles of caskets, techniques of restoration and embalming, and the style of homes for funerals themselves, there is little that is new and unique to the activities of the present-day American undertaker.

After the casket and/or vault have been selected, a contract for goods and services is signed and arrangements for payment are made. The funeral director then attends to many of the backstage details of putting on a funeral: notifying newspapers for obituaries, securing the burial permits and arranging for the grave opening, organizing the religious ceremony and contacting relevant clergy, and preparing the dead body for viewing. This last process includes embalming, washing and cosmeticizing the corpse, setting hair, trimming nails, restoring visible bodily structures that may have been decimated by accident or chronic illness, dressing the body, placing it in the casket, and arranging it in the reposing room among the flowers and specially designed lights.

After negotiation of the contract and preparation of the body for the wake, the stage is set for visitation. During specified hours, such as 2:00 to 5:00 in the afternoon and 7:00 to 9:00 in the evening, the family of the deceased gath-

ETERNAL ACCOMMODATIONS

The casket showroom, which most people visit infrequently, is the business heart of the funeral service profession. Reflecting the American values of individualism and materialism, a wide range of caskets are typically available to the modern consumer.

ers in the reposing room of the funeral home to receive guests who come to pay their last respects and to offer their condolences. The people who come will range from close relatives and friends to distant relatives, acquaintances, and the deceased's friends and work colleagues, some of whom may be complete strangers to the grieving immediate family. This period of visitation generally lasts for one to three days, with the present tendency being toward the briefer duration of one day.

On the morning of burial, close friends and relatives typically convene at the funeral parlor for a final viewing of the deceased. The closing of the casket and the procession to the church and then on to the gravesite for a brief ceremony are all standard features of the contemporary funeral. It is also common for the burial to be followed by lunch at a restaurant (some major cities have restaurants located adjacent to cemeteries that do a very large funeral-related business) or by a reception with food and drink at the home of the deceased.

In a period of two to four days, the funeral process has usually been completed, and formal rituals of grief and bereavement have come to a close. After the ending of formal rituals, the family and close friends are thus left to their own resources and to personal adaptive strategies in coping with the death of a loved one (as we shall understand more fully in the next chapter).

Apart from the fact that the role of the undertaker in many ways has changed very little, the present-day American funeral is a dramatically different affair from the traditional funeral. Community involvement and participation have been replaced by the presence of a diverse group of people, many of whom have little in common with each other. Community involvement has also been replaced by strangers and professionals who have become responsible for the care of the dead and direction of the funeral proceedings. Simplicity has been superseded by the moderate lavishness of the modern funeral parlor. The importance of death itself as an event to be recognized and celebrated has decreased, and attendance at the funeral has assumed secular meanings associated with a transient sense of social obligation. Finally, the solemnity of the traditional funeral has disappeared. Indeed, the contemporary funeral is not a very solemn occasion and often assumes the character of an ordinary social gathering unrelated to death. These changes are consistent with those in the broader American culture during the twentieth century, as stable patterns of community life have given way to specialized professionalism and individual self-expressiveness, and as religious meanings have been severely challenged by secular definitions of reality. Again, I hasten to comment that in reflecting the folkways of American life, the prevailing patterns of the American funeral affirm the theme that the way one dies is a reflection of the way one lives. Let us now turn our attention to the various meanings that have been established for the funeral in American society.

The Funeral as a Social Rite

In his classic study of religion as a social fact, Emile Durkheim points out that religion is composed of two essential ingredients: belief and ritual. His definition of religion indicates how these are integral to the very nature of religion: "A religion is a unified system of beliefs and practices relative to sacred things, that is to say, things set apart and forbidden—beliefs and practices which unite into one single moral community called a church, all who adhere to them."[7] Moreover, beliefs and rituals are not only *both* essential, in an important sense, they are *equally* essential.

The special nature of religious activity, as opposed to everyday behavior, is that it takes place in a context relative to things that are sacred. Thus, drinking wine has an ordinary meaning at a dinner party but assumes religious extraordinary significance at a Catholic Mass. Religious ritual, then, may be defined as the way one is supposed to behave in specific circumstances while in the presence of sacred things. In this way, ritualistic behavior is predetermined and becomes a habitual form of conduct. For example, if one walks into a Catholic church when the Host is present, one automatically genuflects in a predetermined manner. In this way, genuflection is an expected and standardized form of behavior upon entering the sacred building that is a church.

In his study of religion, Durkheim describes many of the beliefs and practices of primitive religions, but more important to his analysis are the ways in which the rituals related to religious beliefs are sociologically useful. The benefits that religion generates for the social fabric of society are many, but three are especially salient. First, as Durkheim observes, rituals provide participants with a basis of common identification. As individuals come together to participate in ritual activity, they become collectively identified in terms of group affiliation. Rituals become, therefore, an active means of attaching an individual to a group and strengthening the group's idea of itself.[8] In a curiously similar way, it would seem that participation in collective cheers and gestures at major sporting events provides the individuals involved with a fleeting, but nevertheless real, sense of belonging to a collectivity. Religious ritual, whether participation in Sunday Mass, a Jewish celebration of Purim, or a primitive snake dance, identifies the participants as being part of a collectivity and engenders a sense of belonging.

Second, rituals function to facilitate a sense of well-being. Anyone who has taken part in a community ritual of prayer and song and has felt strengthened—perhaps even gained a sense of elation from doing so—knows precisely what Durkheim is referring to when he says:

> It has its share in the feeling of comfort which the worshiper draws from the rite performed. . . . After we have acquitted ourselves of our ritual duties we enter into the profane (ordinary, everyday) life with increased courage and order, not only because we come into relations with a superior source of energy, but also because our forces have been reinvigorated by living, for a few moments, in a life that is less strained, and freer and easier. Hence religion acquires a charm which is not among the slightest of its attractions.[9]

The individual is able to transcend, at least temporarily, his or her fears, vulnerabilities, and anxieties. This state is achieved by becoming part of something greater and more vital than oneself, and by absorbing the energy and vitality that are generated by the group rituals of worship.

A third important function of rituals is that they reaffirm the legitimacy of the broader social order. Rituals, by definition, are connected to the past, and to its traditions and authority. By drawing on the past, the present is given a sense of legitimacy and viability through its connection to authoritative symbols, practices, and meanings. In this way, not only is ritual a moral force related to the beliefs of a particular religion, it is a social force related to the everyday folkways and practices of social living. Accordingly, religious ritual serves to strengthen the social nature of human beings[10] and gives individuals a greater sense of confidence in being a member of their society.

Durkheim's model of the social functions of religious ritual is relevant to understanding responses to the deaths of community members and loved ones. In fact, Durkheim makes the connection himself. In defining them as the ritual care given to the corpse (the way in which it is buried, and so forth),[11] he indicates that funeral rites are an important means of reaffirming the group that has been shaken by the death of one of its members. Rituals provide comfort to those most deeply affected by death, and they are a means of reassuring all involved that the legitimacy of society remains intact.

The death of someone in a community deprives it of a member. In addition, each member of the group is reminded of his or her own mortality. A natural reaction of those personally and socially touched by death is to assemble together and find comfort in their collective strength. As Durkheim observes:

> When someone dies, the family group to which he belongs feels itself lessened, and to react against this loss, it assembles. A common misfortune has the same effects as the approach of a happy event: Collective sentiments are renewed which then lead new men to seek one another and to assemble together. . . . Not only do relatives, who are affected the most directly, bring their own personal sorrow to the assembly, but the society exercises a moral pressure over its members to put their sentiments in harmony with the situation. To allow them to remain indifferent to the blow which has fallen upon it and diminished it would be equivalent to proclaiming that it does not hold the place in their hearts which is due it; it would be denying itself. A family which allows one of its members to die without being wept for shows by that very fact that it lacks moral unity and cohesion: it abdicates; it renounces its existence.[12]

Thus, funeral rituals enable the group to affirm its own existence and offer solace to each member, especially to those most deeply affected by a death. The absence of funeral rites would leave the group diminished by death, and its individual members would feel confused and vulnerable. These feelings could jeopardize the order and stability of the broader society, as well as the individuals' ability to perform therein.

Of course, the model of social rituals that Durkheim formulates is derived from his observations and analysis of primitive religions. The rituals described by Durkheim therefore took place in social settings characterized by strong moral and communal bonds. Yet in contemporary funerals, individualism and fragmentation are more apparent than a sense of community. For this reason, the model of rituals that Durkheim provides must be revised before it becomes fully applicable to the study of modern funerals.

Many of the societal forces of community, tradition, and stability that gave the Tame Death its shape are no longer present in contemporary society. Indeed, instead of being committed to predetermined beliefs and ritual practices, modern Americans are dedicated to the idea of self-expressiveness. In many ways, individualism has become so dominant in American society that relationships—from acquaintanceship, to friendship, to love, to marriage—are unstable and often temporary. And, as Americans become detached from a pattern of enduring relationships, role expectations, and community norms, individual self increasingly becomes the prevailing moral guide to what is right and wrong. Robert Bellah and his colleagues effectively make this point when they note that the meaning of life for most Americans is to develop their own sense of self—almost to give birth and rebirth to new selves continually. This process involves loosening the ties that connect us to family, community, and inherited ideas. In social situations, as well as in the course of planning and living one's life, the individual self must be its own source of moral guidance, and each individual must always know what he or she desires and feels.[13] In fact, for individuals who are ambivalent or inadequate in their self-perception, an entourage of therapeutic and self-help services has emerged during the past two decades to assist them in strengthening their sense of self.

Randall Collins pushes the point even further when he observes that the individual self, and the way in which it is brought into interaction with other individuals, becomes a modern form of a sacred ritual.[14] In both a metaphoric and a real way, individualism has become a modern type of religion. The styles, mannerisms, and uniqueness that individuals bring to their role behavior and social interactions become ritualized in that self-expression in social situations is dependent upon the cooperation of others in creating the opportunity for that expression. This cooperation of others may be rooted in some form of manipulation, acceptance, tolerance, social protocol, interpersonal power, or dominance; but regardless of the reason, the result is an unfolding of a web of social interaction that might reflect the temporary or ongoing expression of harmony, conflict, or manipulation. Collins effectively puts the rise of individualism as a social ritual into perspective:

> Small, isolated, homogeneous groups put a very strong pressure upon the individual, and that is what generates the feelings expressed in religious beliefs about the omnipresence of supernatural spirits. For those individuals in modern society whose social experiences consist of a great variety of different encounters in the large-scale networks of acquaintanceship, the rituals of interaction take quite

a different form. They remain rituals, nevertheless, and produce a distinctively modern type of "secular religion," the cult of individualism. . . .

As Goffman points out, one is not only allowed to be an individual, one is actually required to be so. . . . And the same social conditions also produce the expectation that we should be self-aware, ironic, detached, and all the rest of the modern style of presenting the self. The modern ideal of the casual, "cool," self-possessed individual is not a reaction against society; it is the very form that social ideals are molded in today.[15]

Thus, people come to a social setting both expecting and required to express themselves as individuals, while simultaneously providing the opportunity for others to do the same. In this way, individualism and its manifestations have become a sacrosanct way of life in American society.

Individualism and the ritual social interactions that stem from it function like traditional rituals of religious belief. The modern rituals of the cult of individualism unite people in a common setting whereby individuals participate in a reciprocal process of self-expression. A sense of well-being is generated by this process in that an opportunity is created to promote, represent, and assert one's self. And these rituals obviously reaffirm the legitimacy of the idea of individualism as a central value of the broader society.

The modern American funeral, as a ceremonial way of responding to death, fulfills some of the traditional uses of ritual that Durkheim associates with religious belief. More important, however, it provides a setting for the ritual of individual self-expression as an adaptive coping strategy in the immediate aftermath of death.

One of the most important functions of the contemporary funeral, as generally defined in the literature, is that it provides for an organized period of grieving. Some have suggested that the traditional funeral, with an open casket for viewing, drives home the irrevocable fact that death has occurred. The idea that "seeing is believing" is relevant here, as it is argued that the physical presence of a corpse provides substance to death that would otherwise be lacking. Viewing the corpse also allows the restorative art of the undertaker to recapture an appropriate "recall image" of the dead person, an image that may have been obliterated through an accident or through the ravages of chronic illness. In addition, the presence of a visible dead body facilitates and focuses attention on the fact of death and allows for greater expressions of sympathy and grief to take place.[16]

Empirical studies on the value of the funeral as a facilitator of successful "grief work" are contradictory in their findings. In a study of 565 widows and widowers, Fulton found that those who participated in a traditional funeral ceremony reported a more positive memory of the deceased, closer ties with relatives, and fewer adjustment problems than those who did not view the deceased's body.[17] A study by Hutchens reports that a significant majority of respondents felt that a traditional funeral ceremony, with open-casket viewing, was useful in their attempt to come to terms with the reality of death.[18] On the other hand, a study by Schwab and his colleagues found that "viewing the body and crying were not associated with the resolution of grief but were reported more frequently by those with unresolved grief."[19] And in a study by Glick, Weiss, and Parkes, 52 percent of the participants felt that viewing the body had a negative effect.[20] Other studies have suggested that the traditional funeral ceremony, with open-casket viewing, was valued by those people who defined themselves as being *very close* to the deceased but elicited a negative response from those who defined themselves as being just *close* to the deceased.[21] The very close group saw the presence of flowers as supportive and functional, whereas the close group felt that the money could have been spent more wisely. These findings may be interpreted as meaning that the traditional funeral ceremony, with viewing of the deceased, is helpful for people who suffered from extreme disorientation as a result of their closeness to the dead person. People less close, who experienced less disorientation, may find viewing the body unhelpful and unpleasant.

Thanatology Snapshot 4.1

An Image of a Dead Man

The dead man lay, as dead men always lie, in a specially heavy way, his rigid limbs sunk in the soft cushions of the coffin, with the head forever bowed on the pillow. His yellow waxen brow with bald patches over his sunken temples was thrust up in the way peculiar to the dead, the protruding nose seeming to press on the upper lip. He was much changed and had grown even thinner since Peter Ivanovich had last seen him, but, as is always the case with the dead, his face was handsomer and above all more dignified than when he was alive. The expression on the face said that what was necessary had been accomplished and accomplished rightly. Besides this there was in that expression a reproach and a warning to the living. This warning seemed to Peter Ivanovich out of place, or at least not applicable to him. He felt a certain discomfort and so he hurriedly crossed himself once more and turned and went out of the door—too hurriedly and too regardless of propriety, as he himself was aware.

Leo Tolstoy,
The Death of Ivan Ilych and Other Stories
(New York: New American Library, 1960), pp. 98–99.

It is therefore difficult to state with any degree of certainty that the traditional funeral ceremony facilitates grief work and initiates the process of resolving grief. In terms of Durkheim's model, however, it can be noted that the funeral ceremony brings together a diverse group of people and unifies them in a common, death-related setting. Regardless of the ultimate therapeutic benefits or nonbenefits of the traditional ceremony, individuals (some close to each other, others not so close or perhaps even strangers) are involved in a social setting that explicitly and collectively reminds them of the finitude of human life. Moreover, the value of the traditional funeral, with open-casket viewing, may be largely variable in different ethnic and social groups and from individual to individual. During the social interactions of the funeral, individuals may impute their own meanings and values to the open casket and other traditional elements. As Richard Kalish simply but effectively observes:

> The first time I attended a funeral of someone I loved, I was disturbed by the cosmetic job that presumed to make him look "just like he did when he was alive." Not only could the embalmer not overcome the ravages of cancer, but I resented the waxy appearance. Some 30 years later I attended a funeral with a closed casket . . . and I found myself resenting the absence of the person within the casket; I was angry because I felt that the empty box did not represent the person I loved. Of course, the box was not empty, but it felt empty to me. . . . Thus in one instance I was upset because the body was on view, and in the other I was upset because it wasn't.[22]

Kalish proceeds to argue that the complexity of modern society and the place of death in the social order make it difficult to assign a collectively shared, singular set of functions to the ceremony of the funeral. He argues that "we need to consider our own value systems rather than pretend to any objective information." The issue that Kalish raises is an important one, namely, that in modern society the funeral is no longer a ritual secured in stable patterns of tradition and community; rather, it has evolved into a forum for individual expression and individually derived meanings. This interpretation is explicitly consistent with Collins' description and analysis of individualism and self-expression as modern forms of ritual activity in everyday American life.

A second function of the funeral has already been suggested, namely, that funerals furnish social support by facilitating the gathering of relatives and friends. Although the contemporary American funeral cannot recapture the dwindling spirit of community in America, it does offer some formal framework of support and a semblance of group identification, even if in a fragmented and temporary way. The funeral provides for a standardized response to death. In this way, in an age of individualism and normlessness, some structurally rooted guidelines (such as attending the funeral parlor during visitation; going to the funeral and burial service; or, as in the Jewish faith, visiting the grieving at home during Shiva) provide for the presence and support of people in the death setting. While the group identity and cohesiveness created by the modern funeral are a far cry from the solidarity that Durkheim describes, the funeral does provide a formalized opportunity to connect with others while responding to the death of a loved one. At the very least, then, grieving individuals are assured that they do not have to face the immediate aftermath of death in isolation from others.

Thus, whereas the patterns of traditional funerals were truly a community affair, modern funerals assemble a network of grievers. A community is a fixed and visible collectivity in which each unit is part of the whole; a network is an independent collection of people who are united not by their attachment to each other, but by their attachment to a common source. In a community, people are bonded together by mutual concerns and shared patterns of living. In a network, individuals are separated from each other, although they are tied independently to one common denominator. A community is a small, stable ethnic village or, perhaps, Trappist monks living together in a monastery. A network is a personal telephone directory: One knows the people in it but, generally speaking, they often do not know each other.

In keeping with our interpretation of Durkheim's model, it is important to recognize that while the funeral does not represent a stable and regular pattern of community ritual, it does provide an opportunity for individual members of the grieving network to express themselves in a collectively organized setting. It also enables the individual to derive personally defined meanings from the funeral scenario. By assembling a diverse group of individuals into a grieving network, the funeral is acting as a modern form of ritualistic individualism.

A third function of the funeral is to provide a base of legitimacy for the broader social order. Thanatologists have discussed funerals of national leaders in terms of their value in reassuring a large and varied grieving population of the continuing stability of the social order. For example, the televised funeral of John Fitzgerald Kennedy was not only a public declaration that the president was dead, but it also gave people all across America the chance to express their personal grief. (These many strangers and fragmented individuals who had so little in common except the tragic death of their president is a vivid illustration of a network, as opposed to a community, of grievers.) Even more important, the funeral served as an opportunity to reassure a shocked nation that social order existed and would persist.

Funerals also provide legitimacy to the social order in that the gathering of the grieving network itself becomes a symbolic manifestation that life goes on. In addition, and perhaps more significantly, the style of the modern funeral reaffirms the authority of the American economic system by its unmistakable connection to capitalism. One of the most obvious and visible things about a funeral is the display of commodities. Funeral homes are typically appointed with luxurious carpeting and furniture. The casket itself is a salient symbol of materialistic display, as are the flower arrangements and limousines in the funeral procession. In a remarkably similar way to weddings, funerals have become an economic statement of social status. And of course, much of the social conversation that occurs furnishes an interactive occasion, not dissimilar to that provided at a cocktail party, to give personalized accounts of how one is faring economically (through the wearing of fine clothes and jewels, talk of significant purchases such as a house or boat, and so forth). In this way, the funeral brings ordinary, familiar, and comfortable patterns of interaction into a setting that is most extraordinary and uncomfortable.

In fact, the type of funeral purchased for a loved one is an important part of the secularized and individualized bereavement ritual. Financing an extravagant display of materialism during the funeral is defined as a culturally approved way of expressing one's sentiments toward the deceased. The ritualistic promotion of the self is also served in that the amount of money spent reflects one's status, prestige, and power in American society. In our advanced capitalistic society, it is understandable that people will associate respect and love for the deceased with spending money. Additionally, individuals will use the occasion to spend money to further their own self-respect and reputation in the surrounding community. In his investigation of this issue, Pine found a positive association between the socioeconomic status of individuals and their desire to spend money on a funeral. According to Pine, there is a sharp increase in funeral expenditures from low-status to working-status groups and a further increase among those who have achieved middle socioeconomic status, at which point there is a tendency for spending to level off. Pine interprets these findings to mean that funeral expenditures have become a ritualistic way to express grief over the death of a loved one:

> The act of buying, receiving and paying for funerals represents a secular and economic ritual of payment formerly performed by more religious customs and ceremonies. . . . Our view is that because people increasingly lack both the ceremonial and social mechanisms and arrangements that once existed to help them cope with death, monetary expenditures have taken on added importance as a means for allowing the bereaved to express (both to themselves and others) their sentiments for the deceased. For with so few modes of expression remaining to the bereaved, funeral expenditures serve as evidence of their concern for both the dead and the conventional standards of decency in their community of residence.[23]

Thus, not only is the American economic system reaffirmed by the funeral, the capitalistic system of funerals augments the modern social ritual of self-expressiveness.

A final salient function of the funeral is that it provides an opportunity for self-promotion and assertion in the face of death. The funeral, especially with open-casket viewing, brings people face to face with the physical realities of death. As you will recall from Tolstoy's great novella (see Thanatology Snapshot 4.1) Peter Ivanovich found himself ill at ease in the presence of the dead Ivan Ilych, but Ivanovich's expression of vulnerability is only one of myriad possible reactions. Responses to the reminder of mortality impressed on us by the funeral are diverse and individualized.

In his study of self-expression and funerals, Rodabaugh discusses the various types of behaviors found at church funeral services and burial services. He examines three types of participants: the Professionals (those ministers who generally direct the ceremony), the Family, and Friends (who are the audience to which the director plays). Rodabaugh describes nine role behaviors of ministers in their carrying out of their professional obligations: the Young Seminarian, the Master Performer, the Political Gladhander, the Eternal Evangelist, the Scripture Quoter, the Harried Professional, the Ebullient Eulogist, the Pessimistic Griever, and the Comforting Shepherd. Among family role behaviors are the Coping Griever, the Bewildered Novice, the Hysterical Performer, the Stoic Spartan, the Party Queen, and the Cosmopolitan Protester. Role behaviors among friends include the Respect Payer, the Party Goer, the Status Accountant, the Family Supporter, and the Professional Griever.[24] Rodabaugh makes the point that these roles are fluid and flexible. According to the situation and one's perception of the behaviors being expressed by others, it is possible for individuals to move between and among various roles. For example, a minister who may act as a Harried Professional during an especially busy week, may assume the character of the Pessimistic Griever during the funeral of a small child, and may become a Comforting Shepherd when a long-time member of the church congregation dies. In addition, some role behaviors complement the role behaviors of other participants, while others clash. For example, the Comforting Shepherd obviously works in combination with the Respect Payer and the Coping Griever, whereas the Political Gladhander is a natural for complementary association with the Party Queen and the Party Goer. As Rodabaugh points out, the Master Performer and the Hysterical Performer compete for attention; however, if the minister moves to the role of Pessimistic Griever, the roles can then mesh.[25]

Regardless of the role assumed, it is chosen by individuals on the basis of what they perceive to be personally valuable. Funerals enable people to express themselves through death-related role performances in a way that elicits positive responses from others and thereby affirms the identity of the individual self. In this way, the absence of rigid or absolute norms to guide behavior at the funeral proceedings allows each individual the opportunity to pursue personal wants, impulses, and meanings. Despite the absence of predetermined, traditional ritual forms of behavior, the modern funeral affords an opportunity for individuals to gather in a death-related setting and experience ritualistic individualism and self-expression. Thus, not only is the broader cultural value of individualism affirmed but, on a more personal level, the individual self is able to triumph over and perhaps even shine in the face of death.

Hence, in order to comprehend the role of the funeral in society, it is important to have an understanding of rituals and their social functions. Yet it is even more essential to understand how the decline in traditional ritual practices is associated with the emergence of the rituals of individualism and self-expression. The modern American funeral offers a semblance of the traditional functions of ritual behavior, but in a way that is more superficial and much more temporary than traditional practices. The modern funeral also fulfills the function of affirming the American commitment to individualism and of providing a social situation whereby the individual self can be asserted in the physical presence of death. It is in this way, in an updated and restyled form, that the American funeral today generates a societal mechanism for "taming" the harsh realities associated with the death of a loved one.

Criticisms of the Funeral

In recent decades, some rather forceful criticisms have been leveled against the funeral and the funeral industry. During the early 1960s, broad societal unease with death, coupled with a growing reaction against excessive materialism, superficial customs, and a generalized critique of capitalistic institutions, made the funeral industry a target of attack and disapproval. Specifically, during this period of general cultural revolt, three primary criticisms of the American funeral emerged. They have to do with cost, superficiality, and the increasing control of the funeral director.

In his analysis of the funeral as a social ritual, Bowman argues that community has been so dissipated by urbanization and individualism that the American funeral can no longer realistically provide social support.[26] In stable rural communities, the funeral and its rituals performed a deeply meaningful and viable function of social support. But as individualism has become increasingly prevalent and as ties to community and to others have become more transient, funeral rituals have lost their viability as supportive mechanisms. Irion asserts that the social processes of urbanization and industrialization have created an impersonal way of life in which meaningful relationships are reduced to the point where loss is minimized.[27] Accordingly, death, unless it is the death of a very close intimate, has merely a superficial impact. Thus, modern funerals have only temporary and minor significance because modern living and dying have become more individualized and impersonalized. Irion explains how hollow the funeral has become:

> Take away the shallow sentiment, the artificial solemnity, the ethereal music, and the emotional expressions vanish with them. The presence of some superficial feelings, successfully limited to a brief period and a specific place, has satisfied the expectation that something should be felt and expressed at such a time. But a truly profound encounter with the meaning of death for life is obviated. The funeral thus may not be permitted to carry out one of its very important functions of enabling the individual honestly to confront his loss as reality while supported by the concern of a community which in part shares his loss.[28]

It has already been noted that contemporary American funerals are not very sad occasions. It has also been argued that the funeral has become an empty, shallow, and increasingly worthless ritual. And it is this superficiality of the funeral, especially of the viewing activities, that has been heavily criticized for creating an image that avoids the reality of death. As Mitford points out, the activities of the embalmer–funeral director are often geared toward generating an illusion of life.[29] And as Irion indicates, the artifacts of death are marketed in a way that gives credence to the idea of the continued physical existence of the deceased.[30] For example, the idea of the deceased comfortably at rest in the casket—luxuriously lined with fine fabric and appointed with a soft, opulent innerspring mattress—is a consoling one. The sense that the loved one is protected from the ravages of weather by means of a vault is also comforting. Both of these images are used as ordinary selling points in the funeral industry and may serve to convey an illusion or fuel a fantasy that physical life continues in and beyond the grave. Minimally, the creation of a comforting memory image establishes a view of death that spares the living from having to confront the harsh and macabre realities of physical decomposition. Thus, while the critics of the funeral may be correct in noting that an aura of unreality surrounds modern practices, it is also useful to note that this delusory process may be helpful in "taming" the emotional impact of the death of a loved one.

Another aspect of the contemporary critique of the American funeral has to do with the increasing dominance of the funeral director. As the process of preparing a body for viewing and burial has become specialized and technically intricate, the funeral director has assumed a more dominant role in the American funeral. In addition to monopolizing embalming practices, funeral directors formally and legally control the entire process of disposing of the dead. And the more funeral directors can ensure that societal responses to death will remain under their jurisdiction and control, the more economic benefit will accrue for the profession and the more family participation in the funeral process will be minimized.

Bowman sees two major consequences of the increasing professional dominance of the funeral director. First is an increased focus on, and centralization of, activity around the body of the deceased. It is the mortician's training that enables him to do what untrained people cannot, namely, embalming, and restorative and cosmetological work on the dead person. In a simple sense, this is the carrot that the undertaker has to offer his clientele. It is logical, therefore, to expect that the funeral ceremony will emphasize that which is unique to the director's training, that is to say, the preparation and presentation of the dead body.[31] Bowman, Irion, Harmer, and others have commented that emphasis on the image of the dead body has aided in the devolution of the funeral from a spiritually centered ritual to a materialistic, self-indulgent ceremony. As Bowman simply states: There is in the contemporary American funeral an exaggeration of the importance of the physical remains in comparison with the spiritual aspects of the funeral.[32]

A second and directly related consequence of the growing professional dominance of the funeral director is the secularization of the funeral. Traditionally, the funeral had significant religious value, made evident a relationship to God, and was largely shaped by American clergy. As society has moved toward secularization, and as the funeral industry has extended its control over the funeral process, the role of the clergy in shaping the funeral and assisting the family has been reduced. A corresponding diminishment of the funeral as a source of comfort and support, and as a ritualistic affirmation of the meaning of life and death, has also occurred. A less important, but still unfortunate, ramification has been friction between the clergy and funeral directors.[33] This conflict is rooted in the struggle for control of the process, as well as in the inevitable clash between the spiritual values of the clergy and the secular concerns of the funeral directors.

Perhaps the most loudly voiced criticism of the funeral has to do with the high cost of death. During recent years the cost of the American funeral, as well as the economic practices of the funeral industry, have been vehemently critiqued. As Harmer states, the cost of death and burial has become one of the most crushing expenses facing American families.[34] Indeed, it is not difficult to see how the funeral of a loved one can be the third largest expenditure that individuals or families will make during the course of their lives. In 1935 the cost of an average adult funeral was about $350. In 1960 it had risen to $1,100,[35] and by 1980 it had reached about $2,200.[36] These figures do not include the expense of cemetery property; grave openings; outer vaults; purchase and engraving of a headstone; fees for clergy, musicians, or singers; flowers; special transportation needs; obituary notices; memorial cards and copies of death certificates; or police escorts. In view of the range of services and items that are available, expenditures of $5,000 or more are not atypical. (Since most estimates of average costs are compiled by the industry, the figures cited here

Thanatology Snapshot 4.2

A Look Behind the Professional Curtain

[Mr. Jones] is whisked off to a funeral parlor and is in short order sprayed, sliced, pierced, pickled, trussed, trimmed, creamed, waxed, painted, rouged, and neatly dressed—transformed from a common corpse into a Beautiful Memory Picture. . . .

. . . [T]he blood is drained out through the veins and replaced by embalming fluid pumped in through the arteries. . . . About three to six gallons of a dyed and perfumed solution of formaldehyde, glycerin, borax, phenol, alcohol, and water is soon circulating through Mr. Jones, whose mouth has been sewn together with a "needle directed upward between the upper lip and gum and brought out through the left nostril," with the corners raised slightly "for a more pleasant expression." If he should be buck toothed, his teeth are cleaned with Bon Ami and colored with colorless nail polish. His eyes, meanwhile, are closed with flesh tinted caps and eye cement.

The next step is to have at Mr. Jones with a thing called a trocar. This is a long hollow needle attached to a tube. It is jabbed into the abdomen, poked around the entrails and chest cavity, the contents of which are pumped out and replaced with "cavity fluid." This done and the hole in the abdomen sewn up, Mr. Jones' face is heavily creamed (to protect the skin from burns which may be caused by leakage of the chemicals), and he is covered with a white sheet and left unmolested for a while. . . .

The patching and filling completed, Mr. Jones is now shaved, washed and dressed. Cream based cosmetic, available in pink, flesh, suntan, brunette, and blond, is applied to his hands and face, his hair is shampooed and combed (and, in the case of Mrs. Jones, set), his hands manicured. . . .

Jones is now ready for casketing (this is the present participle of the verb "to casket"). In this operation his shoulder should be depressed slightly "to turn the body a bit to the right and soften the appearance of lying flat on the back." Positioning the hands is a matter of importance, and special rubber positioning blocks may be used. The hands should be cupped slightly for a more lifelike, relaxed appearance. Proper placement of the body requires a delicate sense of balance. It should lie as high as possible in the casket, yet not so high that the lid, when lowered, will hit the nose. On the other hand, we are cautioned, placing the body too low "creates the impression that the body is in a box."

Jones is next wheeled into the appointed slumber room, where a few last touches may be added—his favorite pipe placed in his hand or, if he was a great reader, a book propped into position. (In the case of little Master Jones a Teddy bear may be clutched.) Here he will hold open house for a few days, visiting hours 10 A.M. to 9 P.M. . . .

Jessica Mitford,
The American Way of Death
(New York: Crest Books, 1964), pp. 54–60.

should be viewed tentatively, if not speculatively. One consumer research group has found that funeral costs typically range from $3,700 to $6,800 or more.) It should also be remembered that these costs are often incurred in the aftermath of exorbitant medical costs. Thus, major expenditure is required not only at a moment of great emotional shock, but also often at a time of financial dislocation. Hence, the charge has not only been that funeral costs are exorbitant, but also that funeral directors are in a position to take advantage of many American families who have accepted the idea that their final duty to and love for the deceased are expressed by the purchase of a "fine funeral."

Not only has the funeral industry been criticized for skyrocketing costs, funeral directors have been accused of unscrupulous and exploitive business practices. First among the charges is that, through an extensive and rigorous public relations campaign, the funeral industry has successfully sold the American public on the idea that the dignity of a funeral is dependent on spending money and displaying commodities. Jessica Mitford cites a relevant passage from the *National Funeral Service Journal*:

> A funeral is not an occasion for a display of cheapness. It is, in fact, an opportunity for the display of a status symbol, which by bolstering family pride does much to assuage grief. A funeral is also an occasion when feelings of guilt and remorse are satisfied to a large extent by the purchase of a fine funeral. It seems highly probable that the most satisfactory funeral service for the average family is one in which the cost has necessitated some degree of sacrifice.[37]

Harmer also comments on how successful the funeral industry has been in selling this idea:

> The public relations campaign has been designed chiefly to give the community the feeling that an undertaker ranks with doctors, lawyers, and even clergymen in status, and to head off any protest about rising costs with cries of "traitor" and "atheist." It has thus succeeded in completely diverting attention from business practices and in impressing many—if not most—persons with the belief that conspicuous consumption is not only the AMERICAN but the SPIRITUAL way to behave when a death occurs. The eagerness with which people have indulged in extravagant display . . . has indicated how successful public relations can be.[38]

Hence, the formula Dignity = Money Spent becomes a driving force of modern funeral practices.

A second major critique involves the fact that the family is at a great disadvantage when "shopping for" and "buying" a funeral. In many ways, the shock of the death of a loved one turns family members into irrational consumers. As Bowman indicates, it is in the extraordinary process of bargaining for the funeral goods and services that customers lose their usual protections and the under taker finds special negotiating advantages.[39] And the situation is obviously not one in which the customer can say to the funeral director, "Well, I have to think it over for a few days. I'll get back to you." The urgency, both presumed and real, that is created by death makes for a unique, high-pressure situation.

And as we noted earlier, one of the major goals of the funeral director in his role as businessman is to sell goods and make a profit. As Mitford reports:

> The funeral seller, like any other merchant, is preoccupied with price, profit, selling techniques. As . . . [the] dean of the San Francisco College of Mortuary Science writes in *Mortuary Management's Idea Kit*: "Your selling plan should go into operation as soon as the telephone rings and you are requested to serve a bereaved family. . . . Never preconceive as to what any family will purchase. You cannot possibly measure the intensity of their emotions, undisclosed insurance or funds that may have been set aside for funeral expenses."[40]

In this vein, funeral directors actively study selling and marketing strategies that range from techniques of conferring about and discussing funerals with customers to the all-important methods of displaying caskets in the selection room. In regard to this last point, it is typical to arrange the caskets in such a way as to tarnish the image of cheaper caskets in relation to more expensive ones. It is hoped that this comparative display will facilitate the customer's choice of a more expensive casket. And, finally, critics of the funeral have loudly proclaimed that misstatements, manipulations, and pressures, coupled with the clients' own lack of knowledge regarding the funeral process and their accompanying emotional vulnerabilities, significantly unbalance the economic relationship between funeral director and customer in favor of the funeral director.

The American Funeral: A Final Reflection

Throughout this book, it has been emphasized that dying in America is a reflection of living. The forces of individualism, normlessness, and technology have given a specific shape and direction to the organizational context of dying in modern America. Indeed, the modernization of American society—with the growth of individualism, the development of technologically specialized professions, and the breakdown of stable patterns of religious ritual—have led to anomie and ambiguity in death-related beliefs and behaviors. This situation is especially evident in the American funeral. Here the diversity of traditions, the myriad strong feelings, contradictory advice about the conduct of funerals, and the relative paucity of research combine to create a climate of uncertainty.[41] Moreover, this uncertainty both reflects and increases the tendency toward self-motivated expressions of individuality in the funeral process.

Thanatology Snapshot 4.3

The FTC Funeral Rule

Each year, Americans arrange more than two million funerals. Many people are not initially concerned about the cost; but a funeral might be the third most expensive purchase you make after that of a house and a car. The Federal Trade Commission has developed a set of regulations, called the Funeral Rule, concerning funeral-industry practices. The regulations, previewed by CR in October, took effect April 30, 1984. . . .

The Funeral Rule makes it easier for you to select only those goods and services you want. Now, for example, you can find out the cost of individual items over the telephone. Also, when you inquire in person about funeral arrangements, the funeral home must give you a written price list of the goods and services available. When arranging a funeral, you can purchase individual items or buy an entire package of goods and services. If you want to purchase a casket, the funeral provider will supply a list that describes all the available selections and their prices.

When you call a funeral provider and ask about terms, conditions or prices of funeral goods or services, the funeral provider must:

—tell you that price information is available over the telephone;
—give you prices and any other information from the lists to answer your questions reasonably; and
—give you any other information about prices or offerings that is readily available, and reasonably answer your questions.

Price List

If you inquire in person about funeral arrangements, the funeral provider will give you a general price list. This list, which you can keep, contains the cost of each individual funeral item and service offered. As with telephone inquiries, you can use this information to help select the funeral provider and funeral items you want, need and are able to afford.

The price list also discloses important legal rights and requirements regarding funeral arrangements. It must include information on embalming, cash-advance sales (such as newspaper notices or flowers), caskets for cremation and required purchases.

The Funeral Rule requires funeral providers to give consumers information about embalming that can help them decide whether to purchase this service. Under the rule, a funeral provider:

—may not falsely state that embalming is required by law;

Thanatology Snapshot 4.3 (*continued*)

—must disclose in writing that, except in certain special cases, embalming is *not* required by law;

—may not charge a fee for unauthorized embalming unless it is required by state law;

—will disclose in writing that you usually have the right to choose a disposition such as direct cremation or immediate burial if you do not want embalming; and

—will disclose to you in writing that certain funeral arrangements, such as a funeral with a viewing, may make embalming a practical necessity and, thus, a required purchase.

The Funeral Rule requires funeral providers to disclose to you in writing if they charge a fee for buying cash-advance items. Cash-advance items are goods or services that are paid for by the funeral provider on your behalf. Some examples of cash-advance items are flowers, obituary notices, pallbearers and clergy honoraria. Some funeral providers charge you their cost for these items. Others add a service fee to their cost. The Funeral Rule requires the funeral provider to inform you when a service fee is added to the price of cash-advance items, or if the provider gets a refund, discount or rebate from the supplier of any cash-advance items.

Some consumers might want to select direct cremation, which is cremation of the deceased without a viewing or other ceremony at which the body is present. If you choose a direct cremation, the funeral provider will offer you either an inexpensive alternative container or an unfinished wood box. An alternative container is a non-metal enclosure used to hold the deceased. These containers may be made of pressboard, cardboard or canvas. Because any container you buy will be destroyed during cremation, you may wish to use an alternate container or an unfinished box for a direct cremation. These could lower your funeral cost since they are less expensive than traditional burial caskets.

Under the Funeral Rule, funeral directors who offer direct cremations:

—may not tell you that state or local law requires a casket for direct cremation;

—must disclose in writing your right to buy an unfinished wood box (a type of casket) or an alternative container for a direct cremation; and

—must make an unfinished wood box or alternative container available for direct cremation.

Unwanted Services

You do not have to purchase unwanted goods or services as a condition of obtaining those you do want unless you are required to do so by state law. Under the Funeral Rule:

(*continued*)

Thanatology Snapshot 4.3 (*continued*)

—you have the right to choose only the funeral goods and services you want, with some disclosed exceptions;

—the funeral provider must disclose this right in writing on the general price list; and

—the funeral provider must disclose on the statement of goods and services the specific law that requires you to purchase any particular item.

The funeral provider will give you an itemized statement with the total cost of the funeral goods and services you select. This statement also will disclose any legal, cemetery or crematory requirements that compel you to purchase any specific funeral goods or services.

The funeral provider must give you this statement after you select the funeral goods and services that you would like. The statement combines in one place the prices of the individual items you are considering for purchase, as well as the total price. Thus, you can decide whether to add or subtract items to get what you want. If the cost of cash advance items is not known at that time, the funeral provider must write down a "good-faith statement" of their cost. The rule does not require any specific form for this information. Therefore, funeral providers may include this information in any document they give you at the end of your discussion about funeral arrangements.

Under the Funeral Rule, funeral providers are prohibited from telling you a particular funeral item or service can indefinitely preserve the body of the deceased in the grave. The information gathered during the FTC investigation indicated these claims are not true. For example, funeral providers may not claim that either embalming or the use of a particular type of casket will indefinitely preserve the deceased's body.

The rule also prohibits funeral providers from making claims that funeral goods, such as caskets or vaults, will keep out water, dirt and other gravesite substances when that is not true.

Consumer Research Magazine,
Vol. 67, June 1984, pp. 28–29.

Of course, not all American funerals are filled with ambiguity and individual diversity. Studies have shown the persistence of some traditional funeral and mourning rituals in American society. For example, Orthodox Jews still practice traditional rituals related to funerals, grief, and bereavement.[42] Afro-Americans use spiritual songs and unashamed emotional expressiveness as ritualistic forms of support during the funeral.[43] Some ethnic groups in American society, though their practices are somewhat redefined by the so-called American way of death, seek to hold on to some of their Old World traditions.[44] In addition, funeral options such as secular-humanistic funerals, cremation, body-organ donation, membership in memorial societies, low-cost, no-frill funerals, and memorial services without a costly viewing-visitation process are all available to American funeral consumers. Some families even conduct their own funerals; before a memorial service, they may prepare their loved one's body for burial, make a simple wooden coffin, dig the grave, and perform the burial themselves. A study of a Protestant congregation in the Midwest found that this alternative type of funeral was being used effectively by members of the church for grieving family members. Three out of four respondents in the study expressed a preference for this alternative, and felt that it provided the congregation with greater sources of support and participation.[45]

There can be no standard or objective delineation of the value of a particular form of funeral over another, or even of the funeral itself. For example, in one study, Colin Murray Parkes found that the funeral did not have a beneficial psychological impact on the widows he studied, but in another study he found that two out of three widows did find value in the funeral of their deceased spouse.

Perhaps the most effective way to interpret the place of the funeral in modern society is through the framework of self-expression and the ritual of individualism that is so important to everyday social life. In this way, with respect to modern American funerals, it may be said: "You pays your money, and you takes your choice."

5

Grief and Individualism:
The Decline of Ritual and the
Emergence of the Therapeutic Model

There is no need to tighten the brow or tense the jaws. We have within us all that it takes to face the immensity of this pain, to let it soak in, to learn to breathe it into the heart, to let it burn its way to completion.

STEPHEN LEVINE

The denial of death has gone beyond the bereaved and the expression of mourning. It has extended to everything that has to do with death, which has become infectious. Mourning or anything resembling it is like a contagious disease that one is in danger of catching in the room of a dying or dead man, even if he is a stranger, or in a cemetery, even if it contains no beloved tomb.

PHILIPPE ARIÈS

The Age of Show Business is upon us.[1] Entertainment in American society has become very big business and a dominant way of life. The lives of entertainment and sports celebrities have become an enormously conspicuous part of the American way of life. Americans seem to have an insatiable thirst for spectator sports, movies, shopping and consuming, television, and other forms of amusement. Indeed, the amount of time, energy, and resources devoted to recreation and entertainment illustrates that fun and pleasure have achieved a prominent place in everyday American life.

Perhaps the most striking example of the rapid growth and emerging dominance of a societal fun and entertainment ethic is found in Huxley's *Brave New World*. The ubiquitous presence of orgy porgy, the centrifugal bumblepuppy, the feelies, and the all-important escape drug, soma, help to make this world a utopia of fun, frivolous enjoyment, and escapism. The inhabitants of Huxley's society become increasingly dominated by and dependent upon the presence of pleasure to the extent that worldly gratifications become the primary motivational force in their lives. One of Huxley's major statements regarding the consequences of the spread of the entertainment ethic is that it makes people less thoughtful and serious. As entertainment and pleasure become dominant symbols of a society, the more incapable its people are of serious social criticism, and the more mindless and trivial they become.

In Huxley's world there is no place for suffering. Moreover, there is no place for death; more precisely, there is no legitimate role for responses to death that are serious, time-consuming, and involve grieving. In this vein, all young citizens of *Brave New World* undergo death "deconditioning"; that is, they are taught that death is inconsequential and is never a reason for pause, sadness, or disruption of ongoing activities.

Certainly, the idea of a suffering-free and pleasure-full life, which Huxley envisioned in *Brave New World*, has not been fully achieved or perfected in contemporary America. Yet there are many ways that modern Americans seek to transcend and escape pain and suffering. Alcohol, aspirin, other drugs, shopping, and television are readily available to disengage suffering and pain from everyday life. On a sociomedical level, pain and suffering have been defined as technical matters requiring technical intervention. As already discussed, the normative medical response to pain is to demand more drugs, doctors, and hospitals, all of which are technological agents of the body social and work together to manage and conquer the problem of pain. Therapy, drugs, and self-help strategies have become salient factors in the quest to remove suffering and grief from the societal record and from the everyday facts of people's lives.[2]

Thanatology Snapshot 5.1

Keep Them Cheerful

The bereaved must carry on and face the future. There is no lack of reverence in facing it with a courageous and cheerful heart. The kindest act you can do is to help those who are left behind to be brave, to look ahead instead of looking back. . . . Talk to them about everyday things . . . get their minds off the sorrow. Urge them to get back to work, get out socially as soon as possible. Public mourning for a year or a month or any stated time is just a relic of past beliefs.

The Clark Grave Vault Company
pamphlet *My Duty*, p. 1.

It needs to be remembered, however, that in reality people are faced with pain and do suffer on a daily basis. Americans also continue to die every day, and, more important, the lives of survivors are touched by a spectrum of grief and suffering precipitated by death. The curious result is that the values, folkways, and institutionalized patterns of American life, which are often organized to promote the pursuit of self-gratification, pleasure, success, and entertainment, are regularly challenged by the ongoing and ubiquitous presence of death, grief, and suffering. The task of this chapter is to explore the nature of grief in modern America, and to analyze how grief and suffering are managed in a society that is neither designed nor equipped for them.

Grief as a Social Concept

In many ways, the identity of a human being is formed and maintained by relationships with others. The role of spouse is created by the presence of a husband or wife, parenthood is achieved by the birth or adoption of a son or daughter, and the existence of a son or daughter is made possible by a parent. In a very important sociological sense, the death of a significant other not only means the loss of that person, but may very well mean that an important part of personal identity is threatened. Grief, then, has to do with the emotional reaction that is elicited by the death of a significant other and its corresponding meanings for the identity and social roles of the survivors.

Bereavement is an essential component of grief. Literally speaking, to be bereaved means to be deprived, to have something taken away. For our purposes here, "bereavement" can be defined as the sense of deprivation or loss generated by the death of another person. An intense emotional response to bereavement that involves sorrow and suffering is grief. Strictly speaking, then, bereavement precedes grief but is not necessarily followed by it; that is, if one is bereaved, it does not always mean that one will grieve, or grieve intensely.

At first glance, it may appear absurd to make such a distinction, but our definition has important implications. As we saw in Chapter 1, the traditional patterns of European and American death encompassed an expansive definition and application of bereavement. Death during these traditional periods was a public affair and involved not just the immediate family but the broader social community as well. Many primitive cultures today also utilize an extended application of bereavement. Some of the small towns of America today, where there is a sense of stability, ethnic attachment, and extended family arrangements, also have an expansive application of bereavement. However, as society has modernized, and as the broad base of social community has been reduced to personal intimacy and friendship networks, a limited application of bereavement has evolved. For example, in a primitive culture or perhaps in a small, stable rural American community, the death of a third cousin may result in a state of loss or deprivation for a person or a family, whereas in the more typical modern, urban settings of America, the death of a third cousin would not be defined as a state of loss. Thus, not only does social environment affect the way in which bereavement becomes relevant, the idea of bereavement itself is a social construct, the application of which changes as society and the social relationships within society evolve and change.

Grief is also defined by social structure. The values and institutionalized patterns of living in society merge and continually define the circumstances under which one should respond to bereavement with sorrow and suffering. Thus, in a coherent rural community, the death of any of a wide range of people is likely to cause sorrow and suffering, but in most urban settings the death of relatively few people is likely to inspire serious emotional expression. Although there are no definitive data to point to, it may very well be (as we saw in the romanticized deathbed of the Victorian period) that as people grieve less often, they may grieve more intensely. Thus, it may very well be that the more narrow and individualized bereavement becomes, the more emotionally intense is the impact of grief. Before concluding this chapter, we will examine the paradox that a society devoted to the denial of suffering may at times actually promote and thereby experience it acutely.

Another dimension of human loss is mourning. Mourning is the behavioral expression of the emotional anguish of grief. The ways in which one is culturally prescribed to behave, such as wearing black clothes, crying at funerals, or taking tranquilizers, are some examples of mourning behavior. Mourning clearly has strong social significance, as particular cultures have specific behavioral requirements, including duration and intensity, in the aftermath of the death of a loved one. Mourning is directly related to grief in that it is the physical-behavioral expression of the feeling of sorrow and suffering. Yet, mourning expressions may not always reflect inner feelings. Zborowski has discussed how various cultures express pain and suffering differently and with varying intensity.[3] Raphael points out these differences by describing how some European cultures, like the Italians, are very expressive and emotional in their portrayal of grief, while the Anglo-Saxon culture is more stoical and behaviorally restrictive.[4] Italians or Greeks obviously do not love their families more or suffer greater grief than people of the WASP culture; dramatic and emotional expressions of grief are simply expected—or even required—of them.

Thus, not only do images of death vary culturally and historically, but responses to death itself are also products of history, social change, and culture. The factors that define the specific social nature of bereavement and mourning are generally obvious and have been extensively reported in the literature.

The social dimensions of grief are more difficult to identify for two major reasons: First, there may very well be some natural, universal, psychologically rooted grief response to death that applies to all humanity, regardless of culture. Second, the scientific study of the social components of human emotions is in its infancy. But it is fair to say, at the very minimum, that bereavement and mourning are directly shaped by culture and society, and that grief feelings are at least indirectly affected by societal factors.[5] Let us now consider how the sociocultural context of modern America shapes and influences the contemporary human experience of grief and mourning.

The Sociopsychological Factors of Grief

To be human is to grieve and to mourn. In modern America, these inherently human processes have taken on new forms and meanings, both of which represent a radical departure from traditional patterns. Historically, periods of extended grief and elaborate patterns of mourning served to assist survivors in coping with the death of loved ones. The spirit of the times during the era of traditional death was supportive of extended grief reactions and of elaborate mourning customs. These included the wearing of mourning clothes and jewelry; significant modification of personal behaviors and lifestyles during mourning; construction of ornate tombs, mausoleums, and gravestones; and an enthusiastic recognition of death, including the ubiquitous representation of death and grief in writing, painting, and sculpture. This traditional relationship of humanity to death and grief, as we saw in detail in Chapter 1, was rich and powerful, especially when juxtaposed to the absence of death symbols, images, and rituals in modern culture. In comparison, grief in modern society takes place in a setting that is "death sterile" and contains minimal ceremonial patterns, folkways, and structures that support the public expression of grief.

As Ariès writes, a great milestone in the contemporary history of death is the rejection and elimination of mourning.[6] Geoffrey Gorer, in a very important study, also discusses the implications of the "denial of mourning" for human life. He comments that people today are without adequate guidance about how to treat death and bereavement, and are without social help in living through and coming to terms with grief and mourning. Gorer proceeds to explain that the lack of rituals, norms, and social-moral guidance is accompanied by "a considerable amount of maladaptive behavior from excessive triviality and busyness through the private rituals of what I have called mummification to the apathy of despair."[7] For the first time in the history of Judeo-Christian culture, according to Gorer and other historians, the overwhelming majority of people lack rituals and social patterns to help them deal with the inevitable crisis of human death and grief.

Indeed, grief and mourning have been redefined in the modern societal context. As American society has modernized, mourning and grieving have been removed from the arena of ordinary social and public activity. The once publicly supported process of mourning has now become defined as the private trouble of the individuals involved, the resolution of which lies in their personal coping and adaptive skills. In this way, then, it may be argued that the entertainment and pleasure ethos of contemporary society is only minimally tarnished by the human realities of suffering. In fact, it is precisely the American preoccupation with fun, pleasure, and egoism which is both a cause and a derivative of the privatization and isolation of suffering, grief, and mourning.

The emergence of a restricted societal pattern of mourning is consistent with and reflective of the forces of individualism and technology that give shape to the modern context of American society. As we have already seen, medical technology has been associated with increased longevity and an increase in the prevalence of chronic illness. In addition, medical technology and public health reforms have successfully contributed to a lower death rate for children and young adults. The general prolongation of life has resulted in an increased

---

Thanatology Snapshot 5.2

The Pornography of Mourning

Giving way to grief is stigmatized as morbid, unhealthy, demoralizing—very much the same terms are used to reprobate mourning as were used to reprobate sex; and the proper action of a friend and well-wisher is felt to be the distraction of a mourner from his or her grief; taking "them out of themselves" by diversions, encouraging them to seek new scenes and experiences, preventing them from "living in the past." Mourning is treated as if it were a weakness, a self indulgence, a reprehensible bad habit instead of as a psychological necessity.

Many people, of course, can adjust to this public attitude by treating it as if it were an extension of modesty; one mourns in private as one undresses or relieves oneself in private, so as not to offend others; and this is probably the best solution now available. . . .

Geoffrey Gorer,
Death, Grief, and Mourning
(New York: Doubleday and Co., 1965), p. 111.

---

number of older people in American society; at the same time, a larger number of older people will die each year. Some 24 million persons are now over sixty-five, and their numbers continue to grow.[8] And about two-thirds of all the deaths that will occur in America this year will be from this age group. As a result, death in the United States is usually the experience of the aged. In addition, 75 to 80 percent of all deaths that occur in America take place in hospitals, nursing homes, or other chronic-care facilities. As most deaths take place in these institutions, death is increasingly vanquished from the everyday life experience of American people and is therefore largely rendered culturally invisible.

In addition to the changing demography of death, individualism is another factor that has diminished the social significance of grief and mourning. As previously noted, social and family ties have been weakened in American society. It is not unusual for adult children to have very separate lives from their parents or even to live in distant parts of the country. In addition, the medicalization of death in America means that most adult children are unlikely to be intimately involved in the daily care of chronically or terminally ill parents. Furthermore, as society ages, the incidence and prevalence of chronic disease in the elderly population increase, and as dying becomes more and more the experience of older Americans, a societal attitude that defines death as normal or acceptable for older people is likely to emerge. Indeed, the death of older people may not only be considered a normal occurrence, it may actually be something anticipated, expected, perhaps even considered overdue. For these reasons, when an aged parent dies, it may very well be minimally disruptive to the emotional, social, and economic lives of adult children. In some cases, especially if there has been a complicated and drawn out dying trajectory in which the terminally ill loved one has become a significant burden, death may be seen as a welcomed relief. A study by Owen, Fulton, and Markusen found that substantial disengagement from social interactions with parents, the advancing age of parents, the corresponding deterioration of their mental and bodily functioning, and often the institutionalization of aged parents have led to the diminishment of grief and mourning on the part of adult children. As they comment:

> It is important to recognize that the mature adult child need not experience the grief reaction or the profound sense of loss that has been documented in the case of a parent or spouse following a death. In modern society, the elderly are often defined differently than are other age groups. They "have had their life." It is only "normal" and "natural" that they die. Death brings "peace to a troubled mind" and "freedom from pain" to a person wracked with cancer.[9]

Thus, in our fast-paced society, chronically or terminally ill older people are often dead socially before they die biologically. The result is that the emotional and social impact of physical death is minimized. Fulton, Owen, and Markusen comment: "We are witness to developments in social change and interpersonal relationships that promise to grow more pronounced in the years ahead. Like the tip of an iceberg only a small proportion of the magnitude of the consequences of modernization for grief and mourning are visible at this time."[10] It is consistent with their view and research data that as society becomes more modernized, grief and mourning expressions will become increasingly diminished. If we extend the theoretical implications of Fulton and colleagues' analysis to their logical conclusion, we are very close indeed to the Huxleyan scenario of death deconditioning.

A second significant implication the work of Fulton and his colleagues is their recognition that grief is shaped by an intricate configuration of societal factors and variables. As the authors emphasize, grief is not a single concept or a uniform experience. On the contrary, these authors emphasize that grief is a highly complex, multidimensional, and individual phenomenon.

One very important factor in determining the meaning and impact of death is the type of death that occurs. As we have just seen, the impact of the death of a socially disengaged older parent is often relatively small. On the other hand, the death of a child or spouse usually has a major effect. Yet, grief and mourning response are also influenced by such factors as quality of the relationship, duration of illness, religious orientation, circumstances of death, ethnic background, sex, and age, all of which combine to make bereavement a unique, personal experience for the survivors. The nature of widowhood and its bereavement will be discussed in greater depth later in this chapter; the impact of the death of a child will be discussed in the following one.

In his critique of the idea that grief is an unvarying phenomenon that goes through predefined stages, Bugen identifies two variables that interact and establish differing intensities and durations to grief responses: centrality of the relationship and preventability of death.[11] Supporting the research of Fulton and coauthors, Bugen recognizes that the quality and closeness of the relationship between the mourner and the deceased are directly related to the intensity of grief. As we might expect, the closer survivors feel to the deceased, the more intense their sorrow; the more peripheral the relationship, the less intense the grief response will be. Additionally, if a death occurs that the survivors believe could have or should have been prevented, the period of grief is more likely to be prolonged than if the death is seen as inevitable and occurring in a context where everything to prevent it had been done.

Bugen proceeds to delineate a four-part model based upon an application of the two sets of variables he considers:[12]

1. Given a central relationship between bereaved and deceased and the belief that the death was preventable, we would expect the grieving process to be both intense and prolonged.

2. Given a central relationship between bereaved and deceased and the belief that the death was unpreventable, we would expect the grieving process to be intense but brief.

3. Given a peripheral relationship between bereaved and deceased and the belief that the death was preventable, we would expect the grieving process to be mild but prolonged.

4. Given a peripheral relationship between bereaved and deceased and the belief that the death was unpreventable, we would expect the grieving process to be both mild and brief.

Bugen's model characterizes grief as a dynamic, variable process—not a static, unchanging one. While this design is useful in identifying some important variables that define the intensity and duration of a person's grief response, it presents only a partial picture. Additional variables such as age, sex, length of illness, ethnic affiliation, availability of support for the survivors, and how much the deceased suffered are all relevant factors that characterize and individualize contemporary grief responses.

The effects of death occurring suddenly versus death following chronic illness have been extensively discussed in the literature about grief. The debate concerns whether or not chronic illness provides an opportunity to anticipate the death of a loved one and if that anticipation lessens the suffering and sorrow of the grief response. The concept of "anticipatory grief" was introduced into the literature by Erich Lindemann, who argued that individuals were able to experience and work through the emotions of loss before physical death occurred. When biological death did take place, the emotional consequences of grief would henceforth be greatly reduced.[13] The beneficial results of anticipatory grief claimed by Lindemann, however, are not fully supported by available research data. Yet anticipation of the death of a loved one does have an effect.

In her study of the impact of anticipatory grief on the grief response after biological death, Sanders identified three types of death: sudden unexpected death, short-term (less than six months) chronic-illness death, and long-term chronic-illness death.[14] As summarized in Table 5.1, Sanders presents some interesting findings and proceeds to comment that survivors of a sudden-death situation exhibited longer-lasting physical repercussions, as well as more anger and guilt, than those who survived a short-term chronic illness. Survivors of a long-term chronic-illness death showed greater feelings of isolation and alienation, which prolonged their grief and gave rise to loss of emotional control. The group making the best adjustment to bereavement had family members who died of short-term chronic illness.[15]

There are two major points to be derived from Sanders' study. First is that the value of anticipatory grief diminishes, and that it ultimately becomes

TABLE 5.1.   Variable Grief Responses

Type of Death		Grief Experience	
		Initial Interview: shortly after death	*Follow-up Interview:* 18–24 months later
Short-term chronic	→	greater social desirability, death anxiety	→ high death anxiety but a generally favorable outcome
Long-term chronic	→	greater social isolation, rumination, loss of vigor, physical symptoms	→ greatest denial, social isolation, loss of emotional control, loss of vigor
Sudden	→	high levels of denial, loss of emotional control, low optimism, depersonalization	→ anger, guilt, depersonalization, physical symptoms

counterproductive as the period of illness is extended from a short-term to a long-term, chronic one. The survivors then often find that their vigor and emotional adjustment are strained by the demands of prolonged dying. The second important point presented by Sanders' study is the vivid and explicit characterization of grief as a complex, varying, multidimensional experience and process.

Yet, anticipatory grief may be even far more complex than Sanders' study indicates. The research literature itself is replete with contradictory and inconclusive findings. Several studies have found that anticipation of death did not affect depression states in older survivors one year after the death of a loved one.[16] In a major study, Parkes found that whether a death was expected or not had no impact on the physical dimensions of grief and bereavement responses.[17] Likewise, Maddison and Walker reported that anticipatory grief had no significant impact on the outcome of grief and bereavement for young and middle-aged widows.[18] However, other studies have found that the shock of sudden death can strain the defense systems of the survivors, so that their ability to cope with intense grief is significantly impaired.[19]

In a very useful study, Glick, Weiss, and Parkes discovered a different value of anticipatory grief than that formulated by Lindemann. Specifically, they found that anticipating a death did not allow a widow to initiate her grief prior to biological death, but furnished an opportunity for the survivors to make the necessary emotional adjustments that would allow them to prepare for the loss:

Thanatology Snapshot 5.3

Therapeutic Death

A particularly striking example of this new "will-to-control" is the activity usually described in the literature as "grief management." This notion, made popular by one of the guiding lights of the American Hospice movement, Marcia Lattanzi, certainly belies the old Shakespearean maxim: "Everyone can master a grief but he that has it." Instead the grief-stricken ones are now urged to "master" and "control" their grief, to "work through" the grieving process—the very words betraying a resolute if not desperate attempt to overcome our vulnerability. As one person, active in the movement, said about his own grief in tones of obvious self-congratulatory relish: "I was surprised how quickly I worked through it." Besides establishing a new battlefield for narcissistic self-enrichment . . . such an attitude denies a fundamental truth about ourselves—that life exists not as a problem that needs a solution but a mystery that continually thwarts our attempts even at understanding, much less controlling, it.

Grief, like any significant human experience, does have its pathological expression, but a criticism of that pathology would lie on grounds other than failure to "work it through." Indeed, what makes "grief management" itself almost pathological is its failure to suffer it through. To undergo grief, to let death be death—in its utter finality—would strike terror in our hearts, leaving us helpless and vulnerable. Yet, not unlike the sexologists who teach their clients how to manipulate their pleasures, the new thanatologists teach us how to get on top of and master our grief and dying.

Preparation for the husband's death seemed to be of great importance for the eventual course of recovery, but not because of the occurrence of what Lindemann called "anticipatory grief." Virtually all the widows and widowers who had known of the impending deaths of their spouses believed that although they had begun to grieve prior to actual death, this did not reduce their subsequent grieving. . . .

Nevertheless, there was a positive correlation between longer advance warning and eventual satisfactory adjustment to widowhood. The value of advance warning seemed to be that it allowed emotional preparation for the loss. Loss without preparation seemed almost to overwhelm the adaptive capacities of the individual. Grief might not be augmented, but capacity to cope seemed diminished.[20]

Parkes and Weiss, in their 1983 study, reported similar findings and emphasized the idea that loss is a different type of experience altogether from the anticipation of loss. The death of a loved one creates sorrow, pain, and despair,

Thanatology Snapshot 5.3 (*continued*)

Such self-conscious manipulation—this feverish anxiety by which we monitor our progress through the various stages—is but an ingenious device for distancing ourselves from the very reality we seek to confront.

. . . [W]hat is so striking [about "grief management"] is the total fixation on the self. Lattanzi expresses it quite well:

> Basically . . . people are mirrors in our lives. We spend our lives getting close to people, and certain people reflect dimensions or parts of ourselves as we'd like to see ourselves. So when we lose someone who is important to us, it's like a piece of ourselves is taken away. And the way we have seen ourselves is no longer the way we are now.

What is mind-boggling in this statement (in which the personal and possessive pronoun is repeated twelve times) is the complete forgetfulness of what occasioned the grief, the death of the other. The death of the unique and irreplaceable other becomes a fissure in ourselves that needs to be therapeutically repaired. The death of the other as an-other is repressed and, ironically, the very devil which the movement sought to expel returns with a vengeance. It is but a short step, then, to deny one's own death, to view it as an illusion, or merely (in Kübler-Ross's quaintly reassuring phrase) "a transition." After all, how could this marvelously self-managed and self-developed person ever face a real break in its existence?

Frances Kane,
"Therapeutic Death,"
Commonweal, 24 February 1984, pp. 110–111.

but also offers the possibility of searching for and finding a new place in the world and, ultimately, acceptance and recovery. Like grief after death, anticipatory grief is characterized by a sense of disorientation and suffering, but it is distinguished by an additional phenomenon: The feelings of attachment to a dying loved one are typically intensified during the period of forewarning. Closeness, tenderness, and a desire to be with each other are common throughout the period of anticipation. Thus, anticipatory grief, as described by Parkes and Weiss, is not just grief begun earlier, but rather is a qualitatively different experience.

In their review of the anticipatory grief literature, Fulton and Gottesman observe that most of the research has focused on the narrow issue of anticipatory grief as forewarning to loss. Agreeing with Parkes and Weiss that postdeath bereavement is different from anticipatory grief, they argue that it is a com-

Thanatology Snapshot 5.4

Widow: A Modern Portrait

. . . And I was shocked. Shocked? Scared out of my wits. Martin looked like death.

To me, that was when the dying began. He had a grayness in his face that had nothing to do with fatigue. My beloved husband had the mark of death. I had seen it before. A dear friend of mine had died of leukemia and he had had the same look.

I didn't say anything. I couldn't say anything. How do you tell the man you love, "I think you're dying. You look like death to me."

Martin's rectal bleeding was getting worse and he was having more pain. His doctor, who was also his best friend, decided that it was time to operate. . . .

"We opened Martin up and it stinks," he said bluntly. "He's riddled with cancer. It has gone through his colon, to his liver."

"Your husband died this morning," he told me.

I don't know what I said. What do you say to the man who tells you your husband is dead? Thank you? I have no idea.

Recently I asked Jonny, "Did I cry?"

"No," he said. "You were acting brave." Acting brave! Children are so aware of what is going on. He knew all about role-playing. Why hadn't he known about death?

So at last Martin died. No more pain. No more hope. No more denial. No more fighting for life.

No more.

It was the end.

A widow. A widow is different. It takes time to realize just how different. There was a transition period when every morning I had to grapple with the fact of Martin's death all over again. Every morning it was new. A raw wound that took a long time to heal over.

---

plex phenomenon that needs to be studied on at least three levels: psychological, interpersonal, and sociocultural. Psychologically, such factors as personal coping abilities, feelings of guilt and anger, and feelings toward the dying patient are all relevant to the nature of anticipatory grief. On an interpersonal level, it is important to realize that grief is not merely a private matter. The manner in which the dying person and his or her family accept or deny the realities of the situation will affect the experience of anticipatory grief and may very well

Thanatology Snapshot 5.4 (*continued*)

The first stage of grief is merciful—a numbness that comes with shock. . . . "I felt numb and solid for a week," said one young widow whose husband had died unexpectedly. "It was a blessing . . . everything goes hard inside you like a heavy weight." She was sure that she could not have managed the children and the funeral arrangements without this numbness. She did not cry during this week either. But then her feelings came flooding back. Other women experience a much longer period of numbness.

I have no social life here at all. What have I done? Cut myself off from everything and everyone I love. Oh, Martin, my dearest darling, you sure screwed me. Left me unprotected. And my despair at your negligence is hurting me something fierce. Well, it is day now so I'd best get on with it.

Helplessness was too much to bear, so I became angry. And after he died, my anger took possession of me.

Sociologist Robert Fulton raised this problem in a conference on widowhood. "Whom can you turn to when you are touched by death?" he asked. His discomforting answer was, "There aren't very many people who are prepared to come to your assistance either socially or emotionally. In fact, it is sometimes hard to find anyone who will even talk to you about your loss."

"Widow" is a harsh and hurtful word. It comes from the Sanskrit and it means "empty." I have been empty too long. I do not want to be pigeon-holed as a widow. I am a woman whose husband has died, yes. But not a second-class citizen, not a lonely goose. I am a mother and a working woman and a friend and a sexual woman and a laughing woman and a concerned woman and a vital woman. I am a person. I resent what the term widow has come to mean. I am alive. I am part of the world.

Lynn Caine,
Widow (New York: Bantam Books, 1981), pp. 71, 90, 107, 112, 181

have broader implications for their postdeath adjustment.[21] The behavior of health care professionals, friends, and relatives may also be crucial factors in shaping the experience. On a broader scale, social and cultural trends such as denial of death and denial of grief and mourning may well have an effect. In addition, the social role of the anticipatory griever is especially confusing and problematical, as there are no guiding norms for appropriate behavior. In what has been called the "waiting vulture syndrome," survivors are left in an awkward, ill-defined position and are compelled to cope with their situation in a way that is consistent with their own environment and living experience.[22]

The essential point to highlight is that, like postdeath grief, the period of forewarning that has been labeled anticipatory grief is highly variable, individualized, and defies categorization into a predefined, uniform process. And this is explicitly consistent with the thrust toward greater individualism in the broader society and the general movement toward the individualization and privatization of grief and suffering in the contemporary American setting. In this way, the social circumstances surrounding and shaping the modern processes of grief and bereavement are unobtrusively, little by little, driving the process into increasingly narrow and personal spheres of experience. The general societal movement toward the personalization of grief and bereavement has led scholars to turn their attention to the usefulness of a personal construct approach. Pioneered in 1955 by George Kelly, construct theory takes as its starting point the idea that people make sense out of the world and their lives

THE PRIVATE WORLD OF GRIEF

Loneliness and anguish, often endured in privacy and isolation, are the plight of the American widow.

by attributing unique personal meanings to human experience. According to Kelly, every individual has personalized patterns of experiencing and viewing the world. For example, people who have optimistically looked forward to and used periods of transition in their lives as opportunities for growth and change will be likely to view the future as a cascade of possibilities for personal enhancement and transformation. In other words, individuals develop a pattern of living and of looking at the world that is a direct reflection of the core constructs of their personalities.

Increasingly, use has been made of construct theory as a theoretical and empirical foundation for the study of death, grief, and bereavement. The death of a loved one not only traumatizes the individual in the performance of social roles, but also may jeopardize and even invalidate the core personal constructs with which an individual maintains his or her identity and existence. Woodfield and Viney, in a very helpful discussion, apply the theory of personal constructs to the emotional and psychological states typically associated with grief in the literature. Beginning with the dislocative states of grief, they define the following:[23]

Shock: a sudden and overwhelming awareness of the need to reconstrue events precipitated by the news of the death of a loved one.

Numbness: the inability to make sense out of what has happened; the partial failure of the person's construct system.

Stress: seeing vital aspects of one's identity and life as being seriously threatened by the death of a loved one.

Anger: recognizing that personal constructs that have been used for years are now invalid, resulting in a feeling of hostility.

Anxiety: recognizing that the circumstances and events initiated by death lie outside the normal range of operation of one's construct system. Recognizing that the old construct system is jeopardized and that a new one is not immediately available to take its place.

Sadness: the feeling that results from the recognition that those parts of one's self that were spawned and illuminated by the relationship with the deceased are also to be lost. The invalidation of one's self and identity, at least in part.

Despair: recognizing that one's identity and construct system cannot do anything to change the death and the impact of the death of a loved one.

Woodfield and Viney also interpret the process of recovery through the personal construct framework. Indeed, the death of a loved one often brings a crushing sadness that makes no sense, especially in our modern cultural context. If recovery is to occur, the people whose lives are devastated by grief must participate in creating a personalized program for making life sensible again and for establishing meaningful reasons to continue living. Behavioral and psychological use of denial and overglorification of the deceased loved one may, bit by bit, be useful in allowing the survivors to assimilate the facts of death into their identities and personal construct systems. Reaching deep inside themselves, through the support of others, survivors may begin to transcend conditions such as depression by actively testing or experimenting with new and untried constructs so as to begin to "find some purpose again to life." If they are able to begin to reorganize their identities by creating a new web of personal constructs and meanings, the process of recovery is initiated.

In conclusion, Woodfield and Viney note:

> The personal construct approach maintains that, after her husband's death, a woman's thinking and behavior are directed by her efforts to anticipate events in her world. This anticipation of events is based upon the widow's system of personal constructs, an individual system of constructed interpretations of herself and the events within her world. The psychological states commonly seen to accompany conjugal bereavement are manifestations of changes in portions of her personal construct system.[24]

The weakness of the personal construct approach is that it narrowly identifies grief as the domain of individual constructions. As we have already seen, despite tremendous individual variation, grief and bereavement are significantly shaped by interpersonal, cultural, and social forces. And the fact that grief is becoming more individualized is a consequence of the social factors of modernization. Personal construct theory fails to account for these myriad sociocultural forces. Its strength lies in its recognition of how the variability of feelings and responses defines the modern system of grief and bereavement within the framework of individualism. Thus, as grief is transformed from a public experience into a private one, the nature of the grief response will be individualized according to the specific circumstances that typify the living patterns of each survivor. It is increasingly the responsibility and obligation of individuals to establish for themselves some set of meanings to the death of a loved one.

As suggested earlier, another very important factor underlying the differences in the American grief response is ethnic identity. Richard Kalish and David Reynolds have comprehensively studied the attitudes, values, and expectations related to bereavement of African Americans, Mexican Americans, Japanese Americans, and Anglo Americans. Some of the salient findings are presented in Thanatology Snapshot 5.5.

Thanatology Snapshot 5.5

Ethnic Variations in Mourning Behavior

Item Number	Question:
022	How often have you visited someone's grave, other than during a burial service, during the past two years?

Response*	Black Americans	Japanese Americans	Mexican Americans	Anglo Americans
Never	71	35	56	59
1–3	26	25	27	26
4–10	2	22	8	10
11+	2	17	9	5

Item Number	Question:
052	[In bereavement after the death of a spouse] Who would be likely to help you with such problems as preparing meals, babysitting, shopping, cleaning house, and things like that?

Response	Black Americans	Japanese Americans	Mexican Americans	Anglo Americans
Family member	50	74	65	45
Friend	42	9	14	45
No one	9	17	21	10

People may have certain expectations of a widow/widower. The next few questions are about someone who has just lost his/her spouse: In general, after what period of time would you personally consider it all right [to do the following]:

Item Number	Question:
110	To remarry?

Response	Black Americans	Japanese Americans	Mexican Americans	Anglo Americans
Unimportant to wait	34	14	22	26
1 week–6 months	15	3	1	23
1 year	25	30	28	34
2 years + other (including never/depends)	16	28	19	7

*Numbers are percentages.

(continued)

Thanatology Snapshot 5.5 (*continued*)

Item Number	*Question:*
111	To stop wearing black?

Response	*Black Americans*	*Japanese Americans*	*Mexican Americans*	*Anglo Americans*
Unimportant to wait	62	42	52	53
1 day–1 week	24	26	11	31
1 month +	11	21	35	6
Other/depends	4	11	3	11

Item Number	*Question:*
112	To return to his/her place of employment?

Response	*Black Americans*	*Japanese Americans*	*Mexican Americans*	*Anglo Americans*
Unimportant to wait	39	22	27	47
1 day–1 week	39	28	37	35
1 month +	17	35	27	9
Other/depends	6	16	9	10

Item Number	*Question:*
113	To start going out with other men/women?

Response	*Black Americans*	*Japanese Americans*	*Mexican Americans*	*Anglo Americans*
Unimportant to wait	30	17	17	25
1 day–1 month	14	8	4	9
6 months	24	22	22	29
1 year +	11	34	40	21
Other/depends	21	19	18	17

Thanatology Snapshot 5.5 *(continued)*

Item Number	Question:
114	What do you feel is the fewest number of times he/she should visit his/her spouse's grave during the first year, not counting the burial service?

Response	Black Americans	Japanese Americans	Mexican Americans	Anglo Americans
Unimportant	39	7	11	35
1–2 times	32	18	19	11
3–5 times	16	18	12	18
6+ times	13	58	59	35
(Don't know, etc.)	11	6	3	19

Richard Kalish and David Reynolds,
Death and Ethnicity: A Psychocultural Study
(New York: Baywood Press, 1981), pp. 106.

Grief as a Disease: Bereavement and the Broken Heart

Traditionally grief has never been considered within a medical framework, but a significant number of grieving people do consult physicians because of disturbing physical symptoms during the process of grief and mourning.[25] More specifically, the literature indicates that there is an elevated risk of death, especially for men, along with other physical health problems in newly widowed persons. In their review of the epidemiological literature on excess mortality in the newly widowed, which appeared in a mainline medical journal, Jacobs and Ostfeld emphasize the importance of understanding the physical consequences of bereavement for comprehensive medical care: Clinicians need to understand the determinants of the mental anguish, morbidity, and mortality of bereavement, especially in newly widowed persons, so that reduction of these effects becomes feasible.[26] Let us review some of the research data that led Jacobs and Ostfeld to the above statement.

In a study of 4,486 widowers over the age of 55 who were followed for nine years, Parkes and his colleagues found that during the first six months of bereavement the rate of mortality was elevated 40 percent, with a return to normal levels thereafter. Primary causes of death included coronary thrombosis and arteriosclerotic and degenerative heart disease. Also included as causes of death were influenza, pneumonia, bronchitis, along with other heart diseases and diseases of the circulatory system. Taken together, diseases of the heart and the circulatory system accounted for two-thirds of the increase in mortality during the first six months of bereavement.[27]

Maddison and Viola studied the health complaints of widowed women in Boston, Massachusetts, and in Sydney, Australia. In Boston they found 21.2 percent of widows sustained a marked health deterioration. This figure was compared to a nonwidowed control group, 7.2 percent of whom suffered marked health deterioration. In Sydney, the figures were found to be 37.1 percent for widows and 2.0 percent for the control group.[28] The following list details some of the specific conditions found in the total group of widows and the total group of controls. The numbers in parentheses refer to the percent of the widows and controls respectively with the condition.

Psychological Symptoms
General nervousness (41.3; 16.1)
Depression requiring medical treatment (12.8; 1.0)
Insomnia (40.8; 12.6)
Trembling (10.4; 2.0)

Neurological
Headache (17.6; 9.0)
Fainting spells (1.3; 0)
Blurred vision (13.7; 7.5)
Dizziness (9.1; 4.5)

Gastro Intestinal
Indigestion (9.9; 4.5)
Difficulty in swallowing (4.8; 1.5)
Excessive appetite (5.4; 0.5)
Anorexia (13.1; 1.0)
Weight loss (13.6; 2.0)

Cardio Vascular
Palpitations (12.5; 4.0)
Chest pain (10.1; 4.5)

Habits
Increase in drug intake (37.3; 11.1)
Marked increase in drug intake (5.9; 0.5)
Increase in alcohol intake (6.7; 2.0)
Marked increase in smoking (11.7; 1.5)

General
General aching (8.4; 4.0)
Reduced work capacity (46.7; 26.1)
Gross fatigue (29.6; 11.6)

From the above figures it is clear that psychological as well as physical health complaint symptoms are greater in recently bereaved widows than they are in the nonbereaved control group.

Other studies also point to the increased risks to health that are associated with bereavement. Maddison has estimated that 20 percent of all widows will suffer from substantial health deterioration during the first year of bereavement.[29] In another study, Parkes found that during the first year of bereavement, widows consulted with a physician at three times the expected rate.[30] Other studies have indicated that widows use substantially more sedatives, spend more time in bed, drink more, take too much medication, and lose too much weight. And a study by Rees and Lutkins found that the risk of death during the first year of bereavement was seven times greater for the bereaved group than it was for their control group.[31]

The composite picture which emerges from the bereavement literature is that widowhood represents a threat to good health and a threat to continued living for the survivor. Jacobs and Ostfeld effectively summarize the basic patterns of excess mortality and morbidity that are found in the newly bereaved:

> A basic pattern of excess mortality in the widowed, especially in males is discernible. . . . The peak risk for men is in the first six months; for women it is in the second year. . . .
>
> Cause [of death] varies by sex. For men there is excess mortality from tuberculosis, pneumonia, cirrhosis and alcoholism, suicides and accidents and heart disease. For women, there is excess mortality from tuberculosis, cirrhosis and alcoholism, heart disease and cancer.
>
> [Overall] Men are consistently at greater risk at all ages than women, and younger widowed have a greater relative risk than older widowed.[32]

Despite the fact that there is a fairly consistent picture in the literature regarding elevated death and sickness risks for the newly bereaved, not all are in agreement with the notion that grief and bereavement are the causes of the increased risk. Alternative explanations for the increase in mortality in recently bereaved widows are discussed in the literature. One salient alternative explanation is that of Joint Unfavorable Environment, where the spouses mutually lived a destructive lifestyle (smoking, drinking, stress, etc.) which killed one partner and therefore is responsible for the death of the other shortly thereafter. Homogamy, or the idea that the physically unfit marry the physically unfit, is also a relevant alternate explanation. Hence when one partner dies, the other is already at high risk of death because of his or her inferior physical condition.

In this way, some question is raised regarding whether the state of widowhood or other factors are responsible for the increased sickness and mortality risks for the newly widowed person. Stroebe and her colleagues in their review of "the broken heart" literature simply and eloquently raise the issue: "While bereavement is undoubtedly a stressful life event and the incidence of fatal heart disease has been associated with the lack of strong social ties, in our opinion, it has not been possible on the basis of evidence collected thus far, to establish unequivocally the existence of a loss effect."[33]

There is, however, as Stroebe herself admits, a wide range of indirect evidence to support the "loss effect" of bereavement.[34] Parkes, in discussing the possible explanations for the "bereavement data" which exist, does not altogether dismiss homogamy and Joint Unfavorable Environment. Rather he indicates that these factors more than likely play only a minimal part in the increased mortality during the first six months of bereavement. He also points out that, while it is unlikely that grief is the sole cause of death among newly bereaved survivors, it may very well be that bereavement acts as an aggravating or precipitating factor for the conditions, especially diseases of the heart, which are the ultimate source of death.[35]

In this way, the psychological stress of bereavement may very well be a factor underlying the emergence of life-threatening physical conditions in the bereaved. As Parkes comments, "All of this leads one to suspect that it may well be the emotional effects of bereavement, with the concomitant changes in psycho-endocrine functions, which are responsible for the increased mortality rate."[36] Frederick discusses the relationship between grief as a stress mechanism and the physiological implications of acute stress, specifically, the effect that grief has on the pituitary-adrenal axis and the hypersecretion of cortisol.

> Normally, the synthesis of corticosteroids by the adrenal cortex is stimulated by the adrencorticotropic hormone ACTH, secreted by the pituitary. The regulation of the cycle is via a biofeedback mechanism. When the level of cortisol circulating in the blood attains a certain concentration, the release of additional quantities of ACTH is inhibited, reducing the biosynthesis of corticosteroids by the adrenal-cortex to baseline levels. However, it has been found that the stress of acute grief brings about a rapid reflexive release of ACTH by the pituitary, which is relatively independent of the cortisol levels in the circulating blood. This release of ACTH can only be partially inhibited by extremely high levels of corticosteroids. The stage is set, therefore, for the hypersecretion of cortisol as a direct consequence of acute grief. With increase in circulating cortisol, there is a resultant decrease in the over-all immunity.[37]

Under normal circumstances, the physiological mechanism at work is controlled by self-regulated, biofeedback systems. An increase in the amount of cortisol in the blood results from the secretion of the ACTH hormone. When the amount of cortisol in the blood reaches a certain concentration level, the secretion of ACTH is inhibited. This concentration will allow the cortisol circulating in the blood to return to normal baseline levels. For this to happen is important because as cortisol levels increase, overall immunity is reduced. Thus, by regulating the secretion of ACTH, cortisol is maintained at acceptable levels, enabling the human organism to maintain a normal level of immune response.

Acute grief, however, facilitates the secretion of ACTH by the pituitary, which abnormally increases the amount of cortisol, the result of which is a depression of the immune response of the grieving person. In this way, because immunity is suppressed, the grieving individual becomes more vulnerable to the pathogens that exist in the everyday environment. As Fredrick explains:

> There is little doubt that the immunological response, both on a biological and on a biochemical level, is directly influenced by corticosteroids secreted by the adrenal cortex under stimulation by the ACTH released by the pituitary. During periods of stress, the level of corticosteroids may be elevated quite a bit above base-line values. It is precisely during these high-stress periods (such as bereavement) that immunological suppression is greatest.[38]

Fredrick goes on to conclude:

> It seems logical that relief from the stress of grief should allow corticosteroid levels to return to base-line or normal values. For the bereaved individual, the resolution of grief would be expected to restore the immunological health of the person.[39]

Thus, while the picture of grief as a disease is at best muddled, at this point there is at least indirect evidence of increased death and sickness associated with bereavement and, at least, a very preliminary understanding of the physiological mechanisms at work in the bereavement–disease process. Perhaps even more sociologically significant, however, is the growing emphasis on viewing grief as a disease and as a precipitator of disease and disease symptoms. It is also important to note that the medicalization of the grief process is a direct reflection of the broader medicalization of death as already discussed in chapters 2 and 3 and of the privatization and restriction of grief and mourning in the broader society.

In the American environment of restricted bereavement and grief, the presence and legitimacy of extended mourning patterns have vanished. In the absence of ritual, ceremony, and cultural traditions to support individuals through their grief, it is understandable that bereaved persons are becoming increasingly dependent on doctors and the medical profession. In a social setting which lacks societal norms to guide the response and behavior of bereaved individuals, becoming sick and assuming the sick role has evolved into a legitimate response to grief. Once a grieving person becomes sick and is defined through the sick role, some norms are then established for piloting the behavior of the bereaved person. Going to the doctor, complying with his or her advice and prescriptions, and cooperating with the doctor in his or her approach to treating the disease and its symptoms become firmly established behavioral expectations for ill-bereaved persons. In this way, then, a motivation to assume the sick role is clearly provided by the lack of societal rituals and traditions to assist the bereaved. Hence, the sick role becomes a legitimate avenue for bereaved individuals to express their grief and receive a modern, technologically based form of support and assistance.

It is not meant to imply that individuals adopt the sick role in a hypochondriacal way. Unquestionably, the illnesses that the recently widowed suffer through are real in a biological-physical sense. Yet, in addition to the biological dimensions of bereavement-disease, there are a variety of social implications to the current functioning of the sick role vis-à-vis grief that are especially relevant for the sociological study of death and dying. More precisely, the objective existence of physical disease in no way diminishes the sociological interpretations of the meanings of the sick role as a modern response to grief and bereavement. Additionally, however, the sociomedical science literature is increasingly making us aware of the societal and psychological factors that underlie human disease and illness. In this vein, a very important question which emerges is how much of bereavement sickness is encouraged by the fact that the sick role is a legitimate way to express the turmoil and the suffering one feels from bereavement? It may very well be that the sick role, because it is a legitimate avenue for grief expression and will provide for some semblance of guidance and support for the bereaved individual, may be an important variable in the precipitation of bereavement-disease. At best, the issue can only be speculatively raised at this point, and we must all await results of future investigations. In any event, although the picture is very fuzzy in ways that have important physical and sociological consequences, "the broken heart" is a metaphor that is aptly relevant to the modern, American bereavement experience.

Widowhood: More Than a Feminist Issue

Many of the life circumstances that face and haunt American widows are specifically relevant to gender role expectations for women. The life situation of a widow also bespeaks issues and problems that are associated with generalized patterns of living in our modern society. In this way, widowhood as a role-state in American society is shaped by factors that define the role of women in particular but also by broader factors that shape family living arrangements and the meaning thereof in the modern societal context.

One of the more unfortunate consequences of the recent thanatology movement and of the study of death and dying is the overintellectualization of the human experience of dying, death, and grief. Any woman who has suffered the death of her husband knows that all of the research findings on grief, bereavement, and widowhood only begin to scratch at the surface of the actual human meanings of widowhood. Regrettably, by turning the experience of grief and bereavement into distinct and compartmentalized research questions, thanatologists (including this author) tend to trivialize the human experience of grief. It is important to recognize that the accumulation of information about an issue does not necessarily deepen our understanding of the meaning of the issue as it affects human life. Perhaps the only way to fully grasp the experience of widowhood is to be living through it. There are, however, some very helpful ethnographic–descriptive accounts of widowhood in the literature. These are not based on scientific methodologies but are guided by human experience and the attempt to understand the meaning of that experience. These accounts are especially useful in providing key insights into the human consequences of American widowhood.

As presented in Thanatology Snapshot 5.4, Lynn Caine has written a moving humanizing account of her experience with widowhood. The death of a spouse, as Caine effectively writes, leaves women with an overwhelming sense of urgency, suffering, and disorientation—much of which the widow is expected to cope with in ways that keep her suffering private and personal, and therefore does not seriously disturb others around her. Her husband's death leaves her feeling lonely, confused, vulnerable, depressed and without the support and presence of the partner she counted on for many years.[40] Additionally, the widow often finds herself in a financially vulnerable position.[41] Her financial vulnerability may result in the loss of her home and of the everyday lifestyle to which she was accustomed. In addition to the loss of companionship occasioned by the death of her husband, a widow typically finds herself sleeping alone, without the warmth and comfort of her husband's body. And the widow often finds herself in a situation where the physical death of her husband results in the social death of her relationships within the community and the loss of friendship ties, especially those that connect her to other couples. Now the "odd-man out," the widow is a reminder of the frailty of life, and the absence of her husband is a constant reminder to her friends that "there, but for the grace of God, go I." Consequently, not only is the widow often stigmatized in society, she often finds that her place in social interaction with friends becomes tentative, ambivalent, or full of pity.[42] Often, in order to avoid their own painful thoughts, recognitions, and associations, friends of the American widow withdraw from her at a time when she is at greatest need of emotional and social support. Justine Ball summarizes the factors that underlie and shape the life predicament of the American widow very nicely:

The Merry Widow image is a myth. Most of the 10 million widows in the United States are grieving or are recovering from the effects of grief. A widow's grief is especially difficult as she is generally an older person in our youth dominated society; she is often very poor in a very affluent society; she is female in a male dominated society. With one out of every six women over twenty-one a widow, there is a need to study all aspects of her life and particularly her grief process.[43]

Indeed the issue of interpersonal support for the widow is especially important in our social setting of diminished community and heightened individualism. Let us consider it a bit more closely. Ironically, despite the overwhelming loneliness that haunts the American widow, the actual presence of support from married friends presents a problem for the widow. During the immediate aftermath of the death of her husband, when the widow is acutely grieving, the presence of support from married friends is inconsequential to her successful adjustment to the death of her husband. On the other hand, if the widow has moved on to some type of transition phase where she is beginning to become reintegrated into the everyday world, support from married friends becomes actually counterproductive to her successful adjustment to bereavement. Simply stated, she is better off in the long run without such support. Bankoff explains:

> Friendships are typically based on common interests and lifestyles. Therefore, when a wife becomes a widow the underlying basis of her friendship relationships with her still-married friends is sabotaged, leaving at best an ambiguous basis for continued friendships. As Blau's research has shown, widowhood appears to have an adverse effect on friendships when it places an individual in a position different from that of most of her age and sex peers.[44]

The need for support, and especially support *from others with shared and common interests*, has led to the establishment of the widow-to-widow program that exists in virtually every American city. The widow-to-widow program is a voluntary association of recently and not-so-recently widowed people, the purpose of which is to provide for an ongoing and overlapping network of support for the widow during bereavement. In an age of urban individualism, the potential for loneliness for the American widow is high. The self-help, widow-to-widow program has grown out of the realities of loneliness and supportlessness that threaten the bereaved widow. In this way, the voluntary and planned creation of a self-help support group attempts to accomplish what would naturally take place in more rural areas and tightly knit traditional communities.[45] And, perhaps most important of all, the widow-to-widow support group program provides a common base of identification for its members and offers widows an opportunity to receive some meaningful interpersonal support and to build up a network of intimate support that may usefully nurture their personal and social identity. This type of intervention helps us to understand why support from married friends is often a problem, as those friends do not share the same worldview or experiential base of the widow.

Despite the existence of support groups and supportive services, the American widow is still very much on her own as she suffers through the process of grief and bereavement. And although particular kinds of support are useful at particular periods of widowhood,* there is little evidence that generalized support, regardless of its type or source, has a positive effect on the well-being of widows, especially those who are in the middle of intense grief.[46] Widowhood represents an extreme and disruptive crisis that occurs within a societal context which is devoid of custom, ceremony, ritual, and norms to support the widow through her suffering. Social support, within the broader social context of denial of grief and mourning, is only partially and in limited circumstances helpful in providing comfort to the widow and enabling her to make some sense out of an otherwise mysterious and meaningless occurrence. And it should not be forgotten that family members, relatives, married and single friends, who may be potential sources of support for the widow, are subject to the same normlessness, ambiguity, and confusion about death and bereavement that faces the American widow. In this way, people generally available to the widow for support are often ill-equipped to provide the support the widow needs. This gap, then, gives rise to the formation of specialized bereavement support groups, such as the widow-to-widow program.

The circumstances that typically face the American widow result from an integration of the specific role expectations of women in society with the broader forces of modernity, particularly in relation to dying, death, and grief. Although independence and autonomy have increased for American women, the social and economic role of women in society is still typically defined through their association with men. In addition, and even more important to the theme of this book, is the role of heightened individualism and diminished community. As the modernization of society has heightened, the value of individualism and intimate relationships as a source of personal well-being has been elevated to an unparalleled high. In this framework, where personal satisfaction and well-being are intimately linked to a relationship with one other person and the role of broader support systems of community has dissipated, the consequences and pain of the death of a spouse have intensified. In this way, individualism be-

---

*Although overall support seems to make no difference in the recently widowed and only minimal difference in those who are in some sort of transition phase, support from parents (especially the widow's mother) is useful during the early and acute experience of widowhood. The value of support from family diminishes during the transitional period, where the widow seeks reintegration into society, and during this period support from an intimacy network becomes important. In this way, support for the widow during bereavement is only partially functional for her and only under very particularized circumstances.

comes the major determining factor of the American widow's experience. Not only is the individual widow primarily responsible for coping with her own grief, the societal forces of individualism have resulted in an intensification of the emotional sufferings of that grief. The irony of modern widowhood thus emerges: As individualism promotes the expectation that grief should be privately and personally resolved, at the same time it increases the difficulty of privately coping with grief through intensifying the emotional pain of the experience. And, in an era dominated by symbols of entertainment and pleasure, where individuals are neither trained nor equipped to cope with death or grief, the expectation that one cope with one's own grief personally and privately is deeply problematic and places an unrealistic expectation on the grieving American widow.

The Denial of Grief: A Summary

In *A Grief Observed*, C. S. Lewis writes: "It feels like being mildly drunk, or concussed. There is a sort of invisible blanket between the world and me. . . . Perhaps the bereaved ought to be isolated in special settlements like lepers."[47] The words of Lewis are aptly relevant to understanding the place of grief in contemporary society. In both symbolic and real ways, the bereaved are isolated like lepers, but not in the sense that they are physically separated from the world and its activity. The isolation of the bereaved is one of emotional imprisonment. The absence of norms, patterns, and rituals for grieving is reflective of a society that seeks to disengage suffering and behavioral expressions of grief from the fabric of everyday social activity.

Unlike Huxley's society of pleasure-full, suffering-free, death-conditioned living, the death of a loved one in varying degrees and intensity does elicit pain and suffering for the survivors. This pain and suffering, although a typical consequence of the death of a loved one, receives very little public recognition and legitimation. Suffering and grief therefore have increasingly become defined as personal experiences, the resolution of which lies in the private domain of the individual self. To assist the individual in his or her personalized patterns of coping, the sick role and a range of self-help strategies have become legitimately available to the bereaved person. The therapeutic model upon which the sick role–self-help approach to grief is based is rooted in the idea that grief should be resolved as quickly and expeditiously as possible, all of which is consistent with the devolution of society's capacity to provide meaning to the human experiences of grief, bereavement, and mourning. Thus, by individualizing the human experience of bereavement, public expression of grief-related suffering is curtailed and the public sanctity of the pleasure and entertainment ethos is preserved.

6

On Dying, Death, and Children

And a young child shall lead them.

CHILD'S GRAVESTONE
ST. JOHN'S CEMETERY,
REGO PARK, N.Y.

In our advanced industrial society, children are not supposed to die. Nor are children to be exposed to the human experiences of grief, suffering, and death. Indeed, a variety of societal factors have emerged and have successfully protected children from the tragedy of dying and death. Improvements in medical technology, along with public health advances, have helped to reduce dramatically the rate of infant mortality in the twentieth century. The removal of dying and death from the home and community to the professional arenas of hospitals, nursing homes, and funeral parlors has contributed to the exclusion of children from regular participation in the processes and experiences of dying and death. Indeed, there is a strong tendency in the collective American idea of death that seeks to shield children from these realities. In this vein, one of the dominant themes of the thanatology literature during the past twenty years has been that death is defined and organized in the modern social setting in such a way as to exclude children.

In the era of traditional death patterns that preceded the modern organization of death, images, artifacts and reminders of human mortality were deeply embedded in everyday activity. From the Tame Death to the Beautiful Death of the Victorian era, death and dying were commonplace, and everyone, including children, was intimately familiar with them. Today, however, things are different. As previous chapters have discussed, never before in the history of humanity have dying people been removed so far from the scenes of social life as today; never before have human corpses been conveyed so odorlessly, and with such technical perfection, from deathbed to the grave. And in this modern social setting, parents are more hesitant to talk to their children about death and to allow them to participate in death rituals. Unlike earlier times, when

135

the sight of corpses was commonplace, modern American children can grow up without ever having seen or touched a dead body. Perhaps it would not be stretching too far to suggest that dying represents a peculiar and unique embarrassment to the technocratic way of life. (Just think of how awkward people feel in the presence of dying people and the embarrassment of not knowing what to say to them.) The embarrassment that dying elicits in the modern context also results in the isolation of dying from ordinary, public, social activity and the shielding of dying from people in the broader community, especially from children.[1]

Ironically, however, in this modern context, where great and diligent efforts are made to shelter children from death, a countertendency, emphasizing the value of full and open participation of children in death and dying experiences, has emerged. Some of the recent thanatology literature has advocated the idea that children should participate actively in the funeral and in other rituals of grief and bereavement. Other writing emphasizes the importance of allowing children the opportunity to discuss their feelings and ideas about death. Additionally, thanatologists have increasingly emphasized the importance of being open and direct with seriously ill children. They have even argued that these children should not be prevented from knowing that their illness is severe and might even result in death. This openness, it has been recently argued, is comforting and also creates the possibility that spiritual and emotional growth can take place.

With respect to children and death, then, it is accurate to say that polarized ideas now coexist. This contradiction and cultural ambivalence are very much a reflection of the dominant factors that generally shape the modern organization of death in the wider society. Thus, the relation between death and childhood in the modern social context is reflective of the normlessness, individuality, and technical reliance that typify both everyday living and dying. This chapter will explicitly discuss how the relationship between children and death is largely a product of the interaction between broader cultural values and social institutions.

Children's Perceptions of Death

As it had been judged culturally inappropriate to associate children with such a morbid topic, very little scholarly attention had been paid to studying children and death until quite recently. During the past decade, a growing body of literature on the subject has emerged, both responding to and facilitating the cultural thanatological trend toward openness and honesty. Additionally, the fact that many of the research findings on children and death are scattered, frag-

Thanatology Snapshot 6.1

The Social Isolation of Children from Death

In earlier times dying was a far more public matter than it is today. This could not be otherwise, first of all, for it was far less usual for people to be alone. Nuns and monks may have been alone in their cells, but ordinary people lived constantly together. The dwellings left them little choice. Birth and death . . . were more public, and thus also more sociable, events than today; they were less privatized. Nothing is more characteristic of the present-day attitude toward death than the reluctance of adults to make children acquainted with the facts of death. This is particularly noteworthy of the repression of death on the individual and the social planes. A vague feeling that children might be harmed causes people to hide from the simple facts of life that they must inevitably come to know and understand. . . . Undoubtedly, the aversion of adults today to teaching children the biological facts of death is a peculiarity of the dominant pattern of civilization at this stage. In former days, children too were present when people died. Where everything happens in large measure before the eyes of children, dying also takes place in front of children.

Norbert Elias,
The Loneliness of the Dying
(New York: Basil Blackwell, 1985), p. 18.

mented, and contradictory reflects the roles that normlessness and individualism play in shaping the modern relationship to death. And, in a curious way, the recent spurt of activity geared toward the scholarly understanding of children and death is itself an important indicator of the modern trend toward isolating children from death. During the era of traditional death patterns, the study of children's concepts of death, stages of death awareness, and so on, would have been moot in terms of the realities of children's lives and their involvement with death. During more traditional eras, when death was experienced as a fact of life within a framework of shared cultural meanings, children were active participants in this cultural framework of death acceptance. The recent proliferation of studies on children and death not only identifies how children have become disenaged from death and dying, but seeks to refashion a relationship between children and mortality. This connection, however, is very different from that which existed during the patterns of traditional death in that it evolves not as a natural part of prevailing arrangements for social life, but is consciously designed and planned in a spirit of therapeutic control and management. In this way, as we shall see, death and dying still remain very much excluded from the ordinary social world of American children.

In order to begin to understand children's relationship with death, one must enter into their private and encapsulated worlds. The most often quoted study of the way children view death was written forty-five years ago by a Hungarian psychologist, Maria Nagy. Nagy asked children from ages seven to ten to write essay compositions about death, asked children from ages six to ten to make a drawing of death, and discussed death with children aged three to ten. In reporting the results, Nagy identified three specific stages of understanding associated with the child's chronological developmental level.

Children before the age of five (the first stage) have an "immature" concept of death. They do not see it as irreversible or even inevitable. In this stage, the child does not see the dead as even being dead; rather, they are regarded as alive but in a different form. Life, as seen by the child under five, changes upon death and begins to take place in the confines of the coffin, where the dead person "sleeps," "eats," "drinks," and so on. As Nagy puts it:

> In the cemetery one lives on. Movement is to a certain degree limited by the coffin, but for all that the dead are still capable of growth. They take nourishment, they breathe. They know what is happening on earth. They feel it, if someone thinks of them, and they even feel sorry for them. Thus, the dead live in the grave. . . .[2]

As they grow closer to the end of stage one, children may recognize death more fully: "There are among children of five and six those who no longer deny death, but who are still unable to accept it as a definitive fact. They acknowledge that death exists but think of it as a gradual or temporary thing."[3]

From the age of about six to nine, children begin to see death as permanent but not necessarily inevitable. Children during stage two tend to personify death, that is, to define death as something that has an external life of its own. In this developmental stage, death is perceived as a force that can unexpectedly sweep down and capture its victim:

> In the second stage of development . . . the children personify death in some form. . . . Either they believe in the reality of the skeleton-man, or individually create quite their own idea of the death man. They say that the death man is invisible. This means two things. Either it is invisible in itself, as it is a being without a body, or it is only that we do not see him, because he goes about in secret, mostly at night. They also state that death can be seen for a moment before, by the person he carries off.[4]

Instead of denying the irreversibility of death, the child now defines death as having an independent personality of its own. In this way, the child may not fully understand the inevitability of death for everyone; the child believes that death may be potentially thwarted by "beating him over the head" or, perhaps, by avoiding "the boogeyman." Thus, in stage two, the child perceives death as being final but not inevitable.

The third stage, which generally begins after the age of nine, occurs when the child recognizes that death is both irreversible and inevitable. During this stage, the child adopts what Nagy sees as a mature understanding of death. By this she means, but never directly says, that the child learns and accepts the definition of death prevailing in the modern social setting, namely, that death means the cessation of physical life for everybody.

Nagy's study has become the standard, most authoritative source of scholarly information on children and death, and it is widely used and accepted by thanatology scholars.[5] Yet, when one carefully reads her study, several substantive and methodological problems become apparent.

First and foremost, Nagy fails to account for the effect of cultural and social variables on children's perceptions. In this vein, it seems fair to ask whether the study of Hungarian children forty-five years ago is applicable to children in the United States today.[6] Indeed, variations due to class, sex, and ethnic, geographic, or religious differences are never even mentioned by Nagy, nor are they mentioned in the popular recipe-type guidebooks on children's grief that rely heavily on her study and its interpretations. The personal and family situation of the individual child is also ignored by Nagy. For example, the possible impact of a child's own death-related experiences on shaping the development of the child's understanding of death is never considered.

On a broader level, Nagy's analysis lacks an historical and cross-cultural perspective. One would assume from her observations that children of the Tame Death, of the Victorian Death, or of the Puritans would have an understanding comparable to that of a child living in New York City in the 1990s. From everything we have learned about the cultural shaping of patterns of human death from Ariès, Illich, and Stannard, it seems absurd to discuss stages of understanding about death that are acultural and unhistorical. Furthermore, what Nagy identifies as "immature" death perceptions of childhood can be found in patterns of death in historical and cross-culturally relevant situations in the experiences of adults. For example, in the South Pacific, the Tetum of Timor practice a ritual of double burial and believe that death is a two-stage process. Upon cessation of the body's functioning, the Tetum believe that death is coming. They ceremonialize the "not fully alive, not yet dead person" by bringing food and drink and entering into conversations with the nonliving, nondead person. (And as cultural folklore tells it, the more alcohol consumed by the visitors during their festive visitation with the nondead person, the more likely that the nondead will speak back to the living visitors.) After several days of this tran-

sitional stage (it used to be months or even years, before the Portuguese gov-
ernment intervened for public-health, hygiene reasons), the second stage of
death—"real death"—occurs, at which time the now fully dead person is committed to the earth for burial.[7] It is fascinating to observe that what Nagy considers a childish, immature definition (the stage one childhood belief that death
is gradual and/or nonpermanent) is a cultural norm for an entire society. One
suspects that Nagy would label the entire culture of the Tetum as confused and
childish with regard to their conceptions and rituals of death.

Moreover, the cultural patterns of the Danse Macabre during the era of
Remote and Imminent Death and the spiritual drama of the battle for the dying
person's soul, which took place during the Death of the Self, are remarkably
similar to Nagy's description of the personification of death in stage two. Like
the "boogeyman" who secretly slips in during the night to kill, the foul, rotten-
fleshed transi could surprise anyone, anytime. In both scenarios, death is per-
sonified: It is seen as a frightening, ugly, and external force. Like the child who
believes that this external force could be seen just prior to the moment of death,
the dying person not only visualized the death process and the spiritual struggle
at the deathbed during the Death of the Self, but also participated in the per-
sonified ritual. Thus, would Nagy repudiate the customs of the traditional death
eras as being infantile and immature? If so, she would have to endorse an eth-
nocentric belief that the modern, rational, scientific conception of death is the
culturally superior one.

More worrying than the imperfections of Nagy's study is the seemingly
mindless regurgitation of her findings, especially in the popular thanatology
literature. This itself is a serious indictment of thanatology and an indication
of general lack of critical reflection in the field. These faults may merely reflect the complexity of the subject, the modern ambivalence toward it, and the
broader cultural need to package the frightening and complex facts of death
into a therapeutically manageable framework. Thus, as we saw in preceding
chapters, the simplification of death, dying, and grief into identifiable stages—
for children as for adults—fosters a comforting, though false, sense of order.
However, if the complexities, vulnerabilities, contradictions, and ambivalences
of the association between modern American children and death are honestly
faced, this comfort is obliterated. In this regard, the damage that may be done
by misrepresenting and oversimplifying modern patterns of death and dying
urgently needs to be studied by thoughtful thanatologists.

Though the research data on children and death are contradictory and incomplete, a major trend in the literature can be identified, a trend that in many ways
is inspired and shaped by Nagy's work. Specifically, there is a tendency to evaluate the child's understanding of death through stages related to age and cognitive development. This approach is based on the assumption that a "mature"

understanding of death is possible only when a child becomes capable of rational, abstract thinking. Inasmuch as the ability to function at higher cognitive levels is related to the ability to communicate "maturely" about death, Piaget's well-known stages of cognitive maturation have been associated with children's knowledge of death.[8] Additionally, Ferguson and White maintain that concrete thought processes, along with logical thought operations, are essential for the child to have an appropriate and mature understanding of human death.[9]

The above discussion alludes to the "mature" concept of death. This term has three components: irreversibility, nonfunctionability, and universality.[10] "Irreversibility" refers to the understanding that once a living thing dies, it cannot become alive again. "Nonfunctionality" refers to the understanding that bodily life functions completely cease at death. "Universality" refers to the understanding that all living things must die.

The idea that death is irreversible, that it results in complete nonfunctionability, and that it happens to everyone is the standard adult conception of death in modern American society. The culture teaches this concept of death to children during the regular, ongoing process of socialization. Yet it does not mean that prior to arriving at a standardized, mature understanding, children have no awareness of death at all. Indeed, children as young as eighteen months old have been reported to have some idea of death.[11] For example, before children come to understand that death is irreversible, some believe that dead things can spontaneously come back to life again. Still others believe that life can be restored by prayer, magic, wishful thinking, or medical intervention and/or hospitalization. (The last belief is particularly fascinating, as it indicates that the broader trend toward the medicalization of death and faith in science and technology is even influencing children's early and immature conceptions of death.) In addition, before children come to learn that death involves the cessation of all physical functions, many believe that dead things function, but not as well or as fully as living things. In this way, one immature perception of death is that it entails the cessation of some functions, but not all. And during the early, immature period, it is not unusual for children to believe that some people can avoid death. Studies indicate that some children believe that death can be avoided by the clever and lucky, by parents and teachers, and by children in general, as well as by the individual child.[12]

In general, there is a wide range of findings on when and how children come to fully understand death. Despite the fact that many studies have sought to associate the mature understanding of death with cognitive ability, it is very unclear what specific cognitive processes are involved. More specifically, it is unclear what enables the child to arrive at an understanding of the components of the mature perception of death. An understanding of children's conceptions of death is further complicated by the fact that the studies use different ap-

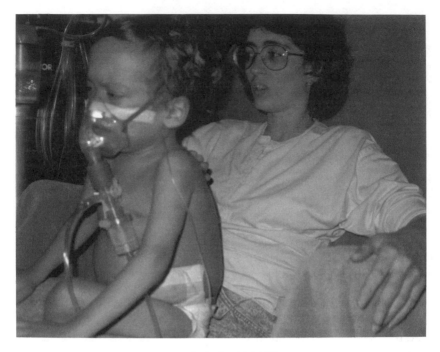

THE BURDEN OF LOVE

The endless medical requirements of a dying child completely absorb the time and energy of the single parent caregiver. This mother, seriously sick herself, provides around-the-clock care for her son, who was born with AIDS.

proaches, methodologies, and populations, a practice that makes them hard to compare. In this vein, it is especially difficult to verify the results of earlier studies using those published later. Additionally, since these studies typically ask children to describe their images and perceptions of death, the responses that children make may well reflect not their understanding of the subject, but their ease or unease with talking about it. In this way, the ability to conceive death, which is measured by the capacity to express the concepts of death, may be readily confused with the level of anxiety felt by the child, which is illustrated by the difficulty of expressing feelings and thoughts about death.[13]

The role that cultural and social factors play in shaping children's perceptions is also unclear.[14] Although gender, race, geographic location, religious background, personal experiences with death, exposure to television, and so on are presumably important variables influencing children's perceptions of death, little helpful research is available to interpret these factors.

It is not difficult to understand that the scholarly literature on children's perceptions of death is sketchy and fragmented. This is not just a reflection of the roles that normlessness and individuality play in shaping the modern circumstances of death and dying, but is also a reflection of the growing complexity of the broader social world and the inherent problems in the scientific study of this world and its myriad complex dimensions. Ironically, the fragmentation and ambiguity of the literature are occurring at a time when the amount of study and information available on children and death have never been greater.

Too Young to Die:
The Tragedy of the Seriously Ill and Dying Child

The life circumstances of the seriously ill, dying child have changed dramatically over the past two decades. Medical gains during these years have altered the course and outcome of many childhood diseases, especially illnesses such as cancer and leukemia. Not long ago, a medical diagnosis of childhood cancer was the equivalent of a death sentence. Indeed, childhood cancer, which was once regarded as a simple, acute, fatal illness, has now become a complex, life-threatening, chronic illness. Childhood cancer used to be simple in the sense that it was typified by an unvarying devolution toward death. It has become complex in the sense that the course of the illness is increasingly characterized by medical and psychosocial uncertainties.

Historically, as these illnesses were nearly always fatal, the diagnosis of childhood cancer therefore meant that the child would be treated as a dying patient from the time of diagnosis. Parental responses would necessarily entail adapting psychologically to the impending death of the child. However, progress in medical science has transformed the diagnosis of childhood cancer from an inescapable death warrant into a situation of uncertain survival.[15] Childhood cancer is thus characterized by the hope of survival, anxiously accompanied by the ever-present possibility of doom. The uncertainty and ambivalence that result from this situation are major factors defining the life circumstances of seriously ill children and their parents.

Ironically, clinical gains in modern medicine have heightened both the medical and the experiential uncertainties of childhood cancer. Not only is the overall prognosis difficult, but the results of particular medical-technological interventions are equally unpredictable. In this way, although survival rates are readily available, the clinical course of cancer for an individual child is uncertain. Largely because of this clinical uncertainty, the psychosocial experience for children and parents is also profoundly ambiguous and anxiety-producing. The struggle for life and against the disease of cancer increasingly transports the child and parents on a roller-coaster journey of hope, diminished hope, anxiety, and ambivalence. The progress that has been made by medical science has replaced certain death with uncertain survival. As medical intervention facilitates prolonged remissions, the value of health and life is affirmed. But this affirmation and kindling of hope take place within a framework of real and potential relapses, which bring to the fore the ever-haunting specter of death. The uncertainty, vulnerability, and stress of the modern experience of childhood cancer, then, are largely consequences of technological development and of the instabilities inherent in living and fighting for health and life in the ever-present shadow of ruin and death.[16]

The search for meaning in the dying and death of children traditionally took place within a context of moral guidance, ceremonial certainty, and community support, which are often lacking in the twentieth century. Within the modern context, it is less sure that the unyielding course of childhood cancer is an inevitable march from diagnosis, to short-lived remission, to relapse, to death. Uncertainty about the medical outcome has become the dominant factor shaping the experience of children suffering from cancer and for their parents. The meaning of this ambiguity is effectively elucidated by Comaroff and Macguire:

> For the most striking feature of the condition is now the unpredictability of its cause and outcome, which turns upon the starkest of alternatives—life or death. In fact, overall improvement in the length of survival of victims dramatically heightens the perceptions of uncontrollable threat in particular cases. Thus the hope of long-term (perhaps complete) remission becomes the preoccupation of all families, despite their awareness that the odds are unfavorable; and this hope is poignantly maintained against counter evidence. The course of the disease now becomes extremely difficult to define and classify. The significance of remission is not easily interpretable at any point in a particular survivor's career. Comprehending clinical predictions and translating them into conventional cultural terms is problematic. While the longer the child survives, the better his chances, relapse can occur at any time. . . .[17]

When a serious illness progresses in a way that becomes decreasingly responsive to medical intervention, the process of living with chronic illness becomes overshadowed by the increasing likelihood of childhood fatality. The fear and possibility of dying begins with the diagnosis of malignancy. But the optimistic biases of doctors[18] and the increased technological capacity to treat the disease initially repudiate the idea that the child with cancer is likely to die. Even as the disease progresses, the perception of the child as sick rather than dying persists. Thus, in addition to the chronic uncertainty that is inherent in the modern medicalized treatment of childhood cancer and other illnesses, another dimension of uncertainty emerges. Specifically, it becomes increasingly difficult to ascertain when the chronically sick child has moved from the stage of living with a serious disease to being in serious danger of dying from the disease. And parents used to living with hope are therefore understandably reluctant and ill-equipped to begin to accept the certainty that their child has become fatally ill.

Yet another dimension of uncertainty is related to patterns of communication with seriously sick and terminally ill children. Some have argued that children should not be told about the seriousness of their disease and of the probability of impending death. Those who believe in the value of protecting children in this way believe also that open disclosure would make the children more anxious and depressed. In a similar vein, it has been argued that young children lack the cognitive ability to understand what is happening to them and the ability to cope with heightened fears and anxiety. Therefore, being honest with children about their disease is both useless and counterproductive, and children should be actively shielded from these painful realities.

On the other side of the issue, there are those who claim, with equal earnestness, that children are able to understand, within a system of meaning relevant to their world, the seriousness of their situation. Additionally, some studies have shown that children are sensitive to the anxiety and emotions of others about their illness and, even at very early ages, come to realize that they are seriously sick and/or dying. The advocates of open communication thus argue that attempts to hide the truth serve no purpose.[19] Protecting children from what they eventually come to intuit and recognize anyway only serves to isolate them in a private, nonsupportive world where fears and traumatic fantasies are actually heightened.

As we evaluate the literature on childhood disease and communication, there is a noticeable trend over the past decade in favor of open communication. As we will see, this emphasis on openness is consistent with the underpinnings of the thanatology-awareness movement in the broader society, the human potential movement, and the therapeutic inclinations of the contemporary social world. Indeed, from the literature (especially popular and quasi-popular thanatology literature), we get the impression that open communication is a simple process of "just being honest" and, moreover, is universally desirable—that is, appropriate for all families with dying children.[20]

However, as we examine the realities, it is clear that no single pattern of communication is typically employed. The range of circumstances and the unique views of the innumerable actors involved in the world of dying children render the patterns of communication and awareness diverse and complex. Indeed, the varied communication patterns among children and their families are still another reflection of the lack of norms involved in the generic experience of dying in our modern society.

One of the most important studies of awareness and coping in terminally ill children is Bluebond-Langner's *The Private Worlds of Dying Children*. In her study of hospitalized, terminally ill, leukemic children, she maintains that by a complex process of socialization, leukemic children come to know about the seriousness of their disease and of their impending death. These children also come to sense and learn that they should keep their awareness secret from others and not talk openly about it. As they live through the disease and the various courses of treatment, children begin to develop knowledge about their condition and an awareness of the meaning of symptoms, procedures, and relapses. (The child's conception of death in this framework is therefore dependent not on age or cognitive ability, but rather on the degree and depth of personal experience.) As children are exposed to more and more experiences of the dimensions of serious and terminal illness, they develop an increasingly sophisticated awareness of their disease. Typically, this understanding evolves through five phases.[21] First, sometime after the initial diagnosis, the child learns that the illness is serious. Second, the child begins to know the names of the drugs that are used and their side effects. Third, children develop some knowledge of their symptoms and of the procedures used to treat the disease. At this time, the child develops an overall idea of the illness and, fourth, begins to see it as a series of remissions and relapses. As the ebb and flow of the disease continues, the child comes to the final recognition: namely, that the series of relapses and remissions may eventually devolve toward death. Table 6.1 illustrates the evolutionary development of knowledge about the disease.

TABLE 6.1.   The Child's Developing Awareness of Illness/Dying

A. *Learning About the Disease*

	1	2	3	4	5
Diagnosis	Illness is serious	Names of drugs and side effects	Purposes of treatments and procedures	Disease as a series of relapses and remissions	Disease as a series of relapses and remissions that ultimately lead to death

B. *Defining the Self in Association with the Disease*

	1	2	3	4	5
Well	Seriously ill	Seriously ill and will get better	Always ill and will get better	Always ill and will never get better	Dying

Source: Adapted from Myra Bluebond-Langner, *The Private Worlds of Dying Children* (Princeton, NJ: Princeton University Press, 1980).

In addition to becoming aware of the clinical facts, the child begins to define the self in association with the disease. Indeed, by witnessing the medical and technological processes that are happening to themselves and to other children, seriously sick children come to recognize their human and medical conditions through the personal experience of being sick, by observing other sick children in the technological confines of the pediatric hospital, and from the reactions elicited from others.

When children are faced with a potentially terminal illness, they and their families are quickly introduced to an overwhelming world of medical activity. The child realizes, in the face of medical procedures such as veripunctures, bone marrow aspirations, and seemingly endless blood tests, and as a result of the reaction of family members (tears, etc.), that the illness is not an ordinary occurrence:

> The children put people's behavior together with the tests and treatments they were receiving and concluded that they were really sick, "seriously ill," "very, very sick." "This is not like when I had my tonsils out" or even "when I cut my head open." "This" was somehow worse, as were they.[22]

In this way, children come to recognize that leukemia, AIDS, or other life-threatening illnesses mean something serious and extraordinary for their lives, and they correspondingly begin to see themselves as being very sick.

As children begin to learn about the treatment procedures, especially about the drugs that are being used, and as those drugs begin to have some beneficial effect, the children begin to see themselves as being "seriously ill but will get better soon." Continuing to learn about treatment and suffering through first and subsequent relapses, children often begin to see themselves as "always sick and will get better."

Additionally, during this period, the child begins to sense that medical staff and parents are reluctant to give a straight answer about what is happening:

> The children were left more and more to their own devices. They sought out peers for information on the purposes of various treatments and procedures. They formulated hypotheses about the relationship between various symptoms and the drugs, procedures, or treatments employed, and checked them out with peers. As one parent said ". . . they know they'll get a straight answer [from their peers]."[23]

As the disease progresses, as side effects occur from medicines, as the child's bodily integrity and functioning are increasingly impaired, and as home and school lives are more frequently preempted by hospital and medical activities, the child begins to understand not only the seriousness of the illness but also its permanence. If and when a peer dies, the realization of impending death also strikes the child.

There are three important lessons to be learned from Bluebond-Langner's study. First, the experience of serious childhood illness is one of suffering, anxiety, and despair, all of which increasingly come to dominate the lives of parents and their dying children. And an implicit finding in her study is that while the medical needs of these children are being adequately addressed in a hi-tech fashion, many others are ignored or inadequately satisfied.

Second, the experience of illness in the modern medical context is a salient factor affecting the child's perceptions of dying and death. Specifically, as the child's life becomes increasingly defined through technological interventions and medical evaluation, this process of medicalization shapes the images and realities of death for children and their families. Thus, an additional variable— namely, the impact of medicalization on the dying process—must be added to the research concluding that age and cognitive ability are the key factors shaping children's ideas about death.

A third important finding is that seriously sick children arrive at an understanding of their medical and human predicament within an environment of closed awareness and mutual pretense. As already noted, physicians, in carrying out their behavioral role expectations as healers, adopt an optimistic bias about the viability of the treatment strategies utilized.[24] In doing so, the emphasis for the physician, in a way similar to the roller-coaster scenario described in Chapter 3, becomes the management of physical symptoms. The issue of dying is averted by relating to the child within the framework of the sick role. Additionally, parents typically seek to withhold information from seriously or terminally ill children in order to avoid distressing and upsetting them.[25] Thus, mutual pretense enables physicians to pretend to safeguard the integrity of their professional role as healers; parents, confused about what to say or how to relate to their children, pretend to protect their role as nurturers, supporters, and guardians.[26] In this closed-awareness environment, the child also develops an understanding that dying and death are not appropriate topics to discuss with staff, family, and other adults. The child who learns quickly not to pursue the issues of dying and death in the presence of adults becomes an active participant in the game of mutual pretense and denial. This, however, leaves the child isolated in a privatized world of awareness. In this world, the child is forced to formulate his or her own adaptive coping strategies for living with the experience of serious illness. The irony is that in seeking to protect children from the tragedy of terminal illness, the modern, medically based conspiracy of silence functions to exacerbate the anxieties and responsibilities of seriously ill and dying children. And, as we see from Bluebond-Langner's study, it is a conspiracy doomed in any event to fail.

In short, the path of serious childhood illness is both torturous and ambiguous. The initial diagnosis in the advanced clinical context of modern medicine

Thanatology Snapshot 6.2

On the Experience of Childhood Dying

Death and disease were by this time constantly on the children's minds. Their world, their thoughts, hence their statements and actions, were permeated with death and disease imagery. In general, as this imagery increased in conversations and play, other topics decreased or were never discussed.

Conversations about the drugs and their side effects declined noticeably, which was not surprising since the children now realized that the drugs were not the answer they and their parents had once thought them to be. Further, when the drugs ran out death became imminent. . . .

In the early stages they had talked about going home, about their progress, in a particular way. When admitted for relapse with complications, they had talked about the procedures. When their condition had seemed to improve and it seemed possible they might go home, they had talked about lab reports and the efficacy of various drugs and treatments. Finally, when going home had been assured, even though they remained in relapse, they had talked about different things than they had in the past (but usually with reference to the disease)—about the way the disease had limited, but not totally curtailed, their activities; about what they could do when they left the hospital—and they had expressed hope they would not have to return for a while.

But, when they knew they were dying, when going home was only a few-week break from hospitalization, they were no longer interested in indications that they might go home (e.g., good counts, absence of infection), or in what they could do when they got there. They were not even concerned about how long it would be before they would have to return to the hospital or what complications would make the return necessary. They feared going home, because they might not be as comfortable there as they were in the hospital: "I want to stay here"; "I'm comfortable here. They can give me more pain medicine here." . . .

This attitude toward life, toward what it had become, was reflected in the children's overall behavior. Previously cooperative during procedures, even when in great pain, they suddenly began to balk, to cry, and to scream before, during and after all procedures. They would argue whenever doctors and nurses wanted to do a bone marrow or just change an IV. Some would also try to talk personnel out of doing routine procedures and complain if a nurse just wanted to make the bed. They did not want to be touched, poked or prodded anymore. If the staff tried to "reason with them" ("It will make you feel better so you can go home"), children often blurted out "You know I can't go home" or "I'm dying."

Myra Bluebond-Langner,
The Private Words of Dying Children
(Princeton, NJ: Princeton University Press, 1983), pp. 191–192.

is often fraught with uncertainty of outcome. The course of the illness itself, and of its medical treatment, is similarly uncertain and unpredictable. Death, of course, is the ultimate source of anxiety in this framework. Death is the thing about which one is most uncertain (in some cases, there is uncertainty as to whether death will occur; and in others, when) and of which one is most fearful. In addition, medical staff, adult family members, and seriously ill children approach the situation from different perspectives. In this way, not only are normlessness and individuality cornerstones of the modern experience of childhood dying, but so are isolation and alienation. And indeed, much of the torment and agony of dying children and grieving parents are directly derived from these modern social facts.

Romanticizing the Tragedy or the Growth of the Therapeutic Model

We should not forget that serious childhood disease comprehensively disrupts the life patterns of the child's everyday, taken-for-granted world. Indeed, not only is the potentially fatal disease a cause for seriousness and uncertainty for the child, it is a source of physical discomfort as well. Terminal illness and its treatment are often associated with nausea, vomiting, traumatic medical procedures, and even mutilation. Furthermore, children are usually hospitalized many times during the course of their illness, and many of the anxieties associated with institutional separation in general then become exacerbated. Not only do hospital stays involve disquieting medical procedures and separation from parents, they also enable the child to see how advanced his or her illness may become by allowing contact with other children who are more seriously ill or dying.[27] In this way, hospitalization not only brings about the experience of isolation through enforced separation from parents, siblings, and friends, but it also illuminates the possibility of a final separation, namely, death. It is not difficult to understand that fear of abandonment, withdrawal, depression, and loneliness are typically associated with the child's experience of hospitalization.[28]

Minimally, the experience of serious or terminal illness creates a threat to the child's bodily image (as a result of invasive, often distressing, medical procedures), a fear of dying, and a sense of isolation and diminished self-esteem.[29] When serious childhood illness was much more short-term and death was inevitable, the necessity to cope with these problems was obviously less. Now, as many diseases have become chronic with an uncertain outcome, the ways in which children come to terms with their life-threatening situations have become increasingly important. Coping strategies have therefore become a core concept in the literature on children and dying.

Because the experience of dying has increasingly become a private process, the attempt to help the seriously ill or dying child handle the stress of illness and its treatment usually begins by emphasizing the need for the child to express his or her feelings therapeutically. Indeed, the idea of openness in communication between staff and dying children, and of the therapeutic benefit of self-expression through direct or indirect means, has become increasingly valued.[30] Particular emphasis has been placed on the therapeutic benefit of emotional catharsis itself and on the value of the child's contribution in designing sensitive systems of care. Self-expression and catharsis have been advocated within a variety of frameworks, including hospice care for children, art therapy, play therapy, and psychotherapy.

One means of facilitating emotional expression has been proposed by Adams-Greenly. Drawing on her clinical experiences with terminally ill children, she identifies seven principles by which the health care professional should encourage communication and self-expression among dying children:

1. Children's cognitive and personal experiences must be taken into account and their perceptions of their situation ascertained.
2. In order to understand the child's perceptions, it is important to understand the child's verbal and nonverbal symbolic expressions.
3. The professional must clarify reality and dispel fantasy for the child.
4. The professional must enlist parental support in addressing the needs of dying children.
5. The expression of feelings should be encouraged.
6. The professional should promote self-esteem through mastery.
7. The professional should make no assumptions about the dying or seriously ill child and his or her needs, but should view each child as a distinct and unique individual.[31]

Thanatology Snapshot 6.3 illustrates the anecdotal vignettes upon which these strategies for facilitating open communication and self-expression are based. It is not difficult to see that rather sweeping generalizations and categorizations are made, based on intuition and personal conviction rather than solid scientific evidence. As we will see, this approach is not unusual among advocates of humanistic therapeutic intervention in the arena of children, dying, and death.

Obviously, as examples throughout this book have shown, human growth and love can indeed be nurtured by the tribulations of suffering. Yet the therapeutic model, as increasingly applied in thanatology, tends to simplify and romanticize the process of dying and death in a way that may trivialize the depth and anguish of human suffering. Indeed, in our contemporary framework of hi-tech, death denial, the death of a child is a haunting tragedy that parents never really

Thanatology Snapshot 6.3

In Pursuit of the Therapeutic Model:
Some Principles of Intervention

Principle 1. Ascertain the Child's Perceptions . . .

Devon, age 6, said when asked why he was in the hospital: "I have Hodgkin's Disease Stage IV B." I said, "Gee, those are pretty big words! What do they mean?" Devon replied, "My mother said if I ate too much candy, I'd get sick. So, that's what happened, and then they cut me open [splenectomy] and took it all out."

Principle 2. Understand Symbolic Language . . .

Kathleen, age 6, was admitted in the hospital in terminal condition when I commented that she looked "kinda tired," she replied, "Oh, yes, I'm so tired. I try to stay awake all night to pray, 'Oh, please, God, don't want me to be an angel!'" After a pause, I wondered aloud what it would be like to be an angel, and Kathleen explained that an angel has wings, wears a white dress and a halo, and flies around, sits on clouds, and eats ice cream. I said that being an angel didn't sound so bad, but I could tell she didn't like the idea, and I wondered why. She started to cry and said, "Don't you see? I wouldn't be with my Mommy and Daddy!"

Principle 4. Enlist Parental Support . . .

When I came to Fred's room the next day, I found that Fred had died. By phone Mr. and Mrs. T. told me that during a long talk the night before, they had realized that for two weeks, Fred had been afraid to sleep and therefore must have known he was dying. "In a way," Mr. T. said, "we had been selfish. We wanted him to hang on for us. For himself, he needed to rest." At the hospital that night, Mr. T. sat on the bed stroking Fred while Mrs. T. held Fred's hand and said "It's okay now, Fred, you can let yourself go to sleep. We are right here with you. Nothing will ever hurt you again." Fred had opened his eyes wide and said, "You mean I can go now?" Feeling reassured, he drifted off to sleep, his breathing relaxed, and he died an hour later.

   Although Mr. and Mrs. T. were sad about their son's death, they treasured this good memory of their last moments with him. They had reached a state of "reconciliation" in which their competence as parents was reinforced. As Mr. T. put it, "We could no longer give him the gift of life, so we gave him the gift of a peaceful loving, death."

Thanatology Snapshot 6.3 *(continued)*

Principle 6. Promote Self-Esteem Through Mastery . . .

When 7-year-old Greg returned to school after having his left arm amputated at the shoulder, his classmates refused to believe that his arm was gone: "It's still there—you just have tied it down to your waist inside your shirt!" Enraged, Greg jumped up and stripped off his shirt, to the great dismay of his teacher. At first, the other children were shocked; then they began to gather around Greg. Although worried, the teacher recognized that to stop the children from investigating would give them the impression that Greg's amputation was too horrible to talk about and make them even more anxious. She also recognized that the situation gave Greg an opportunity to master his own trauma and to educate his classmates about cancer.

As Greg proudly displayed his incision, the class looked closely with awe and respect, counted every suture mark. They asked questions such as, "Why did this happen?" "How did they cut it off?" "What did they do with it?" "How will you tie your shoes?" The teacher helped Greg answer some of their questions, and Greg described life in the hospital in great detail. The teacher reinforced his positive feelings about his experience and knowledge and then developed a teaching plan about cancer and its treatment and about the many ways in which handicapped people can lead normal lives.

Principle 7. Make No Assumptions . . .

When we approach the child with cancer, we must be free of assumptions about what the situation will entail and be open to what each encounter can teach us.

Devon teaches us not to assume that because we think we have explained something clearly, the child will understand. . . . From Kathleen we learn that an idea that is comforting to us is not necessarily reassuring to the child. . . . Fred and his parents teach us that even in the midst of such sadness, there is room for a peaceful, loving memory. . . . And, finally Greg teaches us not to underestimate the child's ability to master life's challenges creatively and with humor and dignity.

Margaret Adams-Greenly,
"Helping Children Communicate About
Serious Illness and Death,"
Journal of Psychosocial Oncology,
Vol. 2(2), Summer 1984, pp. 61–72.

expect to happen to them and is therefore one with which they are never prepared to deal. In this way, the gloss of optimism and opportunity for personal
growth with which Adams-Greenly and others coat their advocacy of the therapeutic, self-expressive models imposes an artificial standard on both dying
children and their families. The image, generated by descriptions of therapeutic intervention, reflects excessively simplified scenarios that often obscure the
complexity and variability of each case. Indeed, when the "advantages" of
bodily mutilation are enumerated and the experiences of the dying child are
emphasized as sources of learning and inspiration for parents and other adults,
the agony, ambivalence, frustration, and alienation felt by the parents and child
are rendered extraneous or inappropriate.

As we observed earlier, there is no more prominent spokesperson for the
thanatological therapeutic model than Elisabeth Kübler-Ross. Fourteen years
after the publication of *On Death and Dying*, she applied the framework she
established in that work specifically to the issue of children, dying, and death.
The following excerpt accurately reflects her idea of the therapeutic model as
a vehicle for joy and inspiration:

> There are thousands of children who know death far beyond the knowledge adults
> have. Adults may listen to these children and shrug it off; they may think that
> children do not comprehend death; they may reject their ideas. But one day they
> may remember these teachings, even if it is only decades after when they face
> "the ultimate enemy" themselves. They will then discover that those little chil-
> dren were the wisest of teachers, and they, the novice pupils. . . .
>
> Thus God who creates us all compensates the little ones as they fail physi-
> cally. They become stronger in inner wisdom and intuitive knowledge. . . .
>
> As the children of the concentration camps at Majdanek scratched little butter-
> flies with their fingernails on the walls before entering the gas chambers, so did
> your children know at the moment of death that they would emerge free and
> unencumbered into a land where there is no more pain, into a land of peace and
> unconditional love, into a land where there is no time and where they can reach
> you at the speed of their thoughts. Know that and enjoy the spring with new
> flowers coming out after the deadly winter frosts, and enjoy the new leaves and
> life bursting forth all around you.[32]

Indeed, the realities of the child's understanding of death in a framework of
closed awareness and mutual pretense, as depicted in the preceding section,
are ignored by this approach and by the idyllic images it engenders. Moreover,
the therapeutic model attempts to establish an inspirational environment that
is inapplicable to the lives of many dying children and their families. Thus,
when families evaluate their complex and profound response to a loved child's
dying in relation to the images promoted by the therapeutic model, fertile ground
for feelings of stigma and inadequacy is being established. Parents may feel

inadequate and inferior if they have been unable to transcend their fears, their tendency toward denial, and the difficulty they have in talking about their child's illness and death. The therapeutic model and its comforting images of "joy" and "growth" become a source of inadequacy for those parents who lack the "virtue" and "courage" to respond to their children's dying with the recommended and required dignity and grace.

The therapeutic model of childhood dying is not wrong in itself. The idea of psychosocial and emotional support for children and their families is distinctively valuable in the modern context, with the alienation, uncertainty, and denial it creates. A problem, however, emerges when mesmerizing and idyllic images counterfeit and substitute for the real experiences associated with the illness, dying, and death of a child. It is not only unhelpful to idealize the difficulties and sufferings of parents and their children, but it is also unproductive from a scholarly point of view. The task of scholarship is to seek truth and understanding, even if the truth is socially unpopular and distressing, not to distort and selectively emphasize particular aspects of reality just because it is convenient, desirable, or profitable to do so.

Given society's focus on self-expression and personal development, the evolution of the therapeutic model for childhood dying is understandable. In fact, the ideas underlying the human potential movement extend beyond the clinical setting of seriously ill and dying children to the more general area of children's relationship with death.

In spite of the societal tendency to deny death and isolate children from death and dying, a countertendency that advocates talking openly to children about the facts and feelings of death has gained a strong foothold in American culture.[33] The idea of educating children about death is consistent not only with the general framework of the human potential movement but also with the therapeutic model. Specifically the goals of education for death awareness are three: (1) to increase the child's knowledge of death; (2) to help the child express his or her feelings about death; and (3) to enable the child to reap the benefits of participating and interacting with others during the course of the dying process. Television specials (*Mr. Roger's Neighborhood, Afterschool Specials*, etc.) have brought the issue to the popular attention of children. Books on death and dying have been written for children,[34] and also for parents to guide them in teaching their children about this topic.[35] The value of open communication about death and dying with children, within the context of their families, is also strongly promoted by the professional therapeutic community.[36]

The exposure of children to the themes of death and dying reflects a general trend toward educating them about controversial social issues: for example, sexual activity, drug and alcohol use, and family instability and divorce. Children thus increasingly confront and even participate in the complex world of the American adult. A number of important recent studies have articulated concern over the rapidly developing trend toward the disappearance of childhood.[37] Indeed, a convincing argument has been made that children are coming to dress, talk, and behave in ways that are traditionally and exclusively reserved for adults. There is little doubt that children, through mass media such as television and films, through music and their peers, and through intensified educational efforts at home or in school, are more exposed to the dilemmas, contradictions, and hazards of everyday adult life. In fact, children are deeply touched by the workings of the everyday world, including its violence, sexuality, drugs, family strain, and sociopolitical corruption.

Some have argued that in the modern world, the opportunities for freedom, growth, and responsibility for American children have never been greater. Others believe that children both need and deserve childhood, and that the cultural arrangements and expectations of modern living are making children grow up much too quickly. The legitimacy of these points of view will not be debated here. Instead, it is sufficient to note that in the context of American society, where the standards, experiences, and situations of adult life are increasingly relevant to the lives of children, the ideas of open death awareness and death participation for children are additional indicators of this trend.

Obviously, however, the competition of social forces to either realize or prevent the "adultification" of the child means that neither goal is accomplished. Likewise, the surging emphasis on open death awareness for children is only partially institutionalized and is seriously hampered by other societal factors that foster denial and exclusion. In this paradoxical situation, we find further evidence of institutionalized normlessness in regard to children and death. Likewise, the emphasis on full and open disclosure, with its complement of therapeutic support and educational programs, is counterbalanced by institutionalized medical realities that seek to protect children from death, and by the social realities that inadequately prepare parents in a consistent and open manner for their child's dying.

In short, and in the absence of consistent norms, a vast difference often exists between the idyllic, mesmerizing images of the therapeutic model and the ambivalent, sometimes agonizing realities of childhood dying and death. In any event, modern families are typically forced to develop their own distinct ways of sustaining themselves through the experience of serious childhood illness, dying, and death. On a personal level, this means that coping with the dying of a child is typically privatized and individualized. On a societal level, it means that the patterns of childhood dying are typically variable and pluralistic.

Parental Bereavement: Attempts to Cope
in the Face of the Unthinkable

Parental bereavement, of course, does not originate at the moment of death. Throughout the period of caring for a seriously ill, dying child, parents not only live with the constant threat of death but also experience major and traumatic disruptions of their lives. In her study of the practical, emotional, and social consequences of caring for a fatally ill child, Judith Cook has identified eighteen psychosocial problems that mothers and/or fathers may potentially encounter during the course of their child's illness: (1) making child-care arrangements for the ill child, (2) arranging child care for other children, (3, 4) encountering disciplinary difficulties with the ill child and siblings, (5) experiencing feelings of helplessness, (6) feeling a loss of confidence in parenting abilities, (7) confronting financial difficulties, (8) experiencing a sense of being avoided by other people, (9) feeling the need to reassure and comfort others, (10) suffering marital strains, (11) feeling excluded from participation in the ill child's life, (12) suffering the loss of religious faith, (13) experiencing the need to protect one's spouse from distressing feelings and upsetting events, (14) feeling that one's family unit is being torn apart, (15) feeling that one's spouse may be excessively preoccupied with the ill child, (16) fearing the inability to cope with the actual death of the child, (17) having to handle the inability of other family members to accept the child's illness situation, and (18) seeing the withdrawal of one's spouse from the family unit.[38] Thus, the chronicity and ambiguity of serious childhood illness create an atmosphere of living in a framework of impending loss, anticipating the possibility or probability of the death of the child, as well as having to deal with a variety of role-threatening losses and strains.

During the process of caring for a fatally ill child, mothers and fathers define the experience of anticipatory bereavement in different ways. Mothers identify the need to protect others from upsetting and distressing information as a prominent concern. Mothers also perceive marital difficulties as being a major factor affecting the course of living with and caring for a terminally ill son or daughter. Indeed, mothers feel resentment from their husbands over the extent to which they themselves are involved with the dying child and his or her illness. Mothers also express concern over sexual problems and the sense of isolation precipitated by the social and emotional withdrawal of their husbands. Mothers believe that existing instabilities in their marriages are heightened by the anticipatory bereavement experience, which often causes problems of drug usage, alcohol intake, and extramarital affairs. Mothers also feel the tremendous burden of keeping up the sick child's morale. And last, as a predictable result of these burdens, mothers believe that the experience of anticipatory bereavement places an overwhelming emotional strain on their lives.[39]

Fathers, on the other hand, feel that a dominant problem stems from the emergence of dual responsibilities—namely, the need to juggle and adjust their work schedules in order to accommodate increased family obligations. Fathers stated that they felt excluded from family interactions, as many of the day-to-day decisions about the care of the ill child were made without their participation. In a related sense, fathers also expressed concern about what they perceived as the overinvolvement of their wives with the sick child and his or her illness. And last—and this too is fully understandable in view of these problems—fathers expressed a longing to be able to spend more time with their wives and ill children.[40]

Parents of dying children face the disruption of personal, marital, and family patterns, and live in the midst of multiple profound changes. Part of this whirlwind of change entails the creation of a new role for both of them: acting as parents of a dying child. As we have seen, anticipatory bereavement is a different experience for men and women, and by studying these differences among mothers and fathers, several valuable insights are gained. First, we learn something important about their respective roles and expectations in society, and how these cultural prescriptions and expectations are shaped by gender. Second, the divergent responses of fathers and mothers illustrate the fragmentation, polarization, and individualization of the modern grief experience in a way that is specifically relevant to children and death. Third, as the mothers and fathers in Cook's study indicated, fragmentation of the concerns and tasks related to caring for a dying child generates special stresses and strains in the parents' relationship.

Judith Cook effectively articulates these considerations in the conclusion to her study of the role of gender and its influence on anticipatory parental bereavement:

> Overall, women seem more pervasively steeped in the culture of the child's illness—a culture in which men feel out of place and uncomfortable. The father's social settings are those of work and of more instrumental activities such as telephone calls and running errands. Thus men may feel excluded from the illness and prevented from taking part in the child's care. . . .
>
> Conversely, since the mothers are more involved in the direct care of the child, they may experience the death as more traumatic than the fathers do because they had a closer pre-death relationship with the child. The same pattern may create difficulties for the woman who is faced with the task of repairing relationships with her husband and her children who resent their mother's lack of attention during their sibling's illness. In addition, the spouses' inability to obtain emotional support from each other may be reinforced if wives continue to shield family mem-

bers from upsetting feelings and if husbands continue to withdraw from involvement in the dynamics of the family. Moreover, after the child's death, the family that previously was oriented toward the dying child may experience a loss of purpose and direction, which may create feelings of ambiguity about already-strained relationships and further impede attempts to resolve familial conflicts.[41]

These conclusions should not be surprising in light of the harsh realities affecting the lives of parents and of their dying children. Not only do Cook's findings instruct us on the influence of gender on bereavement, they identify some of the stresses caused by the dying process for parents in the same way that Bluebond-Langner's study identifies them for dying children.

Coping with the unrelenting turbulence and stress of caring for a terminally ill child becomes imperative for the parents. Of special interest is the fact that these stresses are not inevitably detrimental to the parents' marital relationship. In fact, in a study of parents with still living, seriously sick children, a majority reported that the quality of marriage and family life either remained unchanged or improved since the diagnosis of their child's illness:

> A majority of parents indicate that their feelings toward each other remain unchanged or changed in a positive direction, that their spouses had been a significant source of support during difficult times. In addition, a majority of parents report that their families have become closer as a consequence of the illness and its treatment.[42]

Most of the literature, however, identifies the death of a child as a major source of stress for the parents and their marriage. It is important to note that while most of these studies of bereavement and its impact on the marital relationship are based on samples in which the death of the child had already occurred, the participants in the above study consisted solely of parents of still-living children. In attempting to explain why the other studies typically report that bereavement negatively affects marriage and this one does not, the following explanation is offered:

> Parents of deceased children may have experienced more stress and have more negative views of family relations. In addition, new and alternative treatments hold out the possibility of long-term survival and even care for children currently living with cancer. Until those alternatives are exhausted, parents may not experience the depths of depression and hopelessness that once characterized the situation of families who had little or no hope of effective treatment. . . .
>
> Because of the threat to the child's survival, the couple may not be attentive to the evidence or sources of discord and, consequently, may continue to evaluate the marriage favorably. As questions regarding the child's survival become settled, the "piled-up" issues rise to the surface, contribute to a sense of marital distress, and increase the likelihood of dissatisfaction and discord.[43]

Several interesting questions are raised by this study. It may very well be that during the phase of living with a child's terminal illness, the full stresses of grief and bereavement are not yet experienced. Although this is a plausible explanation, the value of anticipatory grief in facilitating postdeath adjustment is seriously challenged if marital distress increases upon the child's death. It would be useful to study parental responses to sudden infant death or accidents in comparison with responses to death following chronic illness. It may also be, as the above extract proposes, that the hope engendered by medical advances may mitigate the maritally destructive dimensions of coping with the stress of childhood chronic illness. But though the discord of the anticipatory death period may be temporarily reduced, the subsequent shattering of hope caused by death may exacerbate the discord and distress of the postdeath bereavement period. In this way, the clinical gains of medical science may temporarily soften the anguish of anticipatory loss, only to intensify the anguish of loss upon death.

The responses of mothers and fathers in postdeath bereavement are similar in some ways and different in others. For example, both mothers and fathers feel an overwhelming and intense sense of loss. Both experience sadness during significant days and holidays. Additionally—and this perhaps is the greatest similarity—both mothers and fathers have great difficulty in seeing other, healthy children, especially during the first year after their child's death.[44] However, fathers describe their sense of loss as a "void," as the feeling that there is "something missing." They also feel a loss of direction in the family and the subsequent need to "regroup," to "reorganize." Mothers, on the other hand, define their sense of loss less in terms of family impact and more in an intimate and personal way. In other words, while both parents experience an intense and overwhelming impairment, the loss of a child disrupts the father's external world, whereas, for the mother, the death of a child is often experienced as an obliteration of self and of personal identity.[45]

Crucial differences in postdeath bereavement are clearly related to the traditional gender expectations of men and women. Fathers, in connection with their largely instrumental world view and the norm of American male inexpressiveness, emphasize their feeling of responsibility for managing the grief of the family unit. Furthermore, fathers usually shoulder the emotional burden of grief in a solitary and private way. Mothers, on the other hand, suffer from repeated visualizations of the deceased child, see painful reminders of the child around the house, and experience an emptiness related to the child's absence from their daily routine. The more personal dimensions of the mother's grief— her deep loneliness and feeling of personal diminishment—make her grief typically more profound and difficult to handle. Mothers also are more likely to express feelings of distance from their husbands and are less likely than fathers

CHILD IN A CASKET

Eric at age four has physically departed from this world. The memory of his suffering and death continues to haunt his loved ones every day.

to be satisfied with the support they receive from their spouses. Fathers, on the other hand, are less likely to be distressed by an absence of support, real or imagined, as they typically cope with the emotional burdens of grief in a self-contained, private sphere.[46]

Mirroring generic gender prescriptions of the society at large, bereaved mothers are therefore more likely to express their feelings and needs than are fathers. In this way, not only are the parents using different coping strategies, but their differing reactions create distinct psychosocial patterns of need that make it difficult for them to find comfort in each other. Moreover—and relevant to the major themes of this book—in the absence of guiding norms for bereavement and in the cultural framework of isolated grieving, men and women understandably rely on gender prescriptions that have influenced them all of their lives to provide some sense of stability and familiarity to the turbulence generated by the death of a child.

We do not mean to imply that the gender prescriptions and the tendencies noted in the aforementioned studies provide unvarying guidelines for parental bereavement. In fact, very little is known about precisely how traumatic the death of a child is in modern society or how difficult the process of anticipatory and postdeath grieving is. Although some studies, such as those discussed above, provide useful information on parental bereavement, our understanding of the impact of a child's death is still sketchy and incomplete. Pointing to the biases of the studies' methodology, their excessive reliance on anecdotal information, and the lack of broadly representative samples and participants, Sanders puts the current state of knowledge into the following perspective:

> It is obvious from the lack of studies in this area that we know very little about the relative intensity or duration of the effects of grief upon parents who have lost a child.[47]

Many variables combine to make the study of parental grief both complex and inconclusive. For example, there are a wide variety of factors relevant to the overall assessment of parental grief and to the particular impact of grief on marital status. Religious faith,[48] community support, ethnic background,[49] level of education, the parents' own anxiety about death and prior experiences with death, the extent of the sick child's suffering and the manner in which he or she died, and feelings of parental guilt during bereavement,[50] as well as the length, stability, and quality of the marital relationship, are some of these variables.

There is another major shortcoming in the contemporary literature on parental bereavement. The studies typically involve parents living in the traditional nuclear family. The state of research on parental bereavement is thus being shaped by a somewhat outdated framework. Obviously, many parents live and grieve within the traditional nuclear family, but to focus exclusive attention on the problem of bereavement from this perspective is to ignore an entire spectrum of other modern lifestyles and living arrangements. In addition to traditional and modified nuclear families, contemporary families are organized around dual-career couples; cohabitation; nontraditional sex roles; single-parent families through widowhood, separation, and divorce; and reconstituted families, with one or both of the spouses bringing children from former marriages to the relationship. In this light, the studies discussed above and the overwhelming majority of others on parental grief simply fail to take into account the plethora of family living situations in which bereavement occurs. And the abundant patterns of family life in which bereavement takes place further testify to the pluralism of death patterns characterizing our age and the normlessness, privatization, and individualization of modern grief responses.

Parental Postdeath Bereavement and Therapeutic Response

The importance of immediate postdeath therapeutic healing for the bereaved is emphasized in the therapeutic thanatology literature. A considerable amount of anecdotal and inspirational-type material stresses the role that guidance, open communication, and professional support can play in eliciting expressions of parental grief and in initiating the flow of grief work that will ultimately speed parental mending and healing.[51] Explicit and immediate therapeutic intervention is advocated whether the child dies at home,[52] in a hospice for terminally ill children,[53] or in a hospital setting.[54] As two therapeutic advocates put their case:

> The importance of the bereavement group experience to hospital administrators and to mental health professionals in hospitals and in the community cannot be overlooked. . . . Bereavement is not a pleasant subject, but it is a part of life. As social workers, the authors have the privilege of facilitating growth by intervening with families in an emotionally healing manner. Those hospitals and agencies that support programs such as the bereavement group experience provide a much needed service to a special population.[55]

The therapeutic orientation has also been fully extended to sudden infant death syndrome (SIDS) and its associated parental bereavement. Two national organizations (the National Sudden Infant Death Foundation and the Council of Guilds for Infant Survival) offer information and counseling services to parents whose children die from this syndrome. The focus of these organizations, and of the general therapeutic approach to SIDS, is based on encouraging emotional expression, providing psychosocial support, facilitating grief work, and promoting recovery and growth. The therapeutic intervention typically involves nurse counselors, pastoral and social work counselors, physicians, and psychiatrists, all devoted to the emotional and psychosocial support of the parents and other family members. The discussion of an interpretation of autopsy findings with bereaved parents is also claimed to be of salient therapeutic value.[56] A special section of the medical journal *Patient Care* offers guidelines for health care personnel "When Sudden Infant Death Strikes." Their recipe for a response to SIDS parental bereavement was drawn from interviews with four medical doctors and is summarized in the following therapeutic prescription:

> Counseling parents immediately after their child's death consists of providing them with the basic facts about SIDS and reassuring them that the death could neither be foreseen nor prevented. It's often helpful to give them written information on SIDS. . . .
>
> Convey autopsy findings to parents as soon as possible. Ideally, they should have the opportunity to discuss the findings with the pathologist.

After the period of initial shock, parents may need extensive psychosocial support. Attempt to assuage their guilt by re-emphasizing what is known about SIDS,
and provide calm support while allowing them to grieve and express their feelings.
. . . Assess the adequacy and individual bent of their coping mechanisms. . . .

Make sure the parents know how to get in touch with a SIDS parent group or
similar organization . . . some parents may benefit from sharing their grief in
group therapy with other affected parents; others, who may be more depressed
or humiliated by it, may prefer individual talks with another SIDS parent, in
person or by phone.[57]

The importance of the funeral is also emphasized in the overall program of
therapeutic support. First, parents are often advised to encourage the attendance
of siblings (and, indeed, the trend has been toward more and more children
attending their siblings' funeral). Second, the therapeutic thanatology literature defines the funeral as a memorable and therapeutically useful tool for grieving parents. Some writers even suggest that active participation of the parents
in the preparation of the child's body for the funeral viewing (washing the body,
dressing the child, combing her hair, etc.) is of therapeutic value.[58] It is important to note, however, that the idea that the funeral is helpful is presented without concrete corroborating evidence that traditional funerals are associated with
improved emotional or behavioral adjustment for parents or siblings. The words
of Kübler-Ross make apparent not just the idealized viewpoint of the therapeutic advocates, but also the biases underlying their definition of appropriate
and successful reactions:

Children, more frequently in recent years, have asked to be able to prepare their
own funeral. Adolescents particularly want to know ahead of time what they are
going to wear, what music will be played. . . . Needless to say, such prepara-
tions demand a cooperating, well prepared family or friends. It requires an acceptance of the impending death and an openness of communications which is
occurring more and more frequently.

Next of kin should at least have the option to wash, dress and comb the hair
of their own child, to rock a baby, to hold a stillborn for a while until they are
ready to let go; to carry a dead child into the car of the undertaker or drive their
own child to the place of the wake or the viewing. . . .

Funerals are often a time when a family can share the poems their child had
written, offer a philosophy of life that they learned from their dying child, and
begin an opening in the awareness of those who participate—a beginning dawn
of knowledge that a ship which disappears behind the horizon is not gone, only
temporarily out of sight.[59]

Although the words of Kübler-Ross may be embracing and comforting, they
fail to take into account our cultural inhibitions in communicating about death,
marital stress and discord, and divergent ways of coping.

Funerals, however, can be associated with at least one standardized response. There is an institutionalized expectation in American society that parents should begin to mourn the death of their child through a funeral. Ironically, although there is limited evidence to support the view that the funeral is therapeutically helpful, participation does seem to provide one base of stability in a situation plagued by uncertainty and anxiety from the moment of initial diagnosis through the bereavement period. It is also important to note that other systems of support, such as assistance of friends, support of a caring physician, or involvement in a counseling program, are available to parents only in a haphazard, irregular way and according to individual circumstances. In the face of this unpredictability and uncertainty, the funeral offers one guarantee of communal participation in the parents' mourning.

A means of overcoming the parents' sense of isolation is provided by bereavement self-help groups. Such groups have become established in response to the overwhelming sense of isolation and individualism that characterize the drift of our times:

> In an age of alienation and anomie, the self-help group encourages people to jointly take responsibility for themselves and others. The self-help group fosters healing by building a community of people who help themselves by helping others.[60]

One such organization, the Compassionate Friends, is a nationally organized self-help group. In his study of the dynamics of the Compassionate Friends as a vehicle for healing parental grief, Dennis Klass has identified three distinct phases of parental involvement: the decision to attend, the process of affiliation, and the transition to helping others.

Initial attendance is sparked by a variety of personal and social factors, and the ways parents become connected to the self-help organization are varied and often haphazard. They may be referred by a wide range of sources, such as friends, clergy, and health care professionals. Often, contact with self-help groups takes place in conjunction with bereavement counseling or psychotherapy. Sometimes, after a period of aimless floundering, grieving parents themselves reach out and initiate contact with the group. At other times, parents become reluctantly involved in order to please someone else. In any event, feelings of estrangement and isolation commonly motivate attendance; parents who make the decision to participate in the Compassionate Friends often feel separated from other sources of support and comfort, such as family and friends.[61] In the face of loneliness and in the absence of social support, the idea of participating in a voluntary planned bereavement group can become alluring to the grieving parent.

Regardless of how or why they become involved, parents typically begin their participation with a sense of timidity and uncertainty. While some parents do feel that attending will be helpful, they are often unsure of precisely what to hope for. Others are simply confused about why they are there and ambivalent about the value of belonging to the group.

The process of becoming actively involved in the group, of making it part of oneself, and of making oneself part of it has been termed "affiliation." The first part of the process involves catharsis. This occurs when the bereaved parent takes emotional energy (sense of grief, pain, or loss) and begins to share it with the group. By telling stories and showing pictures of the deceased child, the bereaved parents initiate a strong and special emotional interconnection and bonding. In a Durkheimian fashion, sharing emotional energy with the group in this way not only provides emotional release, but also builds collective unity and support. Thus, during a time of crisis and devolution in the parents' life, a collective vibrancy and energizing begins to emerge. Although Klass is remiss in making the connection to a Durkheimian framework, his words are very much cast in this mold:

> The sense of unity with those whose lives have been shattered, the sense of hope at seeing that others have made it, the sense of finding an appropriate object on which to attach the energy formerly given to the child, the sense of family in a supportive community, and the special relationship with someone very much like the self but further along are all part of the cathectic dimension.[62]

The second part of the affiliation process is called the "experiential dimension." Whereas "catharsis" means the sharing of emotions, the experiential dimension involves the sharing of experiences and information about how individual parents solved specific problems and survived watershed events. As a variety of individuals reveal their own strategies for survival and coping, newcomers are presented with a wide array of situations and alternatives. Thus, involvement in the Compassionate Friends does not entail adhering to fixed guiding norms to facilitate grief work. Rather, it provides an awareness of some possible options and encourages individualized responses appropriate to each parent's personal needs and circumstances.[63]

The final phase of the self-help process is termed "transitions": the movement from being helped to helping others. It entails a forward motion toward the investing and reinvesting of the energies of the self in relationships with others and involves a reaching out to others, specifically the newly bereaved, to provide support and help in the immediate aftermath of their postdeath trauma.[64] It is important to recognize that this transition does not diminish the therapeutic value of the group for the helper. Indeed, not only does the helping role give the parents an opportunity to continue a symbolically meaningfully relationship with their dead child, it enables them to continue to participate in the cathectic and experiential aspects of the group process.[65]

The Compassionate Friends is thus a voluntary self-help community of be-reaved individuals who seek the support of others as they try to cope with the death of a loved child. The paradox of the Compassionate Friends' organiza-tion is that it simultaneously emphasizes community and individualism. Clearly, the Compassionate Friends exists to respond to the needs of parents in the ab-sence of guiding norms and bereavement rituals, but participation in the group strongly affirms the value of self-expression. In this way, the Compassionate Friends does not establish collective norms and rituals, but rather focuses on personal, private assertions of grief. The institutionalized structure of the Com-passionate Friends may reflect an absence of and a thirst for community, and its operation may facilitate a temporary semblance of community, but it is grounded in the ideology and framework of American individualism.

The emergence of the therapeutic model is understandable in light of the American emphasis on individualism and self-development. The specific effi-cacy of its application to the area of parental bereavement has yet to be estab-lished. Many questions remain to be asked and answered. For example, it is important to evaluate who is included and served by the therapeutic bereave-ment effort. Is there, as there seems to be, a white, middle-class bias in the organizational realities of the therapeutic model and self-help groups? Are African Americans, other minorities, and the poor unlikely to become actively involved? Does therapeutic and self-help intervention generally work? In what ways does it fail or succeed? How can success or failure be measured, identi-fied, or predicted? For whom is it likely to fail? Does the need for therapeutic, self-help support end at a particular point or is the lifetime inevitability of parental bereavement something that creates a need for perpetual therapeutic assistance? How does therapeutic, self-help participation affect the gender-specific response to bereavement? Is divergent coping, and its potential for mari-tal disruption, ameliorated by the therapeutic process? At present, there is a surfeit of ideologically biased advocacy of the therapeutic, self-help movement (even Klass' studies are biased inasmuch as he serves as a professional con-sultant to the self-help group he studied). In the absence of unbiased studies, the meaning and benefit of the prevailing cultural response to the trauma of parental bereavement can only be anecdotally and speculatively affirmed.

In Summary

The nature and images of childhood are a central part of the cultural and social fabric of any society. Perhaps there is no greater illuminator of the dominant issues, images, or themes related to society and death than the relationship that exists between children and the processes of dying and death. Thus, the epi-graph that began this chapter also conveys a very special secular meaning—

that the changing circumstances surrounding children and death are a prime illustration of the modern shift in the way death is generally treated and understood. Thus, it is useful to study the nature of children's involvement with death and dying as a signal indicator of the place of death in the broader social context. Indeed, as we survey the literature on children and death, discomforting images of normlessness, individualization of response, unresolved tension between open versus closed approaches to communication, an excessive role of technology in shaping the human meanings of the terminal experience, the absence of cultural rituals, and an overreliance on therapeutic response become readily apparent.

The significance of normlessness is clearly evident from the conflicting and contradictory studies on children's perceptions of death. The collective tendency to protect children from death, in conjunction with the countermovement toward openness to death, bespeaks norm dissensus. In this framework, individuals are increasingly left to their own adaptive resources in facing up to the tragedy of childhood dying and death. Individual sculptures, that is to say, personalized carving out of adaptive coping strategies, are a common reflector and a focus of the individualized response to children and death. Additionally, and in conjunction with the competing orientations toward openness and mutual pretense, this facilitates the development of pluralistic death patterns on a societal level. Furthermore, movement toward privatization and pluralism is not only a consequence of thanatological normlessness, but also serves to affirm the social realities of this normlessness.

As we have seen consistently throughout this book, the gains of modern medical technology have dramatically altered the course of the serious illness and dying trajectory in modern society. The impact of technology and the evolution toward chronicity in the experience of terminal illness are never more evident than in the case of dying children. In fact, anyone who has visited a pediatric intensive care unit or pediatric oncology unit will quickly recognize that technology is changing the way children die, and that a revolution in childhood death and dying is taking place. The uncertainties and ambivalence that typically define the course of childhood dying not only enmesh parents and children in an insecure and bewildering world of technological intervention and psychosocial doubt, but are an additional testament to the instabilities and absence of norms that characterize childhood dying. The inability to predict medically how individual children will respond to particular clinical interventions also contributes to the uncertainty of the process of dying, which, in turn, contributes to the growing reality that the paths of childhood dying are increasingly individualized in personal, psychosocial, and medical modes.

The emergence of the therapeutic model, and its application to dying children and bereaved parents, are an attempt to fill the need for support in the

WHEN A CHILD DIES

Regular, taken-for-granted assumptions about life and the world are severely challenged by the death of a child.

absence of supportive rituals of dying, death, and bereavement. The tendency toward overzealous and ideologically biased advocacy of the therapeutic model needs to be evaluated in this light. The growth of this model, and of self-help groups for dying children and bereaved parents, also reflects a cultural commitment, largely internalized by the participants, to the value of self and self-development.

In conclusion, although more information on children, death, and dying is available than ever before, surprisingly little can be said with certainty about children's death perceptions, the process of childhood dying and associated parental bereavement, or the efficacy of the prevailing thrust toward therapeutic intervention. The study of children and death therefore becomes even more important, not just for its value in potentially ameliorating the suffering of dying children and bereaved parents, but also as a rich research area for scholars. The specific study of children and death can provide general insights into shifting patterns of death and dying in the wider society and into their implications for the lives of human beings today.

7

The Death of Humans by Humans, Part One: Violent Deaths of Suicide

> There is but one truly serious philosophical problem, and that is suicide. Judging whether life is or is not worth living amounts to answering the fundamental question of philosophy.
>
> ALBERT CAMUS

It has been said that suicidal behavior reflects a pathological state of being human. In popular terms, this means that if a person either attempts or commits suicide, he or she must be "sick." On a more professional level, suicidal behavior has become defined as a psychiatric and medical abnormality requiring active therapeutic, clinical intervention. Thus, the professional response to suicidal behavior is not to perceive it as immoral, but rather to define and respond to it as a sickness requiring therapy.

The popular and professional views of suicidal behavior define the problem in terms of individual factors of pathology and deviation. This approach is partially correct. Indeed, there are profoundly personal and deeply individual motivations for suicidal behavior. However, when suicide becomes a source of duty or even honor, as in the situations of kamikaze pilots, harakiri, a soldier sacrificing his life to spare the lives of his comrades, a captured spy who ingests a cyanide tablet, or a group of 900 individuals who commit suicide collectively, motivations become very much cultural and social. Additionally, when certain kinds of social factors affect the incidence and prevalence of suicidal behavior, social arrangements and social organization become key factors in explaining human suicide.

In exploring suicide as a form of violent death in this chapter, special attention will be paid to the importance of identifying and integrating the deeply personal and highly social motivations of suicidal behavior. In seeking this integration, the themes of individualism, normlessness, and the technocratic organization of society will once more emerge.

Killing Oneself

An understanding of why human beings kill themselves may never be fully attained. The personal and sociocultural factors of suicide, taken individually and in combination, are so complex that it would be impossible to arrive at a definitive explanation. It is possible, however, to assemble a collage of understandings and explanations that, when viewed as a whole, provides an overview of the problem. Despite many limitations in our knowledge, sociomedical and psychiatric research, ethical analysis, philosophical reflections, and artistic, dramatic portrayals have generated many valuable and instructive insights into the web of human suicide.

Thanatology Snapshot 7.1

Epidemiological Portrait of Suicide

According to official figures, there are nearly 30,000 suicides a year in the United States; the real number is obviously higher, some think as high as 100,000. . . . More people die by suicide than homicide. It is now the tenth leading cause of death in the United States and accounts for at least one percent of all deaths.

Suicide rates are highest in old age: 20 percent of the population is over 60, but 40 percent of suicides are over 60. After age 75, the suicide rate is three times the average. . . . The suicide rate at ages 15 to 24 has gone up 150 percent and is now near the average for the population as a whole. This age group . . . once accounted for 20 percent of male and 14 percent of female suicides. . . .

Three times as many men as women commit suicide; among the elderly the proportion of men may be as high as 90 percent. Blacks commit suicide less often than whites, although the rates are now converging, and young black men (ages 20 to 35) already have a suicide rate twice as high as young white men. American Indians have a higher suicide rate than whites at all ages.

The Harvard Medical School
Mental Health Letter,
Vol. 2(8), 1986, pp. 1–2.

The modern study of suicide has been pioneered by Emile Durkheim's land-mark work, *Suicide: A Study in Sociology*. Durkheim was writing at a time when the discipline of sociology was in its infancy. He therefore sought to establish explicitly, through his study of suicide, the value of the sociological perspective. As is evident from the first several chapters, Durkheim believed that if he could take one of the most private and personal experiences of human life, namely, the act of killing oneself, and demonstrate the prominence of the social element in suicidal behavior, sociology would be established as a legiti-mate scholarly discipline.

Durkheim's explanation for the social basis of suicidal behavior is rooted in his conviction that the cultural arrangements of society constitute social facts that have a coercive influence on the lives of individuals. As Durkheim observes, specific individuals may perish but social facts will persist. Thus, the rate of suicide in various countries, cultures, and circumstances is not connected to the death of individuals but exists in conjunction with the relevant social facts. Rates of suicide, therefore, are not dependent upon individual psychological or emotional states. Durkheim directly expresses his argument, repudiating in-dividual explanations and advocating social explanations for suicide:

> The conclusion from all these facts is that the social suicide-rate can be explained only sociologically. At any given moment, the moral constitution of society establishes the contingent of voluntary deaths. There is, therefore, for each people a collective force of a definite amount of energy, impelling men to self-destruc-tion. The victim's acts which at first seem to express only his personal tempera-ment are really the supplement and prolongation of a social condition which they express externally.
>   . . . It is not mere metaphor to say of each human society that it has a greater or lesser aptitude for suicide. . . . Each social group really has a collective incli-nation for the act, quite its own, and the source of individual inclination, rather than their result.[1]

In establishing the social dimension of suicidal behavior, Durkheim identi-fied four types of suicide: egoistic, altruistic, anomic, and fatalistic. Durkheim organized his fourfold typology around two independent explanatory variables: integration and regulation. In this framework, each type of suicide is associ-ated with the degree of either social integration or social regulation found in a particular society.

"Egoistic" suicide is suicide that occurs as a result of the individual's height-ened sense of isolation and social detachment. It is the absence of social ties that is directly responsible for suicidal behavior in the egoistic framework. In this category, suicide varies inversely with the degree of integration of the social groups to which the individual belongs.[2] Egoistic suicide is thus framed by the loosening of social bonds that tie individuals to each other. On a societal level,

DURKHEIM'S FOURFOLD TYPOLOGY OF SUICIDE

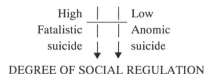

QUALITY OF SOCIAL STRUCTURES

High | | Low
Fatalistic | | Anomic
suicide | | suicide

DEGREE OF SOCIAL REGULATION

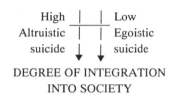

QUALITY OF INDIVIDUALS

High | | Low
Altruistic | | Egoistic
suicide | | suicide

DEGREE OF INTEGRATION
INTO SOCIETY

this detachment from others is linked to the excessive growth of individualism. Egoistic suicide thus results from a person's strong sense of individualism and personal ego, as well as his or her weakened sense of integration and social ego.

Specifically, Durkheim advanced three propositions regarding the relationship between individualism (or the absence thereof) and suicide in the modern social context:[3]

1. The Level of integration into family inversely corresponds to the incidence of suicide.
2. The Level of integration into religious beliefs and rituals inversely corresponds to the incidence of suicide.
3. The Level of integration into political values and activities inversely corresponds to the incidence of suicide.

In effect, Durkheim's thesis emphasized that religious commitment, a stable family life, and political involvement have a prophylactic effect on suicide rates.[4] The integration of individuals into collective values, sentiments, traditions, beliefs, and activities gives meaning to individual life for the members of the collectivity and serves as a protection against suicide.

According to Durkheim's analysis (which is largely correct, as we will see shortly), egoistic suicide is characteristic of modern societies. The fall of mechanical solidarity and its collectivist values, the growth of individualism, and the weakening of the moral and social fabric of society are inherent in the processes of modernization and are facilitators of suicidal behavior. Indeed, the main components of egoistic suicide are the key themes of the sociological landscapes of present-day American life: social isolation, excessive individualism, mobility, emotional isolation, and absence of stable, meaningful social ties.

A second type of suicide is "altruistic" suicide, which is related to excessive social integration. When an individual is integrated into a group to the degree that its welfare becomes paramount to the welfare of the self, the influence of collective forces becomes the source of an individual's suicide. Contrary to egoistic suicide, altruistic suicide occurs when an individual's personal ego is weak and when social ego and integration into the group are strong. Durkheim's account of altruistic suicide identifies three specific types: acute, obligatory, and optional.[5] Acute altruistic suicide is most often related to spiritual belief systems that hold that the individual, physical self is of no value, especially when considered in relation to spiritual forces of the cosmos. Thus, renunciation of the self in deference to spiritual forces becomes a source of joy and satisfaction: The individual commits suicide purely for the joy of sacrifice, because renunciation in itself is considered "praiseworthy."[6] Optional and obligatory altruistic suicide are two sides of the same coin. Obligatory suicide is self-sacrifice mandated or imposed by society.[7] The individual is required to kill himself or herself for the good of the social group. Optional altruistic suicide occurs when the individual responds (voluntarily) to the pressures of the norms, mores, and folkways of society that prescribe for him or her the act of suicide.

In altruistic suicidal behavior, a strong social conscience dominates the individual. This conscience impels the individual to sacrifice his or her life for the good of the group. Whitney Pope provides an effective summary of the societal basis of altruistic suicide:

> For Durkheim these different beliefs arise from the same source as the low valuation placed upon the individual, namely, the underlying social reality. . . . In the case of obligatory or optional suicide, social necessities are translated as duties or options for the individual in the form of beliefs about duty, honor and religious sanctioning in the afterlife. In the case of acute suicide, the relevant belief is that the principle of action is external to the individual. These somewhat different beliefs reflect the same reality, *the crucial aspect of which is that the society directs the individual.*[8]

The individual therefore internalizes the society's belief that the individual self is worth little in comparison with social and collective values.

Durkheim defined altruistic suicide as typifying primitive society, with its strong sense of mechanical solidarity and moral regulation. Vivid examples of altruistic suicide can also be found in the modern military:[9] the acts of Japanese kamikaze pilots or the heroic act of jumping on a live grenade to save the lives of comrades.[10] Although Durkheim is correct in identifying it with primitive society and the military, altruistic self-sacrifice occurs in other circumstances as well: individuals who starve themselves to death for a cause, the well-known mass suicide of the People's Temple in Guyana, the Buddhist monks who immolated themselves in the streets of Saigon to protest the Vietnam War, or the suicide of a Japanese student who performed poorly on important school examinations.[11] Suicides among other groups—the Hindu wife who kills herself after the death of her husband or the modern housewife who commits suicide because she feels she is worthless to and a burden on her family[12]—are also examples of motivation deriving from commitment to the principles and welfare of the social group. While these examples illuminate the nature of altruistic suicide as envisioned by Durkheim, the self-involvement and social isolation dominating modern social activity make altruistic suicide, especially in America, a somewhat rare and isolated occurrence.

The third type of suicide is "anomic suicide." Unlike egoistic and altruistic suicide, anomic suicide is not related to integration or individualism but is associated with social regulation. Similar, however, to egoistic suicide, anomic suicide is a dominant form of self-destruction in the modern social setting.[13]

Anomic suicide is rooted in the normlessness, confusion, and disorganization that result from the acute or chronic disruption of traditional lifestyles. It is a consequence of the breakdown and diminishment of the controlling and regulative power of norms over individual behavior.

Durkheim consistently argues that moral regulation provides a sense of certainty and security. It enables the individual to recognize readily his or her place within the economic, domestic, and social spheres of human living. For example, a peasant, who has a whole tradition of guiding norms to rely on, knows exactly who she or he is. Norms thus provide for individuals in different social positions guidelines or standards for behavior. Their presence promotes orderly patterns of living by providing relatively fixed rules to live by. Their absence leaves the individual disoriented, anxious, and floundering, without the security of fixed and established needs, desires, aspirations, or goals.[14]

Durkheim's discussion focuses explicitly on domestic and economic anomie. Considering specifically widowhood, Durkheim argues that the death of a spouse leaves the survivor disoriented and unprepared to adapt to his or her new social status:

> The suicides occurring at the crisis of widowhood . . . are really due to domestic anomy resulting from the death of a husband or wife. A family catastrophe occurs which affects the survivor. He is not adapted to the new situation in which he finds himself and accordingly offers less resistance to suicide.[15]

Economic normlessness erases old boundaries and standards that defined what was feasible for an individual to aspire to and achieve. Economic needs, which were rigidly defined by social conditions and circumstances, are no longer controlled and restrained. In the absence of regulation, individuals seek to achieve more than they can ever realistically hope to accomplish. The result of these unmet expectations is an intensified and heightened sense of dissatisfaction and unhappiness, which greatly increases the risk of suicide.[16] And as Durkheim observes, economic anomie can be a consequence of acute or abrupt changes in the economic sphere of society and/or of slowly evolving or chronic changes in socioeconomic structure. Again, Pope effectively interprets Durkheim on this point:

> Society normally makes a rough determination of how much the individual may legitimately aspire to, a determination that not only effects a proportionality between means and needs (by scaling down the latter) but is also accepted as just. However, during times of sudden change people find themselves thrown into new situations to which prevailing rules are no longer applicable. Since new rules cannot be immediately established, the individual is temporarily unadapted to his situation. During depressions the individual's means are no longer adequate to his needs, while in times of prosperity the lack of guiding rules permits expanding needs to surpass means, thus creating the means-needs disjunction marking anomie. In contrast to acute anomie, chronic anomie denotes situations in which the gradual erosion of social control permits needs to expand and outstrip means.[17]

Durkheim provides a useful summary of the above three types of suicide:

> Egoistic suicide results when an individual no longer finds a basis for existence in life; altruistic suicide occurs because this basis for existence is situated beyond physical life itself (in a principle or perhaps spiritual existence). The third sort of suicide, which we have just considered, results from human activities lacking regulation and the commensurate sufferings which result. This is, by virtue of its origin, anomic suicide.[18]

Durkheim's scheme is completed by a fourth category: "fatalistic" suicide. Excessive social regulation is its premier quality:

> The opposite of anomic suicide, just as egoistic and altruistic suicides are opposites, . . . is the suicide deriving from excessive regulation, that of persons with futures pitilessly blocked and passions violently choked by oppressive discipline.[19]

Citing the suicide of a slave as an example of this fourth kind of suicide, Durkheim goes on to state that since excessive regulation has little place in the modern world, fatalism likewise is irrelevant as a source of suicide in modern society:

ECONOMIC ANOMIE AND SUICIDE

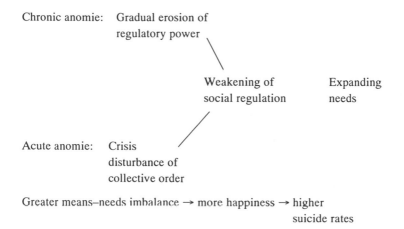

Chronic anomie: Gradual erosion of regulatory power

Weakening of social regulation

Expanding needs

Acute anomie: Crisis disturbance of collective order

Greater means–needs imbalance → more happiness → higher suicide rates

Adapted from Whitney Pope, *Durkheim's Suicide*, Chicago: University of Chicago Press, 1976, p. 29.

But it has so little contemporary importance . . . that it seems useless to dwell on it.[20]

Although this category has tremendous analytic value, fatalistic suicide as defined in this way may appear to be a rare occurrence in the modern social setting. Yet, the suicides of prisoners, of prisoners of war, of concentration camp inmates, and among the inmates of the Soviet Union's gulags seem to be related to the excessive moral and physical despotism characteristic of Durkheim's concept of fatalistic suicide. The relevance of fatalism to modern suicide may pale in comparison to egoism and anomie; nevertheless its use value, especially in total institutional settings, has not been fully explored in the psychiatric, sociomedical, or sociological literature.

The framework established by Durkheim is essential to any understanding of suicide. Of particular relevance to suicide in contemporary America are the forces of egoism and anomie, which dominate the social fabric of this society. It is not surprising, then, that themes of isolation, alienation, social disorganization, loss of role, social disintegration and turmoil, fragmentation, powerlessness, and helplessness run throughout the contemporary suicide literature. Durkheim's *Suicide* teaches us an important lesson: The sociohuman qualities associated with modern suicidal behavior do not exist as isolated components of individual lives but are structurally rooted social facts typifying the organization of modern society.

Obviously, however, suicide is more than a social fact, inasmuch as it is relevant to the private, personal lives of human beings. Durkheim ignores both psychological variables and the ways in which social forces may be internalized by individuals. For example, though Durkheim dismisses fatalistic suicide as inconsequential to modern society, it is certainly relevant to suicidal individuals, especially to their state of mind (as we shall see later). Furthermore—and perhaps this is one of the most salient problems in Durkheim's work—social factors and emphasis on social facts cannot explain why specific individuals in given situations do or do not kill themselves. Thus, while the modern social environment of egoism may be an important risk factor for suicide, Durkheim's analysis fails to tell us which individuals in this high-risk social setting will kill themselves. In this way, by excluding nonsocial variables, Durkheim provides an incomplete picture of suicide.

Because of Durkheim's insistence on social facts as causal factors, his work has been severely criticized in contemporary suicidology literature. Yet there are some mitigating considerations. First, let us remember that Durkheim's dominant purpose in writing *Suicide: A Study in Sociology* was to establish the legitimacy of sociology as a discipline so that it could achieve parallel academic status with other scientific disciplines, such as biology and physics. Thus, Durkheim builds restriction into his purpose, and he puts this straightforwardly:

> Each society is predisposed to contribute a definite quota of voluntary deaths. This predisposition may therefore be the subject of a special study belonging to sociology. This is the study we are going to undertake.[21]

Moreover, rather than repudiating the importance of nonsocial factors, Durkheim clearly acknowledges that his analysis offers only a partial explanation:

> We do not accordingly intend to make as nearly complete an inventory as possible of all the conditions affecting the origin of individual suicides, but *merely to examine those on which the definite fact that we have called the social suicide rate depends.*[22]

The weakness of Durkheim's analysis lies not so much in his actual formulations as in the fact that the restrictiveness of his categories and approach only partially explains the problem of human suicide in all of its complexity. Suicide has a deeper existential significance than can be established by defining it solely in terms of social facts and social rates. Thus, the personal dimensions of human suicide, while not considered by Durkheim, are very relevant to our understanding of the modern collage of suicidal behavior.

In contrast to Durkheim's social realism, the psychological or psychiatric explanation for human suicidal behavior locates the cause of suicide within the individual's instinctual, psychological, and emotional makeup. The psychological explanation for suicidal behavior was pioneered by Sigmund Freud.[23] Freud not only identified the locus of suicidal activity as residing within an individual's instinctual inclinations, but in doing so, he located the explanation for self-destruction within the parameters of psychoanalytic theory. In discussing the dynamics of depression and the instinctual drives toward violence and aggression, Freud saw suicide as a form of internalized hatred of another, that is, a desire to kill someone else that is introverted and directed toward the self. In this framework, depression originates from the hatred of another that is turned back toward the self. Suicide, correspondingly, results from introverted hate, as aggressive urges and the desire to kill are directed away from the hated person toward the self. For example, anger toward a former lover may become a source of depression and a cause of suicide as the desire to destroy the other is transmuted into self-destructive activity.

Others since Freud have adopted and advocated the psychoanalytical explanation of suicidal behavior. In an important study, *Man Against Himself*, Menninger accepts the existence of an inherent death instinct. In considering the dynamics of self-destructive behaviors, Menninger grounds his discussion of suicide (and other forms of self-destruction such as alcoholism) in three dimensions of a human death instinct: the wish to kill, the wish to be killed, and the wish to die.[24] As Menninger observes:

> Merloo modified and elaborated my proposals regarding the unconscious motivations for suicide. . . . He subdivides it into suicide as being killed, suicide as

---

Thanatology Snapshot 7.2

On Psychological Drives Toward Suicide

Destroy ourselves? Surely this is a strange objective. Do we not all endeavor to save our skins and better our living? How can one hold the view that our aggressive nature, which is necessary for self-protection (and in small amounts for amusements!)—how can anyone say that this same aggressive, destructive energy will be turned inward upon ourselves? Are we our own enemies?

Karl Menninger,
The Crime of Punishment
(New York: Viking Press, 1968), p. 164.

killing, suicide as communicating, suicide as getting revenge, suicide as escape, and suicide as magic reversal or transformation.

Thus, suicide as a way of being killed magically is sometimes a self-offering to the gods, sometimes a negative magic gesture: "See, they killed me!" As a communication it sometimes says . . . "See how I have suffered. . . ."

Suicide is often revenge, i.e., an aggressive or partial killing, on those who remain. The survivors all feel to some extent guilty, as no doubt the victim expected. "Now you will be sorry that you were not kinder to me. Remorse will follow you to the grave. The dead exert great power and my death will haunt you. . . ."

Suicide may be even more aggressive, even an actual murder . . . as [evidenced] by some individuals who take bombs into planes and blow themselves and others to pieces. In fantasy, suicide is probably always a psychological murder, a killing of the world or some important person in it—an unfaithful lover, a resented parent, an internalized tormentor, which may be the conscience.

Suicide may be a flight from pain, incurable disease, the threat of helpless senility, or violent death on the gallows. It may be a flight from anticipated rejection. . . . Or it may be a flight from self-depreciation, feelings of sexual or general inadequacy, humiliation, unknown danger.[25]

Menninger thus seeks to explain suicide within Freud's framework of unconscious drives toward self-destructiveness, instinctual inclinations of aggression toward others, and impulsive, aggressive outbursts that seek a magical transformation of an individual's life circumstances. In any event, an individual's hatred of another, despair over life circumstances, or hatred of self become behaviorally expressed in a framework that merges aggressive impulses with unconscious motivations for self-destruction:

. . . there is in the suicidal determination an exposition of that deepest and most incomprehensible yet inevitable characteristic of man, his self-destructiveness. Freud repeatedly emphasized that the manifestations of the self-destructive instinct were never nakedly visible. In the first place, the self-destructive instincts get turned in an outward direction by the very process of life, and in the second place they get neutralized in the very process of living. Self-destruction in the operational sense is a result of a return, as it were, of the self-destructive tendencies of the original object. It is not quite the original object as a rule, because the object of redirected aggression, *aggression reflected back upon the self*, is usually the body.[26]

The psychoanalytic framework, as established by Freud and Menninger, has become a dominant theoretical explanation for suicide. For example, Zilboorg argued that a loss during the formative years of an individual's life predisposed that person to suicide.[27] Palmer also identified the early childhood formative period as being the cause of suicide in later adult life. Specifically, he argued, the basis for suicide is laid in the early years by the absence of crucial role models, the result of which is arrested psychosocial development.[28] Indeed, in the scholarly literature, classic psychoanalytic concepts are used to explain the causes of suicide: "unconscious dramatization," "aggression against guilt," "paranoid schizophrenia,"[29] "frustrated dependency," "spiritual rebirth," "depressive personality system," "veiled aggression," and "hateful personality."[30]

The psychological explanation for suicide also includes a wide array of theories that expand the narrow parameters of Freudian psychoanalytic thought. Indeed, the weakness of the Freudian-Menninger explanation of suicide is that it ignores a range of relevant psychological and social factors. These factors include deeply personal and psychologically based motives such as dependency, shame, guilt, hopelessness, pain, excessive boredom, and absence of self-esteem. And, as we shall see, many of the personally relevant psychological factors, such as hopelessness, and problems in coping with pain or shame, are also explicitly related to the fabric and organization of the broader social structure. Indeed, the psychological approach, with its emphasis on the individual, fails to take into account a number of social and structural issues; rates of suicide also vary according to cultural differences, regional location, family structure, social integration, age, race, sex, urbanism or ruralism, social change, and other social factors. Therefore, whereas Durkheim's social realism cannot explain why individuals kill themselves, psychological interpretations cannot explain the persistence and variability of the social suicide rate. As sociomedical scientists and suicidologists try to develop an expanding and more complete explanation, sociological perspectives must genuinely be integrated with psychological and personal explanations of suicide, and vice versa.[31]

Perceptions and Contexts of Suicide: Individuals and Individualism

Despite the importance of multi- and interdisciplinary perspectives, the dominant societal style is to consider suicide as the private trouble of individuals and to define it as a psychiatric and therapeutic issue. *The Harvard Medical School Mental Health Letter* characterizes this approach in the following way:

> The mental health system is the institution most often called on to deal with suicidal thoughts, threats, and behavior, and the emotional consequences of suicide. From this point of view it [suicide] appears not as a moral or metaphysical issue but as the most common serious psychiatric emergency—one of the most difficult situations in psychiatry, not only for suicidal persons and their families but also for therapists.[32]

There is a logical consistency in defining suicide in this way, especially in light of the widespread trend toward greater therapeutic intervention, as well as the medicalization of morality and social issues in the broader society. In this manner, the management of the human problem of suicide, like the management of the dying process or the grieving process, becomes rooted in a technological-therapeutic paradigm. Additionally, as the focus of attention in the therapeutic model is the individual, the problem of suicide becomes personalized. And, as much of society's response is based on psychologically rooted treatment plans, the relevance of social facts and social structure is obviated. Yet, this psychologization of suicide profoundly amplifies the value of individualism, which mobilizes the broader context of modern American society.

Specific responses to suicidal behavior include forced psychiatric hospitalization, individual counseling and therapy, specific forms of medical management to reverse the physical symptoms of self-injured suicide attempters, psychiatric intervention with possible chemotherapeutic treatment, and preventive and postventive efforts in the community, such as crisis lines and suicide self-help groups. Similar therapeutic efforts are also exerted in support of the family members of a completed or attempted suicide. These include individual counseling, family therapy, and self-help support groups.[33]

Indeed, in the study of suicide, we can clearly see that the medical model and the social value of individualism converge. The psychiatric, medical, and psychological literatures, as well as active therapeutic intervention into suicidal behavior, strongly emphasize the individual design and nature of the process. It is the prevailing assumption that suicide motivations are so diverse and variable that suicide therapy must be similarly individualized. Perhaps suicide therapy does not differ radically from other kinds of psychotherapy, but it may very well accentuate the individualized and emotional dimensions of the therapeutic perspectives.[34]

The individualization of the problem of suicide, however, goes far beyond the therapeutic characterization and response. Individualism, as both an abstract and a practical value, is reflected in the growing societal trend toward personalized conceptions of the morality and justification of a suicide. There is no question that suicide is still strongly stigmatized in our society, and strong legal, medical, and social sanctions against it continue to exist. However, over the past decade or so, groups and perspectives advocating the right to suicide have emerged. Indeed, the ideas of rational suicide, free choice, and self-determination became a part of medical, academic, and even popular discourse in the 1980s. During the past decade, a change in social attitudes toward suicide has created a more tolerant attitude toward the idea of individual freedom of choice, especially in extreme contexts such as serious, painful, chronic, or terminal illness:

> While suicide rates are increasing, denunciation of the act is diminishing. We read and hear more about "rational suicide" and the right-to-suicide. In Califor-

nia, an organization called Hemlock has been established whose purpose is to promote public discussion of rational suicide. They also publish a suicide guide. A Society for the Right to Die lobbies for state legislation giving people the right to refuse extraordinary life preserving measures. Another organization, Concern for Dying, has mailed out millions of copies of the Living Will, a document which enables individuals to designate when in the case of a serious illness they would want treatment to cease. Others have proposed schemes for regulating and "sanitizing" the act in the hopes of presumably destigmatizing suicide or removing the fear of death.[35]

One of the most prominent contemporary spokespersons for the right-to-suicide movement is Thomas Szasz. As Szasz articulates, rational suicide is based upon the principles of individual self-determination and self-control. In order to protect these principles of autonomy, the state must not forcibly intervene to prevent an individual's freely chosen suicide. Szasz, a psychiatrist, puts the point simply:

> A man's life belongs to himself. Hence, he has a right to take his own life, that is, to commit suicide.[36]

Rational suicide is therefore anchored in the moral argument that taking one's life is a fundamental right rooted in the natural law of human freedom and autonomy.[37]

Support for the right to commit suicide is not isolated within a circle of ideological and philosophical supporters. There is a growing pattern in the American public, and even among health care professionals, of greater tolerance and acceptance of suicide, especially when related to serious chronic and terminal illness. Studies have also documented a growing acceptance of suicide among youth and have found that college-age students, in particular, have a generally accepting attitude:

> The majority of students demonstrated liberal attitudes toward suicide. They agreed that people have the right to decide their own fate, that it is not morally wrong to commit suicide.[38]

Youth's general acceptance of the right-to-suicide argument represents a substantially different view of suicide than that held by the preceding generation. Specifically, it has been found that, even within the same family, people of the younger generation had a distinctly more liberal attitude than members of the older generation. Menno Boldt summarizes the findings of one such study:

> The major findings that emerge from the comparison between the true intrafamilial generations indicate that there are important generational differences in conceptions and valuations of suicide and death. The youthful generation, in contrast to the parental generation, consistently viewed suicide in terms that were less judgmental of the individual, less stigmatized, and less calamitous after death.

They were also less inclined to think of suicide in religious-moral terms, and they placed greater emphasis on the individual's right to suicide. Taken together, as a complex of attitudes, the observed generational differences support the hypothesis that the subculture through which youths perceive suicide is more accepting of the act than is that of the parental generation.[39]

While young people's attitudes are clearly more tolerant, this fact should not be interpreted as an across-the-board repudiation of suicide by the older generation. To the contrary, national studies (conducted in 1977, 1978, 1982, and 1983) are beginning to show that, increasingly, adults are recognizing suicide as a rational alternative for those suffering from incurable, terminal illness. This view, which legitimates situationally specific suicide, however, should not be confused with the idea of supporting suicide as an ethical and legitimate option of free choice. Additionally, acceptance of suicide in other, specific crisis situations such as bankruptcy, family dishonor, or just being tired of living has remained extremely low.[40] Specifically, the data indicate that there has been a steady increase in acceptance of suicide in cases of terminal illness during the years 1977 to 1995. Suicide in the face of devastating and incurable illness is currently approved by about 50 percent of the general adult population. Recent data indicate that the drift toward approval of terminal-illness suicide is so strong that even health care professionals, who have an explicit professional code of ethics regarding patient suicide, are more tolerant of suicide when a terminal, chronic illness is involved:

> The reasonableness of an intentional death over a sustained but torturous existence is becoming acceptable in physical medicine. . . . The ability of nuclear medicine to forestall death at the cost of a painful or vegetative existence has forced the question of the quality of lives being saved. There are many medical patients whose conditions make a prolongation of life synonymous with unnecessary torture.[41]

In this view, health care professionals are less likely to sanction strongly and judge negatively the suicide of an incurably ill patient.[42] And I hasten to add that the slow but emerging lenience of the medical profession toward euthanasia is related to the fast-paced growth of technology and its corresponding indignities of dying, as discussed in earlier chapters. In the face of chronic and severe pain, vegetative and lingering states, and severe physical devolution, some physicians are accepting the idea that it may be merciful and humane not to prohibit a patient's freely chosen suicide. A select few, most notably Jack Kevorkian (the "suicide doctor"), are even at the forefront of assisting critically sick patients with their suicides.

The change in general societal attitudes concerning the individual's right to influence the circumstances of his or her death are rooted in a variety of broader

social factors. The value of individualism, coupled with the specific orientation of the human potential movement, underlies this trend. For example, Betty
Rollins' biographical account of the suicide of her mother (Thanatology Snapshot 7.3) following the diagnosis of metastatic ovarian cancer has received
much popular acclaim and has been assessed favorably by newspapers and
magazines:

> "It is about people who discover a depth of love that they never knew."

> "A poignant and loving story."

> "A compassionate message you must not miss."

> "Moving. . . ."

> "The book is not really about death. It is, instead, a loving tribute to Mrs. Rollins'
> spirited progress through life."

> "It embraces you and makes you want to embrace the writer for her courage in
> both the act and its recounting."[43]

When a book that advocates suicide as demonstrating dignity, courage, and
love receives such popular acclaim and makes several national best-seller lists,
it is evident that its message appeals to the sensibilities of many American readers. Additionally, its themes are so consistent with the mainstream of American perceptions that the book was recently presented as a made-for-television
movie. It is certain that the style and substance of the book and film have been
deeply influenced the prevailing human potential–therapeutic emphasis. It is
in this vein that the acceptance of suicide, especially therapeutic suicide in the
situation of incurable illness, is a reflection of the societal emphasis on individualism, autonomy, and self-expression.

The second important factor in the developing acceptance of the right to die
is rooted in the increased prevalence of seemingly endless chronic illnesses and
the corresponding societal reaction to technology. As we saw in Chapters 2 and
3, the plight of dying patients—pain, a sense of meaninglessness or hopelessness, stigma, and increasing psychological suffering—is largely unresolved by
the technological management of dying. As discussed in Chapter 2, the hospice
movement is one cultural response to the indignity and dehumanization that result
from the technocratic management of dying patients. Still another response to
the technological extension of human life is rational suicide. Thus, in reaction
to the unnatural prolongation of human existence by machines and other technologies beyond the point where life itself is emotionally and socially meaningful, many individuals have become supporters of the right to suicide.[44] In
fact, thoughts about terminal suicide, as well as concern about the indignities of
dying and the decimation of personhood, are so salient in contemporary society
that people have become interested in learning, in specific detail, how to go about
the business of killing themselves. This is reflected by the fact that the founder

Thantology Snapshot 7.3

Euthanasia Suicide as Romantic Death

Two hours before my mother killed herself, I noticed she had put on makeup. This shocked me but it shouldn't have. Whatever the occasion, my mother liked to look her best. That was her way. Just as it was her way to die as she did—not when death summoned her, but when she summoned death.

She would have preferred to do her summoning alone, without help. But she had to ask for help. And when she did, I gave it.

The illness that made her want out of life was ovarian cancer, one of the sneakiest malignancies a body can house. That's why ovarian cancer is so deadly. Unlike a breast tumor, an ovarian tumor can't be detected except by gynecological examination and by that time it's often too late. That is, a metastasis—a spread—has already occurred. . . .

This is going to happen. This is really going to happen. . . .

My mother props up the pillows, then eases herself under the covers, smoothing down her nightgown as she lies back. She looks dreamily out of the window and sighs. "I would have liked to die with my hair."

I sit on the edge of my bed and take her hand and kiss the back of it, not trusting myself to speak.

"Is it time now?" she asks. . . . "Yes," she says, as if the end of the evening has come and the party is over, "it's time."

She sits up and takes the bottle of Nembutal in her hand. She opens the bottle and carefully taps its contents onto an indented place on the blanket. She puts the bottle down. The windows are tightly shut and, except for the small flapping sound of the digital clock passing from one minute to the next, the room is silent. She turns back toward the table and grasps the bottle of soda water in her left hand and the opener in her right. She tries to open the bottle and fails. . . .

. . . She leans over in the direction of her knees and looks down at the small pile of shiny capsules lying like candies on the yellow blanket. She picks up two or three of them in her fingers, places them on her tongue, and lifts the glassful of soda water and, with a short swig, swallows them, and three more. Now three more. Soon she falls into a rhythm—pick them up, toss them on the tongue, lift the glass, swallow; pick them up, toss them on the tongue. . . .

---

of the Hemlock Society wrote a how-to-do-it, recipe-type book on suicide techniques, and its brisk sales catapulted it to the *New York Times'* best-seller list. The stunning popularity of this book, entitled *Final Exit*, certainly does not reflect a widespread cultural desire for self-destruction. Rather, it is indicative of

Thantology Snapshot 7.3 (*continued*)

"You're doing it, Mother," I whisper. "You're doing it. You're doing great."
I am on my feet now. My hands are crossed on my chest as if to keep my heart
from bursting through my chest wall. The play on stage has turned into a sports
event. My mother is performing the decathlon.

I want to jump out of my stone encasement and cheer. I want to wave a flag,
blow a horn, scream, weep, cry, shout. Hooray for you, Mother! You're doing
it, Mother! You're doing it just right! You're—

. . . The room has darkened, except for the pool of light from the bedside lamp,
which shines on her as if she were an oil painting. We enter the light, my husband and I and we take our places on either side of her as if in a ceremony. We
each take one of her hands and for a moment we are all perfectly harmoniously
still. Then my mother begins to speak in a way that sounds like a chant:

"I want you to know that I am a happy woman. I made a man happy for forty
years and I gave birth to the most wonderful child, and late in life I had another
child whom I love as if I had given birth to him too. No one has been more blessed
than I. I've had a wonderful life. I've had everything that is important to me. I
have given love and I have received it. No one is more grateful than—"

She stops suddenly and opens her eyes. "I'm not sleepy." "You will be,
Mother," I hear myself say, as my spine straightens and my blood turns to ice. . . .
Please let it work. Please. Please.

"You'll soon be asleep," whispers Ed. "You'll sleep peacefully."

"Oh yes, I'm starting to feel it now. Oh, good. Remember, I am the most happy
woman. And this is my wish. I want you to remember . . ."

"I love you, Mother!" I call to her, "I love you," and when she does not answer, I lower my face into the soft flesh of her neck. Now the cement that has
been holding me together begins to crack. I stumble backwards, collapse into a
chair and with both hands clamped over my mouth, I sob. I sob heavily, but
not for long. Because when I look up and see how still she is, I know that she
has found the door she was looking for and that it has closed, quietly behind
her.

Betty Rollin,
Last Wish
(New York: Warner Books, 1985), pp. 5–6, 233–236.

a great cultural unease with the potential suffering and horror that can characterize the modern process of medicalized dying.

A third factor influencing this right-to-die movement is the growing awareness of the feminist idea of right of ownership over the body. A central argu-

ment of the women's rights position on abortion, one that was legally and formally supported by the decision in *Roe v. Wade* and is supported by a sizeable majority of the American people, is that a woman has sole responsibility for her own body. The idea that the individual owns his or her own body has been a popularly discussed social issue during the past two decades. Rational suicide is based on a similar notion, namely, that since one owns one's body, there can be no "theft of life" if one kills oneself. In this way, the seeds for the expanding idea of self-determination regarding suicide have been sown by the feminist revolution. Within the sociopolitical platform of feminism, the general idea of ownership of one's own body had already received wide public commentary and legitimization. This has helped to pave the way for a growing acceptance of suicide based on the principle of autonomy of self and body.

Another relevant factor is the decline of religious tradition and sanctions. Throughout the development of American society, religious tradition has consistently acted as a force of prohibition against suicide. In a society where a secular world view has become increasingly prevalent, the ability of religion to influence the thinking and actions of individuals has correspondingly diminished. Religion and its prohibitions against suicide have not disappeared, but they are often ignored, and the values and doctrines of religion are often in direct competition with secular values and folkways. It is fair to say that while religious sanctions against suicide are relevant for the lives of some individuals, they are less relevant and even irrelevant for the lives of others. A good indicator of the effect of ignored religious taboos can be found in Boldt's study of cross-generational attitudes on suicide, where secular, ethical world views of American youth were positively associated with greater acceptance of suicide.

We do not mean to imply, of course, that the right-to-suicide position has become the dominant or unchallenged one in American society. Indeed, while social conditions are right for the growing support of the right to suicide, strong negative sentiments and sanctions against suicide continue to exist. Societal reactions to suicide attempters, and to family members of a completed suicide, typically result in stigma for the survivors as well as the attempters. As we shall see later, the reactions of others to attempted and completed suicide can be a significant source of stress for survivors.[45] Suicidal behavior would not elicit such strong societal reactions invoking guilt, shame, stigma, and psychological stress if suicide were not perceived as a form of deviance by the majority of Americans.

Additionally, those who are critical of the right-to-suicide position (in other words, the majority of the American public and of health care professionals) argue that the concept of rational suicide is misleading and inaccurate. Suicide as a response to life crises, it is argued, emerges not from rational, conscious thought processes but from the subjective and deeply irrational side of the human being:

> Indeed, one of the inherent problems of the rational suicide position is its exclusive focus on thought processes. The urge to suicide, however, usually originates in the realm of feelings and emotions and not in the domain of evaluative reasoning in which the individual can objectively assess his situation.[46]

Karen Siegel amplifies the point:

> Rational suicide implies not only that the reasoning of the individual is in no way impaired, but that his or her motives would seem understandable, if not justifiable, to the members of his/her community or social group. It should be pointed out that the tendency to think that the value of one's life can be measured on a balance sheet is a characteristic of suicidal people. Their thought processes often seem molded to narrow possibilities. Depression rigidifies their thinking, making them unable to envision alternatives. It restricts and distorts their view concerning possibilities for the future. Their thinking reflects a twisted sense of values, impoverished insight. An objective, rational view of life is rarely present. Furthermore, part of the problem with the right to suicide position is precisely its emphasis on thought processes. It is in the realm of emotions and feelings—about one's own self worth, about one's relationship with significant others—that the urge to suicide usually originates and can be stemmed.[47]

Overall, as has been emphasized throughout our observations, moral certainty regarding appropriate forms of human death, which was present during the eras of traditional death patterns, has been replaced by growing normlessness, ambiguity, and pluralism. The social, moral, and medical dissensus on styles and responses to modern dying finds similar expression in the area of suicide. Yet, despite the pluralism of attitudes toward suicide, a common basis for contemporary perceptions is emerging in the form of the therapeutic model.

Psychosocial and Sociological Issues

The range of medical, psychiatric, psychological, and self-help responses to the problem of suicide are increasingly defined within a framework of active treatment and therapy. In fact, the usual societal response to suicidal behavior actively employs a therapeutic program of prevention, intervention, and postvention. In practice, this program entails intervening in order to stop suicide attempts and developing an individually designed program of treatment that responds to the psychological crisis underlying the suicidal activity. Additionally, postventive care for the survivors of a loved one who has successfully completed suicide focuses on stabilization, rehabilitation, and personal renewal.

Thanatology Snapshot 7.4

It's Over, Debbie

The call came in the middle of the night. As a gynecology resident rotating through a large, private hospital, I had come to detest telephone calls, because invariably I would be up for several hours and would not feel good the next day. However, duty called, so I answered the phone. A nurse informed me that a patient was having difficulty getting rest, could I please see her. She was on 3 North. That was the gynecologic-oncology unit, not my usual duty station. As I trudged along, bumping sleepily against walls and corners and not believing I was up again, I tried to imagine what I might find at the end of my walk. Maybe an elderly woman with an anxiety reaction, or perhaps something particularly horrible.

I grabbed the chart from the nurses station on my way to the patient's room, and the nurse gave me some hurried details: a 20-year-old girl named Debbie was dying of ovarian cancer. She was having unrelenting vomiting, apparently as the result of an alcohol drip administered for sedation. Hmmm, I thought. Very sad. As I approached the room I could hear loud, labored breathing. I entered and saw an emaciated, dark-haired woman who appeared much older than 20. She was receiving nasal oxygen, had an IV, and was sitting in bed suffering from what was obviously severe air hunger. The chart noted her weight at 80 pounds. A second woman, also dark-haired but of middle age, stood at her right, holding her hand. Both looked up as I entered. The room seemed filled with the patient's desperate effort to survive. Her eyes were hollow, and she had suprasternal and intercostal retractions with her rapid inspirations. She had not eaten or slept in

---

Again, it is important to recognize that societal values that give rise to this response to suicidal behavior are related to our cultural commitment to life, dignity of the individual, and fulfillment of human potential.

In the social circumstance of anomie, however, commitment to cultural values and principles may often assume divergent and even contradictory paths. Thus, intervention, prevention, and postvention are not the only common therapeutic responses to suicide. Indeed, the right-to-suicide position is rooted in the therapeutic model and in the pursuit of the values of respect for human autonomy, dignity, and self-determination, all of which are central to the human-potential inclination of contemporary society. On a more concrete level, suicide in the shadow of terminal illness is increasingly defined as a therapeutically legitimate human choice. In the context of pain, suffering, physical deterioration, and mechanical maintenance of life, suicide has become one possible way

Thanatology Snapshot 7.4 (*continued*)

two days. She had not responded to chemotherapy and was being given supportive care only. It was a gallows scene, a cruel mockery of her youth and unfulfilled potential. Her only words to me were, "Let's get this over with."

I retreated with my thoughts to the nurses station. The patient was tired and needed rest. I could not give her health, but I could give her rest. I asked the nurse to draw 20 mg of morphine sulfate into a syringe. Enough, I thought, to do the job. I took the syringe into the room and told the two women I was going to give Debbie something that would let her rest and to say good-bye. Debbie looked at the syringe, then laid her head on the pillow with her eyes open, watching what was left of the world. I injected the morphine intravenously and watched to see if my calculations on its effects would be correct. Within seconds her breathing slowed to a normal rate, her eyes closed, and her features softened as she seemed restful at last. The older woman stroked the hair of the now-sleeping patient. I waited for the inevitable next effect of depressing the respiratory drive. With clocklike certainty, within four minutes the breathing rate slowed even more, then became irregular, then ceased. The dark-haired woman stood erect and seemed relieved.

It's over, Debbie.

Name Withheld by Request

"A Piece of My Mind,"
Journal of the American Medical Association
Vol. 259(2), 1988, p. 272.

of asserting the dignity of human life. This is especially true in the face of the many circumstances of chronic illness trajectories that threaten to degrade and dehumanize.

Ironically, the perspectives on suicide do not stem merely from the pluralistic patterns of living and dying in the broader society. They also reflect inherently divergent and contradictory manifestations of the central value of individualism and the resulting emphasis on human potential and therapeutic orientations. In this way, a new scenario of suicide is on the horizon of American society. The realities of changing cultural attitudes toward suicide are just being discovered, and their impact has yet to be determined. All kinds of intriguing questions remain to be answered. For example, will growing acceptance of suicide increase the social rate of suicide? Will an obligation to commit suicide be established for those who are a drain on society's resources, such as the chronically sick elderly or terminally ill AIDS patients? Will the role of health care professionals evolve from prohibitors of suicide to facilitators of therapeutic, euthanasia-type suicide? What will be the impact on individuals, especially adolescents, of this greater tolerance? Will technologies and perhaps even fashions or styles of suicide be established in American culture? Will the American judicial system be affected? The questions themselves are fascinating. But of even greater significance will be the impact of this societal evolution on redefining American values and on *our understanding of the meaning and value of human life.*

In the contemporary cultural framework, suicide has increasing relevance for both the sociological and psychological dimensions of human life. The individual, in relation to suicide, has greater leverage in making choices than ever before. Suicide prevention and postvention have a strong psychological component. In this way, the cultural context and social attitudes have created a situation where the psychological and individualized nature of the suicide is accentuated. However, it is precisely this psychologization of the suicide problem, of responding to suicide as a private trouble, that is rooted in the social structure and organization of contemporary society. The social facts of egoism, individualism, and anomie have created a cultural context where psychologizing and individualizing the suicide problem have flourished. In this way, when the place and perceptions of suicide in American society are examined, it is clear that psychological and sociological factors are not just both relevant, they have merged to establish the frame of reference that defines modern American suicide. The following discussion on the sociopsychological causes of suicide further establishes the inseparable confluence between the sociological and psychological points of view.

Explanation for Suicide: A Psychosocial Collage

The impulse to commit suicide typically arises from private anguish and personal turbulence. Powerlessness and feelings of guilt, depression, hopelessness, loneliness, and despair are psychological symptoms often associated with suicide. Indeed, the psychological portrait of the suicidal individual is one of deep personal disturbance, and the degeneration of psychological well-being is frequently accompanied by drug and alcohol abuse. For these reasons, suicide is often defined exclusively in terms of individual, psychological factors. However, as just noted, the feelings of emotional turbulence, confusion, and self-abnegation that compose the psychological condition of the suicidal individual increasingly are associated with the patterns and organization of contemporary social life. The etiology of suicide, therefore, is explicitly related to a range of psychological variables and to a societal context. In searching for the reasons for suicide, it is therefore essential to transcend traditional disciplinary boundaries and emphasize a multifaceted, interdisciplinary approach. As one surveys the relevant literature, the web of psychological and sociological factors is readily apparent.

Unfortunately, the literature is fragmented and often restricted by disciplinary perspectives and methodological requirements. Nevertheless, when viewed from a distance, the suicidology literature provides a perspective rich in both sociological and psychological insights. A useful way to gain this perspective is by examining the issue of teenage suicide. Suicide among adolescents is second only to accidents as the leading cause of death. And even many of the adolescent deaths due to accidents may have had latent or even explicit suicidal intentions. Adolescent suicide has been identified as a national problem, and some have even gone as far as to label it an epidemic.[48] The epidemiological data on suicide reveal that 5,000 young people commit suicide each year in America; the actual number is probably higher. Estimates of the number of teenagers who attempt suicide each year range from 500,000 to 2 million.

The increasing rate of suicide among American youngsters is clearly disturbing. Equally disturbing are the spread and growth of those forces that render contemporary youths more susceptible to suicidal thoughts and behaviors. The factors that increase the vulnerability of American youth to suicide are multifaceted and often difficult to identify—one reason why so many completed or attempted suicides are perceived as unexpected and shocking to the survivors. In addition, the path to suicide is long and usually meandering. Suicide is seldom a totally impulsive, unpremeditated act. Rather, the road to suicide passes through many preliminary options and alternatives. The young person turns to suicide when other attempted solutions fail.[49] In addition, the combination of factors that drive individual young people to suicidal behavior is highly complex. Typically involved in every adolescent suicide are some of the following: psychological state of being, family context and history, interpersonal relationships, patterns of coping with stress, traumatic life experiences, and lifestyle. Another important factor is the social milieu, not just as it relates to the family, but as it affects the lives of youngsters and is transmitted to them by peers, schools, and the media.[50]

Studies of suicide increasingly emphasize the roles of instability and social isolation. Instability and isolation within family life are especially important factors:

> [M]any experts feel that a child's environment is the most significant factor in the proneness to commit suicide. Helplessness and vulnerability always make a teenager more prone to suicide—and factors such as family violence, intense marital discord, or loss of a parent through death, divorce, or separation can sig-

nificantly increase a child's sense of helplessness and vulnerability. In some families, a child is made to feel "bad"; as a result, he feels rejected, and he wants to die.

Teenagers who are more prone to suicide tend to have had difficulties relating to one or both of their parents. . . .[51]

Family background can be a salient precipitating variable of youth suicide. Broken homes, family disorganization, sudden family crisis, lack of communication, negative relationships, absent relationships, and abuse are family environmental factors related to youth suicide.[52] Thus, family conflict, parental problems, and a dysfunctional family context are explicitly associated with adolescent suicide:

> Parents of suicidal adolescents have been shown to have more overt conflict, more threats of separation and divorce, and more medical and psychiatric problems. . . . In contrast to other parents, mothers of suicidal adolescents have been found to be more anxious, to more frequently abuse alcohol, and to be experienced by the suicidal adolescent as less interested in them. Fathers, also, have been shown to be more depressed, to have lower self-esteem and to abuse alcohol and drugs more often. . . .
>
> Affection is given less often, and discipline in these families is less often by reasoning and explanation. Suicidal adolescents, therefore, understandably tend to hold more negative views of their parents. Time spent with the family is described by the adolescent suicide attempter as less enjoyable, and parental marriages are perceived (correctly) as more in conflict and families as more maladjusted.[53]

The adolescent's negative evaluation of family life leaves him or her feeling unloved, unworthy, rejected, and maladjusted.[54] Other factors that have been identified as causes of teenage self-destructiveness include social isolation, sexual confusion and maladjustment, social alienation, mobility, loneliness, emotional isolation, loss and grief, depression, alcohol or drug use, media attention to suicide, and a feeling of hopelessness.[55] As the adolescent increasingly feels that the stresses and turbulence in his or her life are insurmountable and unmanageable, the likelihood of suicide is increased:

> Many troubled teenagers see death as an attractive alternative to stresses. For some, suicide is a way to find peace and escape, while for others it becomes a statement of protest and rage. As society becomes more violent, this is reflected in our children. We are seeing a dramatic increase in depression and self-destructive behavior among adolescents.[56]

The central point is that growing numbers of young people are unable to find a release from emotional pain and loneliness. In other words, their perceptions of the world have become increasingly desperate and fatalistic. "Fatalism" as a sociopsychological concept means that an individual feels unable to change or improve important circumstances that affect his or her life. In this way, the adolescent feels excessively controlled by external circumstances, situations, and forces. In an era where the human potential movement has glamorized and emphasized the virtues of being autonomous and in charge of one's life, psychological fatalism has become a symptom or indicator of personal failure and inadequacy.

As David Matza states, the adolescent's mood of fatalism

> refers to the experience of seeing one's self as [an] effect . . . fatalism is the negation of the sense of active mastery over one's environment. It is likely to culminate in a sense of desperation among persons who place profound stress on the capacity to control [their] surroundings.[57]

In the absence of personal control and in the presence of perceived hopelessness, despair, and neglect, youthful suicide may be viewed as a response to the stresses of personal impotence:

> Youthful suicide may be viewed as the life-terminating behavior of young people who, though they have only begun to live, react to stress engendered by feelings of helplessness, hopelessness, neglect, the loss of significant others, who develop a sense of perdition, and who view the world through a fatalistic lens.[58]

Adolescents who find themselves snared by alienation, isolation, family turbulence, loneliness, emotional desolation, and so on, and who perceive that their life situations cannot be changed or improved, are at high risk for suicide. Suicide is judged, within this framework, as a viable or desirable alternative to an inescapably torturous existence:

> A sense of hopelessness about the future separated the [suicide] attempters from non-attempters. . . . External locus of control was more apparent in suicidals than controls, a finding which supports previous studies in which depressed or suicidal youngsters tended to feel a lack of control over their environment. . . .[59]

Moreover, the significance of fatalism may be especially relevant to the adolescent's family context:

> These relationships suggest that adolescents who feel little control over their environment may experience their families as unavailable, rejecting or overprotective. Escalating problems could then lead to an increasing sense of hopelessness and impatience about effecting solutions within the family, eventually ending in a suicidal mental set.[60]

Thus, attempted adolescent suicide must be interpreted as more than a "cry for help." It is a response to the awareness of painful life circumstances and a fatalistic view of them. A suicide attempt, therefore, is an attempt to restore a sense of personal autonomy and power in the face of perceived powerlessness. Completed adolescent suicides may be similarly interpreted. The adolescent, in committing suicide, not only reacts to oppressive and fatalistic circumstances, but also makes a final and dramatic assertion of personal strength. The final act of his or her life is ironically life-affirming—that is, it is motivated by the adolescent's thirst for a life that is meaningful, stable, and autonomous.

This view of adolescent suicide is especially disturbing. If adolescents increasingly cannot reap meaning from their lives, this is a problem produced by a society that has failed to provide stability, security, and autonomy, which are essential to productive and healthy life experiences. External forces that shape the role of adolescents in society and the formation of the individual self are major factors underlying the contemporary problem of teenage suicide. Adolescents are increasingly vulnerable to events, pressures, forces, and circumstances that are conducive to suicidal behavior. Indeed, the isolation and loneliness characterizing suicidal adolescents are consequences of the structural organization of modern social living, especially the organization of American family life. In this way, while the immediate causes of specific suicidal behavior are rooted in the individual, the reasons adolescents are increasingly distraught and prone to suicidal behavior are rooted in social and cultural forces:

> Three major changes stand out when you consider the external factors that affect teenagers: the changing structure of families, the weakened tie between children and their parents and the increasing importance of the peer group. . . .
> What have these changes produced? Greater behavioral freedom for teenagers without having to be accountable; greater demands for social competence without systematic instruction; heavier peer pressure, with less self-esteem to resist it; more pursuit of pleasure and escape through drugs and sex. . . . But, most importantly, these changes have resulted in a profound loneliness that is different, more abnormal and more pervasive than the kind that should be expected during adolescence.[61]

In this portrait, the suicidal adolescent is distraught, hapless, hopeless, and excessively isolated, with a psychological state of being that is pathologically impaired. But, again, it must be emphasized that the factors responsible for the stresses that adolescents are increasingly confronting are directly related to patterns and norms of social living.

Changes in the nature and organization of the American family—greater mobility, divorce, intergenerational conflict, breakdown of traditional norms, the rise of single-parent families and dual-career couples—are a major factor related to the increasing suicide rate among young people:

> The motives for suicide attempts of adolescents appear to be largely directed toward effecting change in or escape from an interpersonal system. Families are

perceived as in conflict, stressful and nonnurturing. The suicidal adolescent develops cognitive and related affects of helplessness, hopelessness, worthlessness, and rage. Without alternative and effective coping models (perceived or at hand), the suicide attempt may be construed as a last-ditch effort to alter in a dramatic way an intolerable situation.[62]

The isolation, breakdown in relationships, and sense of meaninglessness and powerlessness that define the psychological and social landscape of teenage suicides in America are not confined to the young. Rather, they are relevant to and reflective of the general problem of suicide in modern society. In a very important study entitled *Suicide in London*, Peter Sainsbury found evidence supporting the hypothesis that where social mobility and isolation are pronounced, community life will be unstable, without order or purpose, and this situation will be reflected in the suicide rate.[63] Specifically, Sainsbury found that isolation and loneliness are significantly related to suicide among individuals over sixty. Feelings of abandonment, loss, rejection, and desolation are frequently part of elderly Americans' motivation for suicide. The increase in solitary living among older Americans, the psychosocial toll of chronic illness, unemployment, economic difficulties, a feeling of uselessness, a sense of social rejection, and weakened family ties are also associated with suicide among older individuals. Moreover, Sainsbury noted high levels of suicide among widowed, unemployed, divorced, and separated individuals—groups that also suffered from extreme social isolation and the deprivation of meaningful, stable social contacts.

Social isolation involves more than just living alone. It includes the social and cultural isolation of the immigrant; the solitude of old age arising from a lack of contemporaries to share one's values and outlook; the unemployed's sense of social rejection; the ostracism resulting from infringement of a social taboo by divorce or a criminal act; or any similar activity that diminishes relatedness to the community. A high suicide rate is found in all of these categories; only the concept of social isolation embraces and accounts for suicide in such diverse groups.[64]

In a similar vein, themes of isolation, despair, and powerlessness have been identified in explaining suicide and suicide attempts among women. Indeed, the typical person who attempts suicide (it should be noted that while women attempt suicide more often than men, men complete suicide more than women) is a young, often poor, woman who has recently experienced divorce, separation, serious marital conflict, or other interpersonal problems.[65] Descriptions in the literature of suicidal women are replete with images of pathological and destructive interpersonal relationships. For example, one study discovered that the highest rates of attempted suicide occurred among separated and divorced women who had a long history of troubled relationships and difficulty in es-

tablishing and sustaining viable, productive, intimate relationships.[66] Another
study found that suicidally inclined women are often trapped in loveless and
emotionally barren relationships.[67] Some suicidal women had excessive and
unrealistic expectations of their marital relationships. Despite the unusual emotional needs of these women, the fact remains that these needs remained seriously unmet by their marriages. But, more typically, suicidal women were involved in situations of spousal infidelity coupled with the absence of the spouse
from the home, physical abuse, and denial of affection from emotionally in-
different husbands. What emerges is a portrait of suicidal women beleaguered
by turbulent and unfulfilling relationships:

> It is undeniable that these are unhappy, even pathological relationships in which
> the women have experienced a lack of love, have been beaten and brutalized,
> have suffered sexual betrayal and have developed deep feelings of worthless-
> ness and despair.
>
> The commonality which all of these women share are their feelings of low
> self-esteem, powerlessness, and worthlessness—self-negating feelings that have
> been exacerbated in their relationships with their partners.[68]

Self-destruction among women is related not just to conflict within marital
relationships, but also to problematic relationships with their parents. The loss
of one or both parents seems to make individuals more prone to suicide and
less adaptable to the stresses of modern living. Even more important, however,
female suicidal behavior has been described as a form of self-destructive coping related to protest and despair over multiproblem, pathological family-of-
origin living situations.[69] In fact, the general picture that emerges from the literature on the families of suicidal individuals is one of relationships that are
often extremely disorganized, emotionally impoverished, and unable to give
psychological nurturance or stability. Such deficits may lead to anger at the
parents, which culminates in guilt and depression. The subsequent suicidal
behavior may be in part a misguided effort to punish or manipulate family
members whom individuals blame for their lack of self-esteem and their emotional insecurity.[70] In the case of women specifically, the most prevalent image
characterizing suicidal experiences is lack of nurturance and emotional neglect.
These are explicitly related to nonloving, absent, abusive, mentally ill, or al-
coholic parents.[71] Disorganization and breakdown in the life situation of the
individual's family of origin are linked to the suicidal woman's depression and
feeling of worthlessness. Depression and worthlessness, in turn, are linked to
suicide, in that suicidal behavior is a symptom of and a reaction to a lifetime of
pathological child–parent interactions.

Attempted suicide by women has too often been interpreted as a hysterical
or manipulative appeal for help. What emerges from recent studies is a differ-

ent picture. The suicidal woman is perennially distraught. She is isolated, lonely, and powerless. She feels trapped and impaired by her oppressive circumstances. Thus, her suicide attempt or her successful suicide is much more than a vengeful, manipulative gesture designed to arouse the affection of her husband or to punish her unloving parents. Instead, it is a response to emotional stress and fatalism grounded in a desire to transcend the realities of her life circumstances, to escape stressful family relationships, and to reach out for life experiences that are meaningful, supportive, and satisfying. For this reason, as was previously indicated, the suicidal woman's behavior is not usually rooted in the desire to die. Rather, it is motivated by a thirst for life at a time when the circumstances of her life are perceived as intolerable and otherwise inescapable.

We should remember, of course, that the behavior of suicidally inclined individuals is explicitly connected to both social and personal factors. It is obvious that the motivations for suicide are deeply rooted in the self, and that suicide behavior is one prominent indicator that the person is deeply distraught. However, the individual's troubled self is not the only cause of suicide. Social and cultural forces also underlie the problem. Thus, in order to begin to understand suicide, social and individual factors must be merged and fully taken into account. In his discussion of adolescent suicide, Ronald Maris articulates the need for psychosocial synthesis:

> Sometimes we act as if we believe that modern social conditions (e.g., divorce rates, both parents working, stress, high unemployment rates and scarce jobs, sex role confusion, competition in schools, "the bomb," etc.) alone were producing the rise in adolescent suicide rates. To be sure, changes in the meaning of work, love, marriage, family, stress, parent–child relations, religion, models of peer suicides, and lack of clear, consensual life goals set the broad context for adolescent suicide. But we must always remember that social factors alone cannot account for young suicides. The overwhelming majority of young people cope with the strains of modern life without killing themselves. . . . Thus, we must also examine the particular individual and situational factors that make some young people's lives not viable. These more personal factors in young suicides might include: use of alcohol and drugs, sexual permissiveness and sexual confusion, repeated depression, hopelessness (including chaos), confusion, disorientation, social isolation and failure of adaptive techniques, anger, irritability, impulsivity, having a gun, and so on. All of these personal and social factors come together in a few basic types of adolescent lifestyles that turn out to be in fact, progressive suicidal careers.[72]

The point is that the psychosocial collage of suicide is comprised of structural conditions, of the psychological state of the individual, and of the manner in which structural or social forces converge with the personal state of the individual and define particular life situations for individuals.

Let us return for a moment to consider Durkheim's study of suicide and social facts. The absence of family and interpersonal integration, heightened individualism, and normlessness are important factors influencing and precipitating modern suicidal behavior. In this way, Durkheim's observations regarding the prominence of egoistic and anomic suicide in modern society are confirmed. But the psychosocial collage of suicide also points dramatically to the incompleteness of Durkheim's analysis. Not only did Durkheim ignore psychological variables, but he overlooked the ways in which social forces may be internalized in a variety of ways and become significant factors shaping the individual's psychological and emotional state. For example, Durkheim defined fatalistic suicide as related exclusively to the structure of society, namely, excessive social regulation, whereby the lives of individuals become choked by oppressive discipline and characterized by despair and hopelessness. Durkheim dismissed fatalistic suicide as being inconsequential to modern society, arguing that the social forces of fatalism are incompatible with the organization of modern societies. Yet the idea of fatalism is very relevant to the psychosocial collage of suicide behavior. It is relevant not in terms of Durkheim's limited view of social dictates, but rather in terms of the individual's definition of reality. In this way, fatalism is more than a social fact of excessive regulation. It is a frame of mind derived from the patterns or conditions of modern life, that is, a psychological state of being that eventuates from a particular living situation in a given social context.

Thus, when social forces that engender powerlessness become relevant to the life of an individual, helplessness is rooted in both personal and social conditions. In this way, fatalism as a precipitator of youth suicide is not related to excessive social regulation alone; it is also related to the psychosocial dynamics of helplessness and hopelessness. As we have just seen, the same applies to women and suicide. Thus, it is precisely the way in which personal and social conditions converge that establish suicide as both a social problem and a private trouble. In any event, it is clear that themes of loneliness, disorganization, interpersonal conflict, and powerlessness are central to the psychosocial dynamics of modern suicide. And these themes are relevant both to the private lives of troubled individuals and to the social organization of modern society.

The Aftermath of Suicide

Suicide may bring an end to the suffering and turbulence that motivate an individual to commit suicide. However, it is the beginning of a unique loss and bereavement experience for the survivors, who are also victims in an important psychosocial sense. As the psychoanalytic formulations of Menninger indicate, suicidal individuals may often wish the death of another and, in effect, bring death to bear upon the lives of the survivors by the suicide act. Obviously, the death brought upon survivors is not a cessation of physical life. Instead, it is a symbolic and psychological one that stigmatizes and torments profoundly. As suicidologist Edwin Shneidman observes, survivor victims of such deaths are invaded by an unhealthy complex of disturbing emotions: shame, guilt, hatred, and perplexity. They are obsessed with thoughts about the death, seeking reasons, casting blame, and often punishing themselves.[73]

Thanatology Snapshot 7.5

Stigma and Social Isolation of Suicide Survivors

A forty-one-year-old woman found her husband's body hanging in the basement of their home. When the police arrived, a crowd of people gathered near the house. She felt exposed and embarrassed. After leaving the house she never returned. She was blamed for the suicide by her mother-in-law. At the funeral, she overheard her eighty-year-old mother-in-law criticize her, "You've never liked me; you've never liked anything I've done; this is the final straw" and slapped her across the face.

A conspiracy of silence developed concerning the suicide. Her son tells people that his father died of a heart attack. . . .

When interviewed six years after the suicide, she expressed a fear that others would find out about the suicide and would think that insanity runs in the family. She mentioned that she thought that the stigma of suicide makes bereavement from suicide more painful to the survivors than bereavement from any other type of death. She also asks herself: "Will the stigma be attached to the children, to the children's children, and to their children in turn?"

Mark Solomon, "The Bereaved and the Stigma of Suicide," *Omega*, Vol. 13(4), 1983, p. 385.

The survivors of a suicide (someone else's) are likely to get stuck in their grieving and to go on for years in a state of cold isolation, unable to feel close to others and carrying always with them the feelings that they are set apart or under the threat of doom.

E. Lindeman and T. Greer, "A Study of Grief: Emotional Responses to Suicide" in A. Cain, ed., *Survivors of Suicide* Springfield, IL: Charles A. Thomas, Publisher, 1972), p. 67.

As discussed earlier, sudden death and the deprivation of anticipatory grief create special grief and particular stress for survivors. In many ways, suicide may be the cruelest possible form of death with which survivors have to cope.[74] Suicide entails not only the trauma of sudden, unanticipated death, but an exacerbation of grief because of social stigma.

"Stigma" is a concept that refers to a deeply discrediting quality or attribute. It connotes not just something unusual, but something shameful and bad about the moral status of the individual.[75] When suicide occurs, its survivors are frequently victimized by stigmatizing societal responses. The suicide itself is often perceived as a discrediting act that leaves a blemish on those close to the person who dies. The response of significant others to the suicide of a loved one is shaped and reinforced by this sociocultural attitude. Indeed, self-blame and guilt are typical responses of survivors, a classic example of guilt by association. For insulation and protection, the survivor often tends to grieve alone and to withdraw from interaction with others; this withdrawal, in turn, engenders a societal repudiation of the survivor. Thus, an expanding spiral of blame, guilt, and emotional turbulence is a painful by-product of the stigma felt by the suicide survivor.[76]

As noted above, one of the major consequences of stigma for survivors is isolation from friendship networks and other sources of support. Internalized guilt and external blame often converge to establish a conspiracy of silence about the suicide, a conspiracy that not only seeks to avoid dealing with the issue but that renders adjustment all the more difficult.

> But perhaps the most crucial of the malignant external agents was the blame frequently heaped on the suicide's surviving spouse by his community, his neighbors and his family—especially his in-laws. . . . often almost no support was provided the bereaved spouse. . . .
>
> Far more destructive, though, was the fact that the shame and guilt typically brought about a massive avoidance of communication regarding the suicide, which in turn virtually prevented the working through of mourning. . . . The conspiracy of silence which tends quickly to surround a suicide sharply limits the bereaved spouse's opportunities for catharsis, for actively checking distorted fantasies against the realities of the suicidal act, for clearing up a variety of gross misconceptions, or for fully dealing with and eventually resolving the irrational guilts and particularly the angry reproaches felt toward the person who committed suicide.[77]

As discussed in Chapter 5, modern grief and bereavement are lonely, isolating experiences, regardless of the cause of death. Privatization of grief and mourning is driven to new heights, however, by the special stress and social stigma associated with suicide. As a result, heightened social isolation is common following the suicidal death of a spouse or child. Parents are often viewed

SUICIDE SURVIVORS AND SOCIAL STIGMA

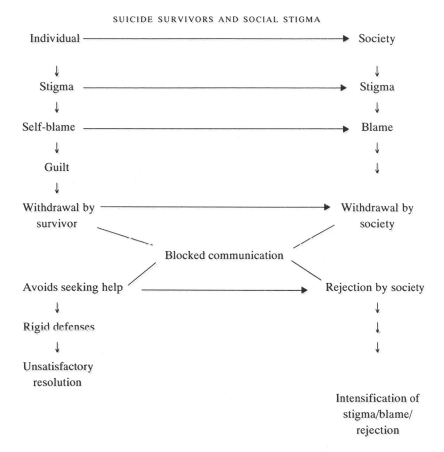

A. S. Demi, "Social Adjustment of Widows After a Sudden Death," *Death Education*, Vol. 8 (Supplement), 1984.

or believe they are being judged severly by others, as responsible for their child's suicide, which inspires deep feelings of shame and embarrassment in them. This kind of parental bereavement is characterized by inhibitions of mourning by the surrounding community, disruption and inadequacy of the usual coping devices, isolation, and a crisis in parental identity:

> Friends or family who avoided the topic or denied the suicide confused the parents and discredited their competence as parents. The hostile world was described as hostile and incapable of understanding their grief. Parents increased their self-blame as they felt blamed by others.

Now [in the case of suicide] parents agreed that past experience with loss was useless. Thus, the usual ways of coping were inadequate. . . .

Even when new coping strategies were developed as grief work progressed, these families still felt isolated, stigmatized and excluded from without and even sometimes within their nuclear families. . . .

Not only were parents unable to control their unpredictable tears and sadness, but now they said they had lost an idea of who they were as parents.[78]

In short, survivors of suicide exhibit more guilt, self-blame, emotional turmoil, diminished social contacts, and personal and social isolation than other survivors. Interestingly, however, studies indicate that though the trauma of death is especially profound for suicide survivors, there is no notably negative impact on the family structure of the survivors. Indeed, it is conceivable that some family units may even be strengthened in the aftermath of suicide. Existing family and personal values may be reexamined, a basis for solidity may emerge,[79] and family members may turn to each other for support and comfort in the presence of larger isolating, stigmatizing forces.

In the isolating web of privatization and social stigma, the goals of therapeutic intervention for survivors of suicide are to provide support, facilitate grief work and mourning, encourage self-expression, and promote self-growth.[80] The emphasis is on self-determination and individualized responsibility, and the ultimate goal is to prompt an active return to life and to block any suicidal inclinations on the part of the survivors. Treatment includes family therapy,[81] individual psychotherapy, and self-help bereavement groups.[82]

Thanatology Snapshot 7.6

Death and Significant Others

There are always two parties to a death: the person who dies and the survivors who are bereaved. Death releases its prey instantly from all further suffering in this world—

. . . [T]he sting of death is less sharp for the person who dies than it is for the bereaved survivor. This is, as I see it, the capital fact about the relation between living and dying. There are two parties to the suffering that death inflicts; and, in the apportionment of this suffering, the survivor takes the brunt.

Arnold Toynbee,
Man's Concern with Death
(New York: McGraw-Hill Book Co., 1969), pp. 267–271.

Although the themes of stigma and social isolation are prevalent in suicidology literature, many questions about suicide bereavement remain. Whether behavioral responses to suicide are different from responses to other types of death is not clear. It is also uncertain whether the psychosocial symptoms of grief, as discussed in Chapter 5, are greater for survivors of suicide. Moreover, the effectiveness of therapeutic amelioration of suicide death, especially in regard to quiet resolution, has been verified only by fragmented clinical studies. And finally, the impact of a suicide on patterns of interaction among family members remains only partially explored.

A Conclusion: Suicidology as the Study of Life

Judging whether or not life is worth living involves much more than merely answering the central question of philosophy. Exploring the question "Are we our own enemies?" involves more than musing about the psychoanalytic relevance of an internalized death wish. Indeed, the problem of human suicide is more than a therapeutic, psychiatric issue. The causes, nature, and incidence of suicide reflect the quality, meaning, and value of life, as engendered by the organization of society and as perceived and experienced in the personal, private lives of individuals. In other words, the way we die is a reflection of the way we live.

8

The Death of Humans by Humans, Part Two: The Holocaust and the Technology of Genocide

> One day . . . I will tell the world.
>
> PRISONER AT BIRKENAU

At first glance, genocide seems unreal and unthinkable. Historical realities of genocide are often dismissed as regrettable but unrepeatable aberrations of unique cultural and social circumstances. Indeed, the idea of genocide is most often associated with Adolf Hitler and his attempt to exterminate the Jewish people. Hitler's holocaust is often explained away as the consequence of an insane, messianic, evil person who was successful in assuming absolute power over German society. In this way, genocide is most often thought of in a specific narrow framework. It tends to be regarded as an individualized and passing circumstance rather than as a social or structural problem. However, the human record of genocide includes far more than Hitler's savage brutality against the Jews. The history of humanity is replete with genocidal thinking, tendencies, and acts. The Spanish Inquisition in the 1500s, the massacre of American Indians during the settling of the United States, the "killing fields" of Cambodia where an estimated 3 million people were brutally killed in concentration camps, the extermination of socially and politically undesirable individuals in the Soviet Union's death camps, and the death and destruction which has taken place in the name of ethic cleansing in Bosnia are painful testimony to the historical record of genocidal behavior among human societies. Additionally, the power and tyranny of Napoleon, Stalin, Mussolini, Bismarck, the Shah of Iran, or Sadam Hussein have been framed by genocidal qualities. The defilement of blacks in South Africa, the starvation of black people in Ethiopia, and the grotesque suffering and mass death in Rwanda also contain disturbing genocidal dimensions. The advent of the nuclear age threatens genocide for the entire planet. In this way, genocidal realities are not necessarily rare, isolated, or unique occurrences. Rather, they are a recurrent part of the historical record of humanity, and in our age of nuclear weaponry, they shadow the entire planet with a perpetual threat of mass death.

The genocidal orientation is thus more than the tyrannical accomplishments of a few despotic individuals in isolated circumstances. Rather, throughout history, genocide has been deeply rooted in social structure and social organization. In this way, whether the motivation of genocide is power, wealth, nationalism, or political ideologies such as "racial purity" or "manifest destiny," the advocacy and legitimation of genocide are derived from social processes and organization. This occurs on a continuum whereby genocide is implemented either as an explicit public policy or as a latent consequence of cultural and social arrangements. It is also important to recognize that the specific manner in which genocidal activities are carried out is shaped by prevailing cultural and social forces.

This chapter will explore violent deaths of genocide by focusing on the Nazi slaughter of European Jews. This is not done to link genocide exclusively with Hitler, but because the crimes of the Nazi holocaust are the most vicious, savage, and incomprehensible violations of humanity ever committed. Additionally, the magnitude of the Nazi atrocities was facilitated by characteristics that typify the organization of modern societies: mass coordination, technological dominance, mechanization, and bureaucratization. In this way, the holocaust is emphasized not to identify genocide solely with the Nazis, but to consider how the social structure of modern technological society is conducive to genocide.

One further introductory remark is warranted. This has to do with the assimilation of suicide and genocide into a common framework. In many ways, suicide and genocide reflect the polar opposites of the voluntary private killing of the self and the mass killing of human beings on a societal level. Clearly, however, despite many apparent differences, suicide and genocide are types of death that are steeped in violence. The commonality, however, runs deeper. As forms of violent death, suicide and genocide have a profound effect on social life in America. The prevalence of suicidal behavior among today's youth and elders is an important measure of the quality of American life. Genocide, as well as being historically significant, demands attention because of the multiplying technological capabilities for destruction of the modern world. In this way, although suicide and genocide appear to be divergent forms of death, they provide important insights into patterns and forms of human life. I have selected these polar extremes of death styles in order to assess the value, sanctity, and meaning of human life in diverse historical and social circumstances. Thus, the overall purpose of the discussion of humanity's hand in the death of human beings is to elucidate how the extremes of suicide and genocide teach us not just about styles of death in modern society, but about styles of living as well.

And this is precisely consistent with the overriding theme of this book, namely, that the way one dies is a reflection of the way one lives.

Let us now turn to the holocaust and explore the Nazi decimation of human life. It is essential to remember, however, that Nazi genocide will be discussed here not just as a situationally specific happening but as a potential for evil that is relevant for all societies, especially technologically developed ones.

Isolation of the Jewish Citizenry

The stage for the mass extermination of Jews was set by a lengthy evolution of anti-Semitic prejudice exacerbated by the defeat of Germany during World War I and the subsequent economic crisis that beleaguered Germany from the 1920s to the 1940s. Careful analysis of the social and economic conditions of preholocaust Germany reveals a growing centralization of capital and wealth in the hands of an industrial elite. The inability of the German middle class to compete and survive in an era of growing monopolization of resources not only disenfranchised German citizens economically but disillusioned them politically as well. The social climate was one of growing scarcity and deprivation, and this context provided a fertile environment for the acceptance of the Nazi agenda of paranoia and hatred. As Martin Gilbert describes in his authoritative account, *The Holocaust*:

> Suddenly events began to favor Hitler and his followers. Inflation began to rise again. Unemployment grew to unprecedented levels.
>
> The internal problems . . . continued to worsen. Unemployment rose yet again, reaching three million by the end of 1929. Both workers and employers were its victims. Small businessmen suffered equally with those on the factory floor.[1]

It was within this social context of political and economic instability that the foundation for a genocidal policy was laid.

Given these circumstances, some Germans, especially the Nazi Party, began to search for a scapegoat within German society in order to mitigate dissatisfaction among the German people. This search led directly to the Jews.[2] Indeed, it was Hitler himself who proclaimed that Germany could become great again if the Jews were eliminated. In fact, hatred of the Jews, central to Hitler's mental state, writings, and speeches, was echoed in the actions of many of his followers. Even before his formal rise to power, individual Jews were attacked in the streets, at public meetings, and at street-corner rallies. Jews were blamed, often in the crudest language, for every facet of Germany's problems, including the military defeat of 1918, the subsequent economic hardship, and the sudden spiraling inflation.[3] Thus, fear and hatred of the Jews, which had existed for many generations prior to Hitler's assumption of power, were cleverly manipulated by Nazi ideologues, facilitating the growth of solidarity and nationalism among the German people.

Germany's nation-state was at the center of Hitler's genocidal policy. The severe political and social crisis that had overtaken Germany after World War I, fed by the harsh economic situation, generated much internal unrest. Hitler correctly sensed the people's collective desire for a cure. And in assessing their urgent need to transcend their economic, political, and social duress, he offered the German people an outlet for their frustration. The outlet was the opportunity to vent their anger on the people who were directly blamed for their ills, the Jews.[4] Additionally, through a propaganda campaign that has now become legendary, Hitler's message of transcendence was deeply embedded in a totalitarian idea of nationalism. His point was unambiguous: Vitality and confidence in German society could be restored only by the total commitment of all Germans to the common good of Germany. In this way, Hitler's vision of societal regeneration was bound up not only with a vicious program of ethnic hatred but also with the identification of nationalism as a supreme value. Hitler's promise to a defeated, desperate society was simple. Omnipotence, perhaps even immortality, could be achieved through the explosive power of Germany as a unified nation. The future of individual Germans could become grandiose and glorious only by deliberately honoring the will and needs of the nation.

Thus, not only did mass killing emerge as the public policy of the Third Reich, it was designed to revitalize and enhance the nation of Germany. In this vein, one of the Nazis' most popular slogans was "Deutschland erwache, Juda verrecke!" ("Germany awake, death to Jews!"). Indeed, the murderous machinery of the Nazis was not implemented as part of an underground movement. Rather, ideological motivations for the mass murder of Jews and other undesirables, along with the strategies and techniques that made it possible, were explicitly part of the organization and structure of Nazi society:

> The German civil service and ministerial bureaucracy had traditionally commanded the implicit confidence of the people. The Nazi regime utilized this confidence and the attitude of the people to achieve its own aims. Immediately after the seizure of power, totalitarianism was injected into the inherited bureaucratic system primarily through the ideological Nazi party and the SS terror. This was accomplished by the fusion of both the new and the old systems, thus ensuring a reasonably smooth functioning of the bureaucratic apparatus. In the party and its formations, especially in the SS and the political police system (Gestapo), this was achieved very expediently, as the odious efficiency of the destruction machinery during the war years attests.[5]

Thus, the mainstream of German society—the political organization of the state, the economic elite as well as the middle class, the professional classes, doctors, lawyers, professors, even the clergy—was forced to submit to the will of the Nazis. Indeed, through the use of terror and persuasion, "trooplike cooperation" was exacted from the professional classes. For example, universities were cleansed of politically unacceptable faculty, all physicians were pressured to accept the Nazi ethos, and medical students were required to integrate their medical training with service to the Nazi ideology. One of Hitler's driving ambitions was to ideologize completely all political, social, and cultural institutions.[6] Thus, although not all professionals fully embraced the Nazi agenda, open dissent was not tolerated. Very quickly, those who disagreed with the Nazi restructuring of society learned to remain silent. In this way, no matter how extreme Hitler's vision of the world may have been, ordinary processes, structures, and interactions within the larger society (such as using the railroads to transport people to the concentration camps) were central to the implementation of his vision. The Nazi program of genocide therefore was implemented through the required cooperation of, and with, all political, social, cultural, and economic institutions of the German nation-state.

Hitler's utopian view of a Jew-free, genetically superior nation was a dominant ideological and political force within the new German society. The evolving genocidal policies of the Third Reich identified Jews as biologically inferior and racially impure. In many ways, the entire totalitarian regime of the Third Reich was based on and motivated by a quest for biological superiority and racial purity. Sterilization and euthanasia programs were technological responses to the Nazi vision of a healthy, viable society. As Hitler wrote in his bible of Nazism, *Mein Kampf*, the mission of the German nation-state was as follows:

> Assembling and preserving the most valuable stocks of basic racial elements in this [Aryan] people . . . [and] slowly and surely raising them to a dominant position.[7]

Illness, according to Hitler, was a social and moral issue that concerned not only the health of individuals but the viability of the entire nation as well. Again, as he expressed it in *Mein Kampf*:

> [A]nyone who wants to cure this era, which is inwardly sick and rotten, must first of all summon up the courage to make clean the causes of the disease.[8]

For Hitler, Jews were the obvious cause of the disease, and they quickly came to be labeled the "misfortune" of the German people. He characterized them as "destroyers of culture," agents of "racial pollution" and "racial tuberculosis," as well as parasites and bacteria causing sickness, deterioration, and death in the host peoples they infected. They were the "external bloodsucker," "vampire," "germ carrier," "people's parasite," "poisoner," and "maggot in a rotting corpse."[9] The cure for the disease, as envisioned by Hitler, was similarly infused with metaphorical, biomedical imagery and focused on the excision of infectious Jews from Aryan society. As one Nazi physician put it:

> [O]ut of a respect for human life I would remove a gangrenous appendix from a diseased body. The Jew is the gangrenous appendix in the body of mankind.[10]

This ideological defilement of the Jewish people isolated them from the mainstream of German life and Nazi society. It also provided the motivation and legitimation of political and social policies designed to expunge Jewish life from German culture. The systematic, physical separation of Jews from German society through ghettos, labor camps, and death camps stemmed from this ideological decimation of the value of Jewish life. This isolation served to provide the Nazis with an essential supplement in their efforts to eliminate the influence of the Jews from every facet of Ayran life. Again, the biomedical metaphor is relevant. The Nazis needed to transform mere words and ideas into a concrete, bureaucratic system that would implement a selffulfilling prophecy of the danger and uselessness of Jewish life. As the spread of contagious disease is halted through quarantine, the Jewish pestilence was to be contained by isolating the Jews from the mainstream of German society.

One of the earliest concrete attempts to separate the Jewish citizenry occurred in Poland in what has become known as the Warsaw Ghetto. Initially, the Nazi plan for dealing with "infectious Jews" in Warsaw centered on deprivation of food. After the Nazi invasion of Poland in 1939, food was rationed to Jews at an absolutely minimum level of subsistence:

> Germans in Warsaw were allotted 2,500 calories a day and could obtain a wide variety of goods at fixed prices in special stores. The Jewish ration came to less than 200 calories a day, for which Jews paid twenty times the price Germans did.[11]

This food deprivation fulfilled several functions for the Nazis. Jews were so weakened by incessant hunger that their strength was systematically drained. Additionally, the deprivation established a built-in system of euthanasia as Jews began to die regularly of starvation. Hunger dramatically increased the incidence of chronic disease and death among the Jewish population:

> Everywhere people were dying of starvation. Daily the newspaper-covered corpses lay in the street. On leaving the house in the morning, it was not unusual to be accosted by the weak voice of a starving man leaning against the wall of a building faintly asking for bread; nor was it unusual on coming home that evening to see the same man lying full length, his breath stilled forever. In the reception centers for refugees, now openly called Death Centers, the death toll still mounted. In the hospitals, typhus patients rose from their beds and fled

from the appalling conditions, ravenous with hunger and eaten by disease. In private homes the story was the same, especially in the crowded, dingy flats of the poor, where relatives of the diseased would hastily drag their bodies into the streets at night and leave them there to be discovered in the morning. The purpose was to avoid paying 20 zlotys to the undertaker and to avoid reporting the death so that they could keep the dead man's ration card.[12]

Starvation haunted the Jews in Nazi-occupied Poland from 1939 on. Living conditions were cruelly inhuman. Filth, disease, suffering, poverty, and death were everywhere. The degradation of Jewish life became so extreme that it is virtually unbelievable and incomprehensible to ordinary human consciousness. Yet to the Nazis, the extermination of the Jews was not happening quickly enough. So the second phase of the Warsaw extermination began. Jews were sealed up behind nine-foot-high ghetto walls and were isolated from the outside world.

The creation of the Warsaw Ghetto took some doing. Jews were relocated in the ghetto in the fall of 1940. They were literally driven from their homes and forced to relocate haphazardly in new accommodations, typically congested with six or seven people per room. The forced relocation brought additional chaos and hardships to Jewish life. As one eyewitness recounts:

> Today [the day of resettlement] was a terrifying day; the sight of Jews moving their old rags and bedding made a horrible impression.[13]

Another eyewitness describes this horror in more explicit detail:

> Try to picture one-third of a large city's population moving through the streets in an endless stream, pushing, wheeling, dragging all their belongings from every part of the city to one small section, crowding one another more and more as they converged. No cars, no horses, no help of any sort was available to us by order of the occupying authorities. Pushcarts were about the only method of conveyance we had, and these were piled high with household goods, furnishing much amusement to the German onlookers who delighted in overturning the carts and seeing us scrambling for our effects. Many of the goods were confiscated arbitrarily. . . .
>
> In the ghetto . . . there was appalling chaos. Thousands of people were rushing around at the last minute trying to find a place to stay. Everything was already filled up, but still they kept coming and somehow more room was found.
>
> The narrow, crooked streets of the most dilapidated section of Warsaw were crowded with pushcarts, their owners going from house to house asking the inevitable question: Have you room? The sidewalks were covered with their belongings. Children wandered, lost and crying, parents ran hither and yon, seeking them, their cries drowned in the tremendous hubbub of half a million uprooted people.[14]

Once the Jews were relocated in this confined area, they were totally separated from the outside world. The rate of starvation accelerated, and formal methods of killing were initiated to complement what disease, hunger, and exposure could not accomplish quickly enough. Extermination squads roamed the streets, killing at will, often at night, and leaving the corpses to be found lying in the streets in the morning. The containment of the Jews and their isolation in the ghetto enabled the Nazis to make sport out of their extermination, and the killing became a regular part of life in Warsaw:

> The relentless killing missed no day and took no rest. The Nazis had cut off each Jewish Community from the outside world, and from all other Jewish life, and using this isolation they worked without respite to destroy. In Warsaw, during a massive round-up on September 6 and 7 [1942], more than a thousand Jews were killed in the streets, including hundreds who . . . were forced to kneel on the pavement and be shot.[15]

Life in the ghetto was brutally cruel, and the Nazis went about their business with murderous frenzy. Death was visible everywhere:

> In January 1941, two thousand Jews had died of starvation in the Warsaw Ghetto. The February toll was just as high. Almost daily . . . people are falling dead or unconscious in the middle of the street. It no longer makes so direct an impression.[16]

And, perhaps even more disturbingly, no one, including children, was spared from Nazi cruelties:

> In every street of the Ghetto, beggars. "Child in arms, a mother begs—the child appears dead." . . . Three- and four-year-old children were begging, "and that is the most painful."[17]

Indeed, in the midst of winter:

> Small children roamed the streets in rags, sobbing and screaming and many actually froze to death.[18]

The relocation of the Jews to the Warsaw Ghetto provided some sense of solidarity and community among the Jewish prisoners. Efforts to repulse the Nazis developed in the form of underground resistance activities. Food, fuel, and clothing were smuggled to those in need. Papers were forged, authorities were bribed, and occasionally, successful escapes were engineered. It was not uncommon for individuals to risk their own lives to try to save their families. Yet, the thirst for life on an individual and a collective level seemed so profound among ghetto residents that resistance, which threatened the life of individuals as well as groups, was stymied. By force of terror and persuasion, most were required to submit to the Nazi reorganization of life in the ghetto. As one survivor's anguished recollection describes:

> The question torments all of us, but there is no answer to it because everyone knows that resistance, and particularly if even one single German is killed, its outcome may lead to slaughter of a whole community, or even of many communities. . . .
>
> . . . So strong is the life instinct of workers, of the fortunate owners of work permits, that it overcomes the urge to fight, the urge to defend the whole community, with no thought to consequences. And we are left to be led as sheep to a slaughterhouse. This is partly due to the complete spiritual breakdown and disintegration, caused by unheard of terror. . . .
>
> The effect of all this taken together is that when a moment of some resistance arrives, we are completely powerless and the enemy does to us whatever he pleases.[19]

Passivity and compliance were sometimes associated with active cooperation with the Nazis. Jewish policemen, under the authority of the Jewish Council, or Judenrat, actively enforced the rules and regulations of the Warsaw Ghetto. They participated in resettlement programs that sent Jews off to labor and death camps. The benefit for the individual policemen was some protection for themselves and their families against Nazi atrocities. However, in the long run, obedience to and compliance with Nazi directives did no good for these individuals, as the Nazis, in Warsaw as well as everywhere else, killed those who served them. Additionally, in this framework of extremity and insanity, there were even some Jews who sought to maximize personal and entrepreneurial opportunities. Those who did this provided a dramatic contrast to the suffering, illness, and death which characterized the ghetto:

> While hundreds of thousands grew steadily poorer, a handful of men were making enormous fortunes: the leading smugglers, bakers, a few contractors working for the German Army, some Judenrat big shots, the ghetto police, and the thirteen. The money so easily, if dangerously, gained was also easily spent. While the ghetto was dying of hunger, champagne toasts were being quaffed, and drunken orgies and banquets held in the night clubs which had recently mushroomed.[20]

Additionally, well-dressed Jewish ladies were sometimes taken to the restaurants and clubs that were then flourishing. On their way to and from elaborate meals, they were walked past starving men, women, and children in the street.[21]

The essential point is this: A visible, privileged class of Jews was being purposefully created by the Nazi planners. On a less dramatic everyday level, the sick were being isolated from the healthy and strong. The rigorous and fit were to be put to work in labor camps, their lives and sustenance guaranteed, at least for the time being, as long as they were productive. The weakest, which typically meant sick people, orphaned children, the crippled, and the aged, were targeted for death. Thus, the Nazis sought to turn Jews against each other by privilege, status, material lures, and the increased possibility of survival.

On an obvious level, then, Jews were isolated from most of Warsaw and from the rest of the world by the walls of the ghetto. By cutting off contact with the outside world, the Nazis were unhampered in their savage brutality against Jewish people. However, on a less obvious but perhaps even more insidious level, the ultimate source of isolation of the Jewish citizenry resided not in physical separation from the outside world, but rather in the isolation of Jews from each other. As Jews were pitted against each other through stratification and vested interests, fellowship and the bonds of community were systematically fragmented. The perverse, sadistic genius of the Nazi plan in the Warsaw Ghetto lay in the attempt to destroy Jewish collective consciousness and solidarity. In this way, conflicting interests and perspectives on life and its future in the ghetto served to divide the Jewish people, often making them collectively helpless and impotent.

Remarkably, however, there were Jewish people in the ghetto who found the will and the resources to resist. Defiance and resistance were initiated through the formation of a complex underground network of Jewish activists. The Warsaw Ghetto was such a closed, tightly controlled area that active rebellion would almost certainly end in death for the participants. In 1942, when the mass evacuation of the ghetto was beginning, a Jewish fighting organization was formed. The Jewish underground accepted the task of securing as many weapons as they could by any means possible. As despair, hunger, suffering, disease, and death swelled in the ghetto, so did the feeling that acquiescence to the Nazis could no longer be tolerated. Indeed, tens of thousands of Jews fought back violently against the Germans. Jewish sources claim that hundreds of Germans were killed and wounded from April 19 to May 15. Official German records of casualties report a significantly lower number.[22] The real measure of Jewish success, however, is not connected to casualty figures of the Warsaw revolt. Rather, the victory lies in the fact that the Jewish people fought back vehemently against the Nazis for several weeks. By the end of the revolt, all Jews had either been deported or killed by the German response to the uprising. On May 15, 1943, SS General Jurgen Stroop proclaimed that the Third Reich had fulfilled its mission: No Jews or Jewish dwellings remained in Warsaw.[23] But the message of resistance spawned by the gallantry and tenacity of the Jewish people spread far beyond the ghetto walls. In fact, the revolt itself became a major event, not only in the history of the holocaust but in the history of the Jewish people themselves.

In the midst of the crises and extremity that characterized life in the Warsaw

Ghetto, many different and often contradictory patterns of Jewish life emerged. Alongside the underground existed those who cooperated fully with the Nazis. Alongside the work of welfare organizations was the existence of apathy and hedonism.[24] The sense of despair and distress over the horror of the ghetto was responded to in different ways by various individuals and groups. However, it is clear that all Jewish people were subjected to conditions of unprecedented stress. The dehumanization, isolation, and segregation of Jews in the Warsaw Ghetto was so extreme that all of the standards by which a society is normally judged disintegrated completely. This was the common fact of life for all Jews in the ghetto, a fact that made the idea of survival on both a community and individual basis all the more urgent for the Jewish people. The ways in which individuals and groups sought and fought to survive may have differed. Yet, the imperative of survival was a matter deeply and consistently felt by many during this time of collective and personal tragedy.

Anus Mundi: Anus of the World

Concentration camps, designed to exterminate the Jewish people, were the ultimate device for implementation of the "final solution." Jews from Warsaw, and from all over Nazi-occupied Europe, were targeted for resettlement to concentration camps. In the camps, the Nazi vision of life perfected was experimented with in the broadest and most frightening ways. Discipline, power, technological and scientific gains, and mass organization, along with systematic killing, were the facts of life in the camps. Indeed, Nazi ideology and mythology were experimented with, literally played with, as the Nazis pursued their dream of a master society and a master race. The Nazi system of terror, as organized and accomplished through the concentration camps, was the most extreme and encompassing system of dehumanization and death ever devised and implemented.

The experience of the concentration camp began for Jews even before their arrival. The journey to the camps via railroad cars initiated the Jewish people into a defiled and debased environment. In many ways, this journey served as a microcosm of the life that would exist later on in the camps. Elie Weisel recalls the degradation of the transport to the death camps:

> [A] convoy of cattle wagons was waiting. The Hungarian police made us get in—eighty people in each car. . . . The bars at the window were checked to see that they were not loose. Then the cars were sealed. If anyone escaped he would be shot.

Weisel continues:

A prolonged whistle split the air. The wheels began to grind. We were on our way. . . .

Lying down was out of the question, and we were only able to sit by deciding to take turns. There was very little air. The lucky ones who happened to be near a window could see the blossoming countryside roll by.

After two days of traveling, we began to be tortured by thirst. Then the heat became unbearable.

As Weisel's journey continued, he recollects a woman in his cattle car who became increasingly hysterical:

She continued to scream, breathless, her voice broken by sobs. "Jews, listen to me! I can see a fire! There are huge flames! It is a furnace!"

It was as though she were possessed by an evil spirit which spoke from the depths of her being.

We tried to explain it away, more to calm ourselves and to recover our own breath than to comfort her. "She must be very thirsty, poor thing! That's why she keeps talking about a fire devouring her."

But it was in vain. Our terror was about to burst the sides of the train. Our nerves were at the breaking point. Our flesh was creeping. It was as though madness were taking possession of us all. We could stand it no longer. Some of the young men forced her to sit down, tied her up, and put a gag in her mouth. . . .

An endless night. Toward dawn Madame Schacter calmed down. Crouched in her corner, her bewildered gaze scouring the emptiness, she could no longer see us. . . .

. . . As soon as night fell, she began to scream: "There's a fire over there!" . . .

The heat, the thirst, the pestilential stench, the suffocating lack of air—these were as nothing compared with these screams which tore us to shreds. A few days more and we should have all started to scream too.

But we had reached a station. These who were next to the windows told us its name:

"Auschwitz."[25]

The transportation of Jewish prisoners to death and labor camps was an important element of growing Nazi terror. As deportees were transported from Warsaw, Lodz, France—indeed, from all of Nazi-controlled territory—an elaborate system of transportation was developed by the Nazis in conjunction with German railroads. Existing resources were utilized with maximum efficiency, and new resources were constructed where the need developed. Indeed, one of the most salient consequences of the development of a complex, yet highly efficient, mechanism of transporting millions of people to their deaths was the expansion of Nazi brutality far beyond the borders of the camps. In fact, the Nazi horror regularly passed through train stations and rolled onward through German towns, villages, and the surrounding countryside.

The relocation of large numbers of Jews through mass, unsanitary transport was a prelude to the life in the camps. As foulness and filth became the dominant environmental condition in the overcrowded cattle cars, filth also became an ordinary and normal way of life for the concentration camp prisoner. If one were to survive, one must live in spite of foul living conditions. In the death camps, dirt, stench, uncleanliness, and human excrement would become permanent facts of existence. It was the journey to the camp that began this subjection to filth:

> The only place to urinate was through a slot in the skylight, though whoever tried this usually missed, spilling urine on the floor. . . . When dawn finally rose . . . we were all quite ill and shattered, crushed not only by weight of fatigue but by the stifling, moist atmosphere and the foul odor of excrement. There was no latrine . . . on top of everything else, a lot of people had vomited on the floor. We were to live for days on end breathing these foul smells, and soon we lived in the foulness itself.[26]

Thus, the conditions in which prisoners were transported to the camps provided a frame of reference for what was yet to come. It also served to weaken the prisoners and make them more compliant. The incomprehensible horror and insanity of the journey established a standard of living in which the extremities of filth, defilement, and death were norms. In fact, the entire process of deportation and transportation transformed the literally incomprehensible journey to the camps into a standardized operation that assumed uncanny regularity and a disturbing sense of normalcy. In this way, the horrible crimes of the Nazis were submerged into a framework of steadiness and efficiency that served to systematize a seemingly unthinkable journey and an improbable outcome.

The time of arrival at the designated camp was another haunting and terrifying experience. Typically, the initial moments of detraining were characterized by confusion, noise, uniformed Nazis, searchlights (as arrival was often at night), guard dogs, and chaos generated by prisoners stumbling out of the cattle cars and lining up under directions of the SS officers. As one survivor recalls:

> Because you arrived at night, you saw miles of lights—and the fire from the . . . crematoria. And then screaming and the whistles and the "Out, out!" "Raus! raus! raus! raus!" [Colloquial German for "out"], and the uniformed men and the SS with dogs, and the striped prisoners—we, of course, at that time didn't know who they were—and they said, "Throw everything out. Line up—immediately!"[28]

As soon as the prisoners were removed from the train cars, they were assembled on the platform. An initial process of selecting those fit enough to work then took place. Those too old, too weak, too sick, or too young were targeted

Thanatology Snapshot 8.1

Initiation to Oblivion

Nearly every cattle car had a dozen human dead by the time the trains arrived at Treblinka, where they were met by SS men and unloaded with fantastic brutality. Those still able to stand were driven into a large square, ordered to strip, and leave all their things on the ground. The purpose, they were told, was for all of them to have a shower. The naked people were then lined up in columns and sadistically driven to a barracks-type building on which a sign read: "Showers." Screams, shooting, and blows created an atmosphere of panic, so that the victims would call out to one another "Come on! Hurry up! Or a tragedy may happen!"

The crowd would literally race into the building to "prevent a tragedy," and once they were inside, guards sealed the doors hermetically. A quarter of an hour later, all the people inside were dead of poison gas. A special Jewish team, working on the double, with its own reasons for panic—equally to "prevent a tragedy" to themselves—dug large pits and buried the bodies in them. . . .

Clothing and personal possessions were loaded on freight cars for shipment to SS warehouses; the few who had escaped and brought us the news had secreted themselves in those freight cars. And yet even those eyewitness reports were hard for us to believe. Surely the Germans, even the Nazis, were not capable of such utter inhumanity?

True or false, the machine was much stronger than we were. We were atomized, blown to bits, scattered to the winds, helpless beyond hope. Some of our own people were helping our mortal enemies to exterminate us. All we could do was find a hole and cower in it, keep as quiet as mice, and wait. Maybe somehow, on an individual basis, survival was still just barely possible.

Alexander Donat,
The Holocaust Kingdom
(New York: Holocaust Library, 1978), p. 73.

for immediate death. This decision, of who would live and who would die, was typically made by a Nazi physician:

When the transport trains came in, the arrivals had to pass before the camp doctor. . . . He pointed his thumb either to the right or to the left. Left meant death by gas. From a transport consisting of about 1,500 people, about 1,200 to 1,300 went to the gas chambers. . . . Those elected for the gas chambers had to undress

IN THE SHADOW OF THE HOLOCAUST

The haunting images of the corpses, both dead and alive, must never leave modern
social consciousness.

in front of the gas chambers and were then chased into them with whip lashes.
After about eight minutes the chambers were opened [and the corpses taken] for
cremation to the furnaces which were burning day and night.[29]

Those not selected for immediate death were processed into the camp. For
those entering the camp, there was a series of initiation rituals designed to dehumanize, humiliate, and degrade the prisoners. (Perhaps it would be more
accurate to note that the humiliation and degradation begun during deportation and transportation was reaffirmed during the initial "welcome" to the
camp.) Victor Frankl recalls his first hours in Auschwitz:

We were driven with blows into the immediate anteroom of the bath . . . with
unthinkable haste, people tore off their clothes.

    Next we were herded into another room to be shaved: not only our heads were
shaved, but not a hair was left on our entire bodies. Then on to the showers,
where we lined up again. We hardly recognized each other; but with great relief
some people noted that real water dripped from the sprays.

While we were waiting for the shower, our nakedness was brought home to us: we really had nothing now except our bare bodies—even minus hair; all we possessed literally was our naked existence.[30]

The process of initiation continued as prisoners were issued uniforms with the Star of David prominently sewn on, were tattooed with an identification number on the forearm, were issued a small wooden bowl that would become an important link to future survival, and were humiliated and beaten by SS soldiers during extended roll-call lineups. The process of initiation, preceded by the horrors of the transport, quickly destroyed normal qualities of humanness in the prisoners. The extremity of this process overwhelmed the prisoner, reducing feelings of vitality and aliveness. Jewish prisoners were typically left numb, helpless, and impotent in the face of this bizarre reality. Unrecognizable as human beings, their only possible human sentiment during all the chaos, confusion, and degradation was the biological thirst for survival, personally and collectively.

The isolation of Jewish persons from the human community was nearly complete at this point. However, for those who were to survive even for a while in the death camps, humiliation and degradation were merely the beginning. Survival meant adapting to conditions that were so extreme and incomprehensible that they seemed unreal. In many ways, prisoners had to adjust to being physically present in a world that lacked comprehensibility, sense, and connection to ordinary human reality. The death camp inmate lived incessantly in a physical world that was humanly unreal. Thus, the world where these humans lived was not a human world. Oddly enough, however, the persistence of human life in these circumstances was based upon human adaptation to this unreal, nonhuman environment.

Daily life was organized around extremes in the death camps. Extremes of hunger and malnutrition, exposure to the elements, lack of sleep, physical exhaustion, and brutality were ordinary facts of life of concentration camp existence. The daily routine began before sunrise each day. A piercing whistle brought the realities of another day into the exhausted consciousness of the brutally awakened prisoners. As many survivors have remarked, the moment of awakening was the most horrible moment of the day:

The most ghastly moment of the twenty-four hours of camp life was the awakening, when, at a still nocturnal hour, the three shrills of a whistle tore us pitilessly from our exhausted sleep and from the longings in our dreams. We then began to tussle with our wet shoes, into which we could scarcely force our feet, which were sore and swollen. . . .[31]

Another survivor observes:

Awakening is the hardest moment—no matter whether these are your first days in camp, days full of despair, where every morning you relive the painful shock, or whether you have been here, very long, where each morning reminds you that you lack strength to begin a new day, a day identical with all previous days.[32]

It is astonishing that the prisoners in the death camps got up at all. Yet, prisoners did indeed get up and did gather the resources necessary to survive another day. In the framework of extremity—of exhaustion, sickness, and frailty—the thirst for survival continued to motivate the decisions of the prisoners. Prisoners either got up or died; they either faced an unbearable world knowing they would have to bear it or gave up.

It has been argued that the organization of life in the death camps reduced the prisoners to childlike inadequacy and dependency. In developing his case regarding the infantilization of prisoners, Bettelheim observes that inhuman "toileting policies" subjected prisoners to childlike helplessness in the face of their own biological functions:

Because of the small numbers of toilets, the brief time available and the large number of prisoners, they were also forced to form in long lines before each toilet. Those waiting, afraid they might not get a chance to use the toilet, nagged and swore at the prisoner using it to hurry up, to get done. Here the waiting prisoners treated the eliminating one as an impatient parent might urge his infant to get off the potty; another camp situation that pushed prisoners into treating each other as incompetent children.[33]

In a childlike manner, normal biological functions were turned into moments of excretory crisis, and prisoners increasingly became obsessed with functioning of the excretory system.

In a sweeping and total effort, the Nazi agenda sought to debase and dehumanize the Jews. This was done by transforming them from self-respecting individuals into obedient children by making it impossible for the Jews to see themselves as adult human beings.[34] However, while the similarities between the behavior of prisoners and children are obvious from these descriptions, what happened in the death camps was much greater than mere infantilization. First of all, unlike that of the child, the prisoner's excremental situation was a permanent condition of the daily routine of camp life. Additionally, unlike a child, excremental imprisonment and defilement was an explicit consequence of an organized system of degradation. Thus, as Terence Des Pres judiciously notes, it is important to recognize that the prisoners' obsession with biological necessities is not a manifestation of their own childish-infantile drives. Rather, behavior in the camps must be viewed in the context of total environmental domination and of a human response to the hideous necessities that were systematically created by the camp environment.[35] Indeed, it was central to the Nazi scheme of things to systematically drain humanness out of the Jewish prisoners. As Des Pres terms it, "excremental assault" facilitated this dehumanization.[36] One survivor recalls the way in which excremental defilement was achieved at Buchenwald:

> At first there were nothing but open latrine pits twenty-five feet long, twelve feet deep and twelve feet wide. Poles accommodating twelve to fifteen men were set up along the sides. One of the favorite games of the SS, engaged in for many years, was to harass and bully the prisoners even during the performance of this elemental need. Those unable to get away quickly enough when the SS put in an unexpected appearance received a beating and were flung into the cesspool. In Buchenwald ten prisoners suffocated in this fashion in October 1937 alone.[37]

Another remembers:

> Urine and excreta poured down the prisoners' legs and by nightfall the excrement, which had frozen on our limbs, gave off its stench. We were really no longer human beings in the accepted sense. Not even animals, but putrefying corpses moving on two legs. [38]

Living with, around, and in their own excrement, in circumstances unimaginable to ordinary human consciousness, human dignity was being systematically crushed out of existence. The Nazis sought to destroy Jewish humanness with efficiency and precision. As in Orwell's *1984*, the humanness of Winston and Julia had to be obliterated. Killing them would not have fully asserted the power of the inner party. Likewise, Jewish prisoners needed to be totally humiliated and defiled, as a matter of policy, in order to fully establish the supremacy of Nazi power. In this way, death is not enough. If a person dies without surrender, if something remains unbroken to the evil, then the power that has destroyed his or her life has not, after all, crushed everything.[39] Something crucial has escaped domination, namely, personhood and humanness.

In their messianic quest to crush the humanness out of Jewish life, the Nazis figuratively and literally forced the Jews to live in the "anus of the world."

Science, Technology, and Mechanized Death

Hitler's sacred doctrine of racial purity, as described in *Mein Kampf*, was based upon the principle that all human beings are not created equal. The strong, dominant Aryan race possessed a moral obligation to eliminate the weaker, "subhuman" parts of the human community. Hitler devoted many pages of *Mein Kampf* to establishing the inferiority of Jews and to defending the idea that violence and bloodshed must be used in the war against the Jewish pestilence. In *Mein Kampf*, the Jews were theoretically debased to subhuman status. In

Thanatology Snapshot 8.2

Daily Routine: Twenty-four Hours Into Eternity

The camp was awakened by whistles, in the summer between four and five o'clock, in the winter between six and seven o'clock. Half an hour was allotted to washing, dressing, breakfasting and bed-making. . . .

A number of camps insisted on morning calisthenics, performed winter and summer at break-neck pace for half an hour before the regular rising time. They consisted mostly of everlasting push-ups in the snow and muck. Because of numerous fatal cases of pneumonia, this practice never persisted for very long.

Breakfast consisted of a piece of bread from the ration issued for the day and a pint of thin soup or so-called "coffee." . . .

Next came morning roll call. . . . Thousands of zebra-striped figures of misery, marching under the glare of the floodlights in the haze of dawn, column after column—no one who has ever witnessed it is likely to forget the sight.

. . . After roll call came a thunderous command . . .: "Caps off!" and "Caps on!" This was the morning salute for the Officer-in-Charge. If it was not executed smartly enough, it had to be repeated again and again, to the accompaniment of such comment as this: "You god-damned ass-holes, if you're too lazy to ventilate your filthy pates, I'll make you practice till the juice boils in your tails, you sons of bitches!"

The next command was "Labor details—fall in!" There was a wild milling about, as the prisoners moved to their assigned assembly points with all possible speed. The camp band, in the winter-time scarcely able to move its fingers, played merry tunes as the columns moved out five abreast. At the gatehouse caps had to be snatched off again, hands placed at the trouser seams. The details then marched off in double time, the prisoners compelled to sing.

Work continued until late afternoon, with half an hour for lunch, out in the open. . . .

In the winter work ended around five o'clock, in the summer, around eight. . . . Then came evening roll call.

---

the camps, Hitler effectively established a self-fulfilling prophecy of subhuman Jewishness. The camps were based on the idea of Jewish subhumanity, and the realities of subhuman existence were forced upon the Jews by the nature of concentration camp life. The camps became ideologically important to Hitler's agenda, visibly demonstrating the theoretical proposition that the Jews were subhuman. Additionally, the death camps provided the organizational structure to carry out the practical goal of annihilating the Jews. In order to achieve

Thanatology Snapshot 8.2 (*continued*)

In every camp this head count was the terror of the prisoners. After a hard day's work, when ordinary men look forward to well-deserved rest, they had to stand in ranks for hours on end, regardless of rain or storm or icy cold, until the SS had tallied its slaves and established that none had escaped. . . .

Everyone had to appear for roll call, whether alive or dead, whether shaken by fever or beaten to a bloody pulp. . . . The bodies of men who had died during the day, either in the barracks or at work, had to be dragged to the roll-call area. During particularly virulent sieges, there were always dozens of dying and dead laid in neat "rank and file" beyond the block formations, to answer the final roll call. For the SS exacted order and discipline down to the last breath. Not until after roll call could the dying be taken to the hospital, the dead to the morgue. . . .

If roll call had been completed with reasonable dispatch, work had to be continued for several hours deep into the night by certain prisoner groups. . . . When taps sounded—between eight and ten o'clock, according to season—everyone except those on detail had to be indoors, half an hour later in bed.

Prisoners were permitted to wear only their shirts while sleeping, even in the deep of winter, when the barracks grew bitter cold and the damp stone walls often coated with ice at the windows and corners. Block Leaders frequently conducted night inspections, ordering all the inmates in a barracks to line up beside the beds or even outdoors, in order to catch those who might be wearing an additional garment. . . .

. . . For a few short hours each night, sleep spread its balm over the misery. Only the aged, the fretful, the sick, the sleepless, lay awake in a torment of worry, awaiting the ordeal of another day.

Eugen Kogon,
The Theory and Practice of Hell
(New York: Berkley Books, 1984), pp. 77–84

this aim, science, technology, and mechanization became essential forces around which death was organized and promulgated.

Science and technology were explicitly utilized in the killing program of the Nazis. The magnitude of the Nazis' extermination activities was directly linked to scientific and technological developments. As Thanatology Snapshot 8.3 clearly illustrates, the success of the Nazi plan for mass murder was linked to the technologization of death. The efficiency of the killing program was enhanced by the use of technical knowledge. Technical advances facilitated the extermination of a great number of prisoners. Additionally, the technologization of death enabled the Nazi agenda to proceed with maximum order and effectiveness. First, technical developments contributed to the organizational efficiency of mass killing activities, making the operation smooth and reliable. Second, technical advances that enabled mass, impersonal killing of the Jews softened the psychological ordeal of the executioners. In this way, the science and technology of killing enabled the Nazis to pursue their goal of annihilation of the European Jews with greater productivity and efficiency. And it was precisely this mechanization and technologization of mass murder that established the holocaust as the most savage, comprehensive, and brutal crime ever committed against humanity.

Initially, the killing of Jews was accomplished by shooting. This was originally performed by squads called "Einsatzgruppen," which were trained by the SS for the explicit purpose of carrying out face-to-face shootings of Jews. These "killer troops" often wandered from city to city. Upon their arrival, they assembled Jews for shootings in the streets or lined them up for death marches to the countryside. There they were shot and never seen. Upon arrival in the countryside, the prisoners were forced to dig large, mass-grave pits. When the pits were dug, the Einsatzgruppen shot each Jewish prisoner individually at the edge of the grave pit. One after the other, Jewish victims were impelled by the force of bullets and the incipience of death into the pits. The result was a pile of dead bodies in a mass open grave that was ironically similar to those found during the Tame Death. As one member of the Nazi Einsatzgruppen recalled:

> We go into the wood and look for a spot suitable for mass executions. We order the prisoners to dig their graves. Only two of them are crying, the others show courage. What can they be thinking? . . .
>
> Slowly the grave gets bigger and deeper. Two are crying without let-up. I let them dig more so they can't think. The work really calms them down. Money, watches and valuables are collected. The two women go first to be shot; placed at the edge of the grave they face the soldiers. They get shot. When its men's turn, the soldiers aim at the shoulder. All our six men are allowed to shoot. Three prisoners have been shot in the heart.
>
> The shooting goes on. Two heads have been shot off. Nearly all fall into the grave unconscious only to suffer a long while. Our revolvers don't help either. The last group have to throw the corpses into the grave; they have to stand ready for their own execution. They all tumble into the grave.[40]

Adolf Eichmann, the SS officer in charge of deportation and emigration, described what he saw on his tour of various sites of Nazi mass executions. This recollection was articulated during his war crime trial in Jerusalem twenty years later:

There were piles of dead people. They were shooting into the pit—it was a rather large one. . . . I didn't think much about it because I could hardly express any thoughts about it—I only saw it and that was quite enough—they were shooting into the pit and I saw a woman, her arms seemed to be at the back; and then my knees went weak and I went away.[41]

Nazi killing was continuous and ubiquitous. The Einsatzgruppen swept through towns, villages, and cities with a plaguelike momentum, destroying Jews and Jewish communities living there:

[At Ponar on Monday, August 2, 1942] Shooting of big batches has started once again. Today about four thousand people were driven up . . . shot by eighty executioners. All drunk. The fence was guarded by a hundred soldiers and policemen.

This time terrible tortures before shooting. Nobody buried the murdered. . . . Many a wounded writhed in pain. Nobody finished them off.[42]

Such executions swept through Nazi-occupied territories. A thousand Jews were executed in the Molcadz Ghetto on July 15, 1942. The same day, another 1,000 were murdered just thirty miles to the south in Bereza Cartusca. At Horodzei, east of Molcadz, another 1,000 Jews were murdered on July 16.[43] In the Mirosk Ghetto, 6,000 Jews were brought to and executed at death pits. The following day, 3,000 more were slaughtered at the pits.[44] No town, village, or city in Nazi-occupied areas was spared the death ravages of the Nazi Einsatzgruppen. Although it is difficult to ascertain accurately how many Jews were killed by the Nazi killer squads, it is generally believed that approximately 1.5 million Jews were killed in this face-to-face, personal fashion.[45]

However, despite the feverish pitch at which the Einsatzgruppen carried out their task, this personalized method of killing was not particularly orderly or efficient. Indeed, the personal involvement of the Nazi death squads with their victims during the execution began to take a harrowing psychological toll on the executioners. To annihilate the entire Jewish population on a one-to-one basis would have been an unbearable psychological burden. Rudolph Höss remarks in the autobiographical account of his experience as commandant of Auschwitz:

I heard Eichmann's description of Jews being mowed down by the Einsatzkommandos armed with machine guns and machine pistols. Many gruesome scenes are said to have taken place, people running away after being shot, the finishing off of the wounded and particularly of the women and children. Many members of the Einsatzkommandos, unable to endure wading through the blood any longer, had committed suicide. Some had even gone mad.[46]

The solution to the problems of inefficiency and psychological torment of the Einsatzgruppen was rooted in advances in the method of killing. Advanced

Thanatology Snapshot 8.3

The Imperative for Mechanized Killing

With the occupation of the Soviet-held half of Poland after Hitler's surprise attack on the USSR in 1941, the Nazis got a chance to put into effect their long-range plan of exterminating all people considered racially and biologically inferior (the so called *Endlösung*, or Final Solution) and removing all incorrigible political opposition. Since millions of persons were involved, the technical problems of secretly assembling and killing them and getting rid of the bodies were staggering. The first exterminations—of the non-Aryan population of the smaller Polish towns and of Communist party members and political commissars among Soviet prisoners of war—were relatively simple: the victims were marched out of town, told to dig a long ditch, and then shot in the back of the head at the edge of the ditch so that they fell in, layer upon layer, until the mass grave was full. . . . But the system was clearly unsuited to mass exterminations, for reasons succinctly summarized by the French writer Jean-Francois Steiner:

> The method of shooting in itself gave rise to controversy among the [SS] technicians, who were divided into two schools: the "classics" and the "moderns." The first were advocates of the regulation firing squad at twelve paces and the *coup de grace* given by the squad leader. The second, who felt that this classic apparatus did not square with the facts of the new situation, preferred the simple bullet in the back of the neck. The latter method finally prevailed, because of its efficiency. It was here that the psychological problems vividly emerged.
>
> With a firing squad you never knew who killed whom. Here, each executioner had "his" victims. It was no longer squad number such-and-such that acted, but rifleman so-and-so. Moreover, this personalization of the act was accompanied by a physical proximity, since the executioner stood less than a yard away from his victim. Of course, he did not see him from the front, but it was discovered that necks, like faces, also individualize people. This accumulation of necks—suppliant, proud, fearful, broad, frail, hairy, or tanned-rapidly became intolerable to the executioners, who could not help feeling a certain sense of guilt. Like blind faces, these necks came to haunt their dreams. Paradoxically, it was from the executioners and not from the victims that the difficulties arose. Hence, the technicians took them seriously.
>
> Thus there arose, no doubt for the first time in the world, the problem of how to liquidate people by the millions. Today the solution seems obvious, and no one asks himself the question. In 1941, it was quite otherwise. The few historical precedents were of no use, whether it was a question of the extermination of the Indians by the Spaniards in South America or by the Americans in the United States, or again of the Armenians by the Turks at

the beginning of this century. In these three cases, no attempt had been made at a new technique, no advance beyond the time-honored hanging and shooting, which, as we have seen, did not satisfy the technicians.

It was necessary to invent a killing machine. With a methodical spirit that is now well known to us, the technicians defined its specifications. It had to be inconspicuous to avoid arousing anxiety in the victims or curiosity in the witnesses, and efficient enough to be on a par with the great plans of the originators of the Final Solution; it had to reduce handling to a minimum; and finally, it had to assure a peaceful death for the victims.*

The killing method that ultimately came to be used was one that had been tried in a tentative way during the euthanasia program: gassing by carbon monoxide. Engine exhaust was piped into the back of a hermetically sealed van loaded with 15–20 victims. The technique was clearly inadequate when it came to really large numbers, such as the 400,000 inhabitants of the Warsaw ghetto (whose fate has been so graphically described in John Hersey's 1950 novel, *The Wall*). That problem was solved by backing the van against a sealed building and piping the exhaust gases into it. Brought to assembly-line perfection by a young SS lieutenant named Kurt Franz at the Treblinka extermination camp near Warsaw, the system ultimately handled 2600 victims in 13 gas chambers with a 200-person capacity in 45 minutes, including the time needed to strip them, cut off the women's hair for use in the war economy, and removing wedding rings and extracting gold teeth from the corpses.

That still left the problem of disposing of the bodies. At Treblinka, the first 700,000 were buried in the usual ditches before the chief of the SS, Heinrich Himmler, decided after an inspection that they would have to be disinterred and burned—a gigantic task that no one had anticipated at the start of the euthanasia program. Even with earth-moving machinery and an unlimited labor supply, it was hard to see how more than 1000 bodies could be handled per day, a rate at which it would have taken 700 days or nearly 2 years to do the job—even if the putrid corpses had not proved so very difficult to ignite. At that point an SS specialist in cremation was summoned, Herbert Floss, who had perfected a new technique at various smaller concentration camps and was itching to try it out on the really large scale that Treblinka afforded. He constructed a giant grill of railroad rails placed on concrete supports about one meter off the ground, on which the bodies were piled in layers and burnt in open air. The giant funeral pyres were an immediate success and the grisly job of excavation and burning began at once. It did not stop for several months.

Charles Susskind,
Understanding Technology,
(Baltimore: Johns Hopkins University Press, 1973), pp. 110–112.

*John François Steiner, *Treblinka* (London: Weidenfeld and Nicolson, 1967), pp. 48–49.

technology would solve these problems if greater numbers of Jews could be killed in a manner that was more mechanical and less personal. The Nazi scientists, leaders, and engineers took this problem seriously and began to search for new methods. This led to the development of Xyklon-B, a poisonous gas that killed swiftly and with certainty. The creation of Xyklon-B provided a fresh method of killing that could be organized into procedures that were efficient, orderly, and psychologically desirable. Indeed, the development of the gas became the inspiration for the design and implementation of mechanized killing procedures in the death camps. Appropriately enough, it was advances in the technology of killing that precipitated sweeping changes in the social and organizational structure of the Nazi killing agenda.

Efficiency, rationality, orderliness, and mechanical precision were organizing principles of the death camps and the gas chambers. As noted earlier, the killing process began with the arrival of prisoners in the transport trains. The prisoners were quickly expelled from the train cars, separated from their belongings, and selected for death or life. Those selected for death by gas, which typically were 85 to 90 percent of all arrivals, were marched directly to the gas chambers. The gas chambers were generally disguised as shower rooms in order to prevent hysteria and resistance among the prisoners. The prisoners were forced to undress in an outer room from the death chamber; this seemed natural since they were going to shower. They were then forced to enter the chamber itself. The doors were locked, and the lethal Xyklon-B was admitted into the room. Death occurred quickly: in as little as three minutes for the weakest and frailest. Death took as long as fifteen minutes for the strongest—who died last and were found lying on top of those who were weaker and had died first. All the while, precision and orderliness were the mainstays of the killing process:

> When they arrived, all had to undress completely in rooms which were made to look as if they were set up for delousing purposes. The steady work detail which worked at these installations . . . helped in the undressing process and advised the skeptical ones to get ready so that the others would not have so long to wait. They were also told to note where they left their things, so they could find them immediately after the bath. All this was done to dispel any suspicions that might arise. After the undressing they went into the next room, the gas chamber. This was set up like a bath, i.e., showers, pipes, drains, etc. had been installed. As soon as the entire transport was in the chamber, the door was closed and the gas thrown in through a special opening in the ceiling. . . . The people became stunned with the first breath of it, and the killing took three to fifteen minutes. . . . After this period there was no more movement. Thirty minutes after the gas had been thrown in, the chambers were opened and the removal of the corpses to the crematoria was begun. In all the years, I know of not a single case where anyone came out of the chambers alive. The hair was cut off the women's heads and any rings or gold teeth were removed by prisoner dentists employed in the detail.[47]

Another eyewitness to the process offers a similar description:

> From the dressing rooms the way led directly to the "bath," where hydrocyanic acid gas was admitted through the shower heads and ventilator outlets as soon as the doors had been closed. Death took as long as four or five minutes, depending on the gas available. During this time the most dreadful screams could be heard from the men, women and children inside, as their lungs slowly ruptured. Any bodies that showed signs of life when the doors were opened were clubbed into quiescence. The prisoners of the service squad then dragged out the bodies, stripped off any rings, and cut off their hair, which was bundled into sacks and shipped to plants for processing.[48]

The sequence of killing was not yet complete. After the gassing and pillaging of the dead for valuable or usable materials, the disposal of the corpses remained to be accomplished. Initially, crematoria were constructed in the death camps and bodies were systematically cremated. However, as deportation, transportation, and gassing procedures became more efficient, the crematoria could not handle the number of dead bodies that were being produced by the Nazi killing machine. For example, at Birkenau, during the height of productivity, the crematoria could handle around 4,400 bodies in a twenty-four hour period. However, during the height of killing efficiency, about 10,000 Jews were killed each day. Thus, a need emerged for greater efficiency in the disposal of the dead bodies.

Again, the technicians responded. With the crematoria unable to handle the load of bodies generated by the gas chambers, disposal trenches were dug and funeral fires were ignited:

> One had to burn . . . great piles—enormous piles. Now that is a great problem, igniting piles of corpses. You can imagine—naked—nothing burns. How does one manage this? . . .[49]

The problem of disposal of the dead bodies that could not be handled in the crematoria was purely technical. The solution resided in finding the most efficient method:

> The gas chambers were sufficient, you see, that was no problem. But the burning, right? The ovens broke down. And they [the corpses] had to be burned in a big heap. . . . The problem is really a large technical difficulty. There was not too much room, so first one thought one would have to take small piles. . . . Well, . . . that would have to be tried out. And everyone contributed his knowledge of physics, about what might possibly be done differently. If you do it with ditches around them, then the air comes up from below and wooden planks underneath and gasoline on top—or gasoline underneath and wood in between—these were the problems. Well, the solution was not to let the fire die. And maintain the cooperation between the gas chamber and the crematorium. When [the fire] reached a certain intensity, then it was just right—[50]

In sum, not only was killing organized according to the highest principles of efficiency and productivity, but the disposal of "human waste" was a routinized, technical activity. The barbarous activities that took place at the death camps made Dante's Inferno look almost like a comedy. [51] Yet the Nazi concentration and death camps were no joke. They were built, organized, and administered around values that were most real and concrete to the Nazi state. One of the ultimate ironies of the Nazi death camps is that there is no way to describe how horrible, inhuman, bizarre, and extreme they were. And yet these incomprehensible extremities of savage brutality were made possible, refined, and even glorified by the most ordinary qualities of industrial society: efficiency, rationality, science, and bureaucratic-mechanistic organization. In a most uncanny way, the abnormal was normalized through everyday applications of ordinary methods of social organization to the unreal world that was the Nazi death camps.

An Interpretive Summary:
Implications for the Fate of the Earth

The structure and the tyranny of the Third Reich have been obliterated. However, the mobilizing forces of Nazi genocide, namely, racial hatred, ethnocentrism, military might, scientific and technologically advanced capabilities of destruction, bureaucratic dominance, and the massification of society have not been similarly crushed. Thus, while this chapter does not present anything like an exhaustive picture of the death camps, I have sought to portray the essential horror that was the holocaust and to illuminate its relevance for evaluating the structural realities and possibilities of our present and future world. Those of us who have never lived in the death camps can never know what it was like to be there. Nevertheless, and perhaps all the more urgently for this reason, the holocaust can never be allowed to be dismissed by a short historical memory. Nor can we allow the intoxicating appeal of the contemporary entertainment ethos, which so successfully expels grieving and suffering from the everyday flow of social life, to dismiss the holocaust as irrelevant to the concerns of present and future generations. Survivors must continue to "bear witness" to the indescribable atrocities that touched their lives directly, and others must listen conscientiously to their disturbing voices. To do otherwise is to render the most horrible of human tragedies meaningless and to cast it undisturbed into a hidden abyss of nonhistory. (One gets the impression that this is precisely what has been done with the killing fields of Cambodia and with the recent genocide in the Middle East, Bosnia, and Rwanda). It therefore must be an explicit aim of the present generation and of future generations to

THE OVENS

The technology of death available to the Nazis is child's play compared to modern capabilities of destruction.

recapture enough of the realities of the holocaust so that it will never be forgotten and can never be repeated.

There is something appealing to the view that the Nazi genocide was the result of a madman and a disturbed ideology of racial hatred. While these factors are relevant to understanding the holocaust, such a psychologically based explanation obscures the social and institutional structures that molded the Nazi agenda of mass killing. It may have been an irrational racial ideology that made Nazi Germany blind. What made it so uniquely dangerous and destructive was the effective mobilization of modern science and technology on behalf of this ideology.[52]

As emphasized earlier, it was the advances in and reliance upon science and technology that enabled the Nazis to kill their victims in great numbers, with expediency and efficiency. Additionally, the technologies of killing developed by the Nazis were put into operation and continued to operate because of the successful use of bureaucratic organization. The Nazis surrounded their program of mass murder with rules, regulations, and procedures that fragmented the process into a variety of specialized tasks. Each person or group of persons had a specific function. Within the social organization of the camps, this function or activity became routine and therefore normal. In this way, the principles of bureaucratic organization promoted a methodological precision that facilitated the regular and continual fulfillment of the duties of those employed in the bureaucracy.[53] Indeed, it was precisely the rational character of bureaucratic social organization that brought impersonalism and matter-of-factness to Nazi genocide. The Nazis fashioned a genocidal bureaucracy that enabled them to utilize scientific and technological developments to kill human beings. They did this successfully by making the needs and operation of the genocide program calculable, rational, predictable, and therefore normal. As Lifton effectively states: "the combination of relative silence and organizational reach puts the bureaucracy in the best position to plan and carry out the details of genocide. That . . . planning contributes in turn to the normalization of a genocidal universe. Mass murder is everywhere but at the same time, through the efforts of the bureaucracy, is nowhere. . . . Bureaucracy, then, does much to render the human killing network into a machine and to deamplify the killing process for all concerned."[54] The structure of bureaucracy, then, ironically creates a dreamlike atmosphere that renders the process of genocide unreal but normal. Hannah Arendt further establishes this important point as she describes the operation of the death camps:

> The human masses sealed off in them are treated as if they no longer existed, as if what happened to them were no longer of importance to anybody, as if they were already dead and some evil spirit gone mad were amusing himself by stopping them for awhile between life and death. . . .
>   It is not so much the barbed wire as the skillfully managed unreality of those whom it fences in that provokes such enormous cruelties and ultimately makes extermination look like a perfectly normal measure.[55]

What must therefore be learned from the experience of the Nazi holocaust is not a reasonably comfortable lesson about madness and irrationality. Rather, we must learn about the dangers inherent in many of the organizational features of our own society that are both institutionalized and normalized by everyday life. As we survey the American condition, we see that racial hatred, tension, and prejudice still abound. Bureaucracy, mass society, and impersonalization have grown by leaps and bounds. And a technology of nuclear genocide now haunts and shadows every human being on earth. If we refuse to admit the holocaust—if we prefer the comfort of pretending it never happened, of ignoring and forgetting it, or of pretending that it wasn't as bad as some would have us believe—a victory for evil is achieved. Hitler and the Nazi version of mass murder may have been smashed, but genocidal principles and their modern organization will have triumphed. By collectively ignoring or denying the structural realities of the holocaust, genocidal possibilities are kept alive. And from Auschwitz and other death camps, we have learned what can happen when the capacity for destruction that exists in modern society is not guarded against with diligence. Since Hiroshima and the advent of the nuclear age, the stakes have become immeasurably greater.

In the final passage of *Night*, Elie Weisel describes how the camps had been liberated and Hitler was being defeated, but how, despite this triumph, the presence and shadow of death would remain forever:

> Three days after the liberation of Buchenwald I became very ill with food poisoning. I was transferred to the hospital and spent two weeks between life and death.
>
> One day I was able to get up, after gathering all my strength. I wanted to see myself in the mirror hanging on the opposite wall. I had not seen myself since the ghetto.
>
> From the depths of the mirror, a corpse gazed back at me.
>
> The look in his eyes, as they stared into mine, has never left me.[56]

The corpses that the Nazis created, likewise, can never leave the vision of the human race. The danger lies in denying and forgetting: Not only are the dead and the survivors of the holocaust betrayed, but our hopes for the future are also compromised. The corpses and the voices of the holocaust are an anguished yet powerful call to life, to faith, and to salvation.[57] To ignore these voices is to side and conspire with the most insidious and evil of all human activities: genocide.

9

Easing Death's Sting:
A Conclusion

O death, where is your victory?
O death, where is your sting?

I CORINTHIANS 15:55

In a biblical sense, the sting of death is sin and eternal damnation. In an ordinary, secular sense, the sting of death is physical debility, pain, suffering, loss, and grief. According to the Bible, death's sting is diminished by the promise of eternal life. But, over time in Western society, human beings have come to fear a sudden or bad death, punishment, loneliness, meaninglessness, the dehumanization of mechanized dying, or loss of control over dying. Humanity also confronts certain inescapable biological truths, including the fact that all living things must die. It is this human awareness of mortality, of individually and collectively knowing that death is inevitable for ourselves and loved ones, that makes its sting especially piercing in a secular society. In large part due to their awareness of death, human beings have established a wide range of cultural and social patterns of dying, death, and mourning throughout history. As we have seen throughout this book, humanity has created ways of confronting death—of easing or taming the sting of death—related to the patterns of living that typify the broader social order. Thus, the styles of death, ranging from the Tame Death, Victorian death, the perceptions of the Puritan Americans, the modern patterns of high-tech death, and the Happy Death movement to suicidal and genocidal death, are reflective of and created by the styles of living in the society. Within this framework, an interesting reciprocal relationship between life and death is thus apparent: The styles of death are a reflection of the ways of life, and the ways of life establish the styles of death.

Death may have not been as naturally and tranquilly a part of life as Ariès and others have suggested during more traditional eras. Individuals may have privately feared dying and death. Nonetheless, private, psychologically rooted issues had little role in guiding human behavior during these periods. The psychological-individualized view of the world so prevalent in today's society had little meaning in eras when community, tradition, and moral authority were dominant cultural values. Thus, the rituals of traditional death patterns, despite vicissitudes of form, were steeped in folkways, mores, and traditions that established death and dying as a communal affair. The individual dying person was, in fact, deeply subordinated to societally anchored traditions, conventions, and ceremonies, and individual expressions or differences in dying were obviated by these social pressures.

During the evolution of traditional death patterns, many interesting variations occurred. Ariès has described the communal ceremony of the deathbed during the Tame Death, the struggle for salvation during the Death of the Self, the savage and vivid images of the macabre during the period of Remote and Imminent Death, and the sentimental affectivity of the Victorian deathbed. While profound differences existed in the particular ways that humanity responded to death during these eras, the relationship between human mortality and everyday life was obvious and notable. Indeed, dying, death, and mourning held a deeply meaningful and visible place in the landscape of everyday life during these times. This regular involvement of death in day-to-day life enabled humanity to mitigate the ordeal of death and to tame its sting in a public and communal way.

During the twentieth century sweeping changes have occurred, not only in the forms of dying and death, but also in the relationship between human beings and mortality. In modern twentieth-century America, dying, death, and mourning have lost cultural meaning and significance. Additionally, death has become increasingly invisible. Thus, the language of death in the twentieth century includes terms like "denial," "avoidance," "fear," and "loneliness." To be sure, these terms, and the corresponding images they convey, are wholly inapplicable to the previous eras of traditional death and express an entirely new societal approach to dealing with human mortality.

Dominant death themes that have emerged during the twentieth century, and are distinct and unique to it, include bureaucracy, medicalization, patient lingering, death as an adversary, emotional neutrality, and technological dominance. These broad themes are most often translated into concrete reality through the processes of high-tech, medicalized death. A high-tech framework, combined with the general unease with death in the broader society, has created a conspiracy of silence around dying and is the basis of its corresponding patterns of closed awareness and evasive interaction. The invisibility of modern death is a direct outgrowth of both the absence of shared cultural meanings and the lack of established norms to guide physicians, families, and patients through their interactional processes. Yet, despite confusion, avoidance, and

TO LIFE

For those willing to listen, the dying can be the greatest teachers about life and living.

normlessness, technology remains a common denominator throughout the medicalized dying process: Whether physicians adopt the save-at-all-costs mode, avoidance-neglect, or detached-sympathetic support patterns, technology is the central factor that defines their view and response to the process of dying.

One of the interesting ironies, however, of the modern organization of death is that, as the problems of dying have become increasingly submerged in society, a counteremphasis on openness and candor has emerged. Pluralistic death is one consequence of this trend and represents a negotiated compromise between the open-awareness, death-acceptance orientation and the closed-awareness, death-avoidance orientation. A related trend has encouraged an expanded role and responsibility for individuals to shape their own deaths. Both of these trends reflect the lack of shared meanings and the absence of cultural consensus about the process of dying.

But the open-awareness, happy-death movement and the conspiracy-of-silence approach have an important feature in common. The foci of both approaches increasingly individualize dying and death. In the latter, individuals are on their own to overcome isolation and loneliness. Similarly, the individual and self-growth are the explicit focus of the human potential, happy-death movement. The thanatology revolution has brought many issues of death and dying to the public arena; death-related themes are appearing with greater frequency in the courts, arts, classroom, and media. There is indeed a greater public preoccupation with dying and death than there was twenty years ago. Nonetheless, the movement toward therapeutic death strongly emphasizes individualism and privacy. When the individual, through personally organized efforts, is able to transform dying into a growth experience, the thanatological ideal has been achieved. This is precisely why seemingly opposite approaches to dying and death are in fact compatible with each other. Despite variations in style and form, both affirm very powerfully the deeply held American value of individualism and the current cultural evolution toward autonomy, independence, and separateness.

Similarly, the professional success of Elisabeth Kübler-Ross is a direct result of the forces of technology and individualism. Indeed, her popularity has been cultivated by the social fact of technological dominance and by a reaction against the indignities of technological control. Moreover, the broader societal value of self-growth and the evolution of the human potential movement have fueled the public acclaim of her therapeutic ideas of death acceptance and death as a stage of human growth. There is popular comfort contained in Kübler-Ross' message, especially its consoling simplicity. Applying the deeply cherished value of individualism, one can ease the sting of one's own death by traveling through the stages she has formulated to the ultimate goal of acceptance.

Yet, despite the appeal of Kübler-Ross' stage theory, the simplicity and singularity that typified styles of death during the traditional eras are difficult to achieve in the modern setting. If simplicity can be achieved in the modern dying process, it is a consequence of heroic individual effort, and always takes place in a cultural context of complexity and normlessness. More than likely, the course of dying in modern America is not a simple matter of courageously accepting the coming of death. Rather, it is characterized by confusion, uncertainty, anxiety, and turbulence. However, in establishing a model for easing the sting of death, Kübler-Ross and other advocates of the happy-death movement dismiss the harsh, torturous realities of dying and replace them with images of peace, growth, tranquility, and acceptance. These images have created a set of behavioral expectations for dying patients that mandate concealment of the wretchedness of dying. Thus, the sting of death is removed from public visibility and remanded to the private worlds of dying individuals and their loved ones.

"Successful dying" thus becomes the responsibility of the individual person and of therapeutic-medical management. And as the process of dying has become individualized, so have the responses of survivors. As we have seen, American funerals have changed notably since colonial days. Perhaps the most important of these changes is not the shift from simplicity to lavishness, or even the expansion of the role of the undertaker; rather, it lies in the significance of the funeral. Historically, the American funeral provided for a ceremonial establishment of the meaningfulness of death, which connected humans and their earthly concerns to sacred and spiritual things, as well as to each other in the ties of community. More recently, however, the funeral has lost its value as an established and prominent community ritual. The funeral has become a rite of individual expression—a personalized response to the death of a significant other. It is therefore not surprising that conflicting views on the role of the funeral abound in contemporary society. Clearly, there are advocates of the funeral who see its traditional form as useful. There are also those critics who view the American funeral as unnecessarily lavish, crassly materialistic, shallow, and exploitive. Stemming from this span of opinions is the emergence of an increasing variety of postdeath funerary folkways. These range from drive-in funeral parlors, to growing acceptance of cremation, to advocacy of alternative funerals, to the traditional funeral ceremony, all of which may or may not include varying elements of religious or spiritual expressions. Thus, the response to death has been removed from the public arena of established funeral norms and now consists of individuals making personalized choices in regard to the death of a loved one.

Not only do the type, style, and meaning of the modern funeral vary, but the funeral itself provides a forum for self-expression. In more traditional eras, the individual was submerged in the broader community. The funeral system in traditional death patterns was therefore geared to reassuring the social community of its viability. Today, the American funeral is typified by a wide variety of roles for all participants. The person adopts those roles that most comfortably reflect his or her individuality. Even within the limits of the traditional funeral, many different role expressions and self-presentations may take place. Thus, in the contemporary framework of extreme individualism, the sting of death impinges not on the broader community but on the lives of individuals, and the funeral is organized to help fortify the individual self against it.

The emphasis on individualization and privacy grows even greater after the funeral, during the grieving period. In the modern era, where society is geared to pleasure, success, and therapeutic enhancement of the self, grieving and mourning are necessarily restricted and diminished. One of the dominant ways in which grief is restricted is by its removal from public interaction and its isolation in the private worlds of grieving individuals. In this way, grief and mourning are corraled and their potential for the disruption of everyday life is minimized. The disappearance of grief from the arena of everyday social interaction has redefined the nature and meaning of grief. Unlike earlier eras, when grief and mourning were publicly and ceremonially supported, modern grieving is brief and private and affects only a narrow range of people. The absence of ritual and the focus on privacy have led to the diminishment of helpful norms, and to the development of confusion and ambivalence.

The therapeutic, self-help response to grief and mourning is also an outgrowth of the broader value of individualism. The cementing of grief, along with its symptoms and associated problems, in the therapeutic, psychiatric, and medical domains also serves to isolate the realities of grief from ordinary social interaction. Additionally, the prevailing self-help, therapeutic approach emphasizes the prompt transcendence of grief and the speedy return to everyday life.

Ironically, however, the restriction of grieving and mourning does not mitigate the anguish caused by death. On the contrary, social isolation and the narrowing of the grieving community intensify the grief experience. As individuals become separated from the general community and as human relationships become more specific, the death of a loved one generates a greater and more piercing sense of loss. Thus, for example, we find widows in America describing such problems as loneliness, emptiness, despair, confusion, hopelessness, and meaninglessness. Though the general social community is "protected" from the sting of death, individuals are isolated from public systems of support, ritual, and meaning. The interesting paradox is that despite the great cultural emphasis on individualism, the sting of death is exacerbated for individuals. The value attributed to individualism is therefore counterproductive to the needs of grieving individuals while serving the interest of the death-avoiding, broader social community.

The meaning of death in modern society is most vividly exemplified by the relationship between children and death. The role of technology and individualism in responding to the processes of dying and death is especially illuminated when looking at the social context that defines the child–death relationship. There is perhaps no greater embarrassment to the technological ethos of modern society than the death of a child. Simply put, children are not supposed to die in our advanced technological era. For this reason, tremendous efforts to prevent the death of children—the development of pediatric intensive care units and the growing sophistication of neonatal units, for example—have recently flourished. We are now undergoing a dramatic technological revolution in the way sick children live and die. Ironically, however, the advances made in the high-tech care of life-imperiled newborns have made the death of children all the more difficult to accept. Whereas not long ago parents had to acknowledge the possibility of infant death as an everyday reality, modern parents, so very reliant on technological intervention and beneficence, find intolerable the idea that their child may die. Thus, technological sophistication and aggressiveness are cornerstone values in the modern response to imperiled children and largely shape the worlds of dying children and their families.

However, while some important medical advances have occurred, ambiguity, complexity, and confusion are the result of this progress. For instance, leukemic children who would have died from their disease thirty years ago are now living longer and substantial numbers of children are cured. But, as they live with their disease, they and their parents continually confront their uncertain fate. The twisting, tortuous course of many childhood illnesses leaves both parents and children anxious, vulnerable, and insecure.

Not only is there an uncertain medical outcome for endangered newborns and critically sick children, there is tremendous psychosocial confusion as well. The confusion is rooted both in the general societal unease about children and death and in the cultural ambivalence between the open and closed approaches to dealing with death. No single, dominant pattern guides people in responding to the possibility of a child's death. Although sophisticated medical activities become an ordinary part of the life of a sick child, systematic psychosocial support is infrequent and irregular. Often, the confusion of parents and medical staff about how to relate to critically ill children isolates those children in a private world of coping.

In response to the child's need for comfort, security, and support, the therapeutic model has developed and been applied. The anguish of uncertainty that grows from the unpredictability of the fate of the child is somewhat eased by therapeutic messages and support. The tribulations caused by the fluctuating course of the child's illness are also supposedly diminished by the happy-death, open-awareness, therapeutic orientation. Yet it is essential to recognize that the suffering of children and parents continues to be great, though it is privately endured; the establishment of these private worlds of endurance only serves to ease the sting of dying children for American society in general.

Self-help groups, like the Compassionate Friends, have been organized to establish a community of parental mourners in an otherwise isolated, and fragmented environment. The psychological orientation of these groups embraces and cultivates the therapeutic thrust. The value of individualism is another core motivational force of these groups. The individual is the source of responsibility, as joining such groups is a personal choice (as opposed to a social norm) and as beneficial results come only from the efforts of the grieving individuals.

The sting of death is eased in two ways by these voluntary self-help groups. First, the psychological and emotional ordeal of the death of a child may be softened. (It is important, however, to remember that these groups have not been shown to be effective on a broad basis. Responsibility for the success or failure of the self-help approach rests with individual parents, not with the program itself. Also, this individualizing of responsibility helps establish the notion that the parents themselves are at fault if the self-help therapeutic approach does not ameliorate their grief.) Second, the sting of death is eased on a broader societal level. Parental mourning is removed from the public arena. The ordeal of grief is contained within the therapeutic community and is worked through in private.

The connection between individualization and modern mortality is also apparent in the violent deaths of suicide. Suicide not only uniquely expresses the relationship between individuals and their society, but also makes a statement about the individuals themselves and about the organization of society. And as we examine suicidal behavior in contemporary America, it becomes clear that the prevailing drift toward radical individualism is defining and redefining the place of suicide in everyday life. It is the creation in individuals of a sense of alienation, isolation, social and moral confusion, helplessness, and meaninglessness that establish the context for egoistic and anomic patterns of suicidal behavior.

The connection between individualism and contemporary suicidal behavior is also evident in the growing acceptance of suicide and the right-to-die movement. Clearly, the cultural values of autonomy, self-determination, and personal choice are major reasons for this acceptance of rational suicide. The right-to-die movement is also closely aligned with the American therapeutic impulse. Active euthanasia is increasingly viewed as a morally and socially acceptable means of relieving suffering. It is also increasingly described as a legitimate and desirable way of triumphing over the indignities of dying and of asserting the self in the midst of the pain, deterioration, and humiliation often associated with dying.

In contemporary society, when an individual finds that life and the world are no longer endurable, suicide becomes a way of relieving suffering. Thus, suicide is an important indication of the ability of society's arrangements to produce a good quality of life for its members. Genocide, on the other hand, demonstrates how the arrangements of society can imperil life for groups of people. And it is precisely for this reason that the tragedy of the holocaust is a salient and dramatic warning of the special dangers that lie in the organization of an advanced, industrial, high-tech civilization.

The genocide of the holocaust grew out of genetic narcissism, racial hatred, and materialistic omnipotence. Hitler's rise to power reflected the social, political, and moral traditions of German society. Yet the extremes of destruction and death during the holocaust were not merely facilitated by totalitarian coordination. Rather, they were made possible by bureaucratic organization and

Thanatology Snapshot 9.1

Death Is Not the Enemy

Karl Barth, a 20th-century Protestant theologian, wrote, "Life is no second God, and therefore the respect due it cannot rival the reverence owed to God." On the other hand, for secularized persons in a secular society, there is no "first God" and thus nothing due more respect or reverence than life itself. Life and its preservation become more than the necessary conditions for the realization of a measure of self-fulfillment and for capacities to contribute to other persons and to society. They become virtually ends in themselves. The pursuit of health and the preservation of physical life seem to have replaced "salvation," the glorification of God, or the beatific vision as the chief end of man. To the secular person, what theologians call "the conditions of finitude," those inexorable restraints and limitations on human life of which the final one is death, seem repressive since there is nothing real or lasting beyond them. A kind of physical fundamentalism comes into being; the practical dogma is to preserve life as long as medically and technically possible. If God is functionally designed as one's "ultimate concern" (to use a term of another Protestant theologian, Paul Tillich), the preservation of life becomes one's God. If one's ultimate object of trust is fundamentally one's God, life becomes one's god—or one's idol.

We are not concerned to argue for the existence of God, or for some form of life after death. We do not claim that a religious outlook is necessary to avoid absolutizing the value of physical life. Secular persons can consent to the conditions of finitude, to the reality of death, to conceiving of death as sometimes friend as well as enemy at least as readily as the religious person. We are concerned, however, to reflect on some of the outcomes of the preoccupation with the preservation of physical life. The intensification of concern to sustain and preserve life is the other side of concern to avoid physical death. These concerns may have obvious benefits in most circumstances—the prevention of many risks through public health measures and educational activities directed toward preventive medicine and personal hygiene and the development of therapies for countless diseases.

An intense preoccupation with the preservation of physical life, however, seems sometimes to be based on an assumption that death is unnatural, or that its delay, even briefly, through medical and technical means is always a triumph of human achievement over the limitations of nature. It is as if death is in every

Thanatology Snapshot 9.1 (*continued*)

case an evil, a kind of demonic power to be overcome by the forces of life, propped up by elaborate medical technologies. Dramatic medical interventions portrayed in the media become living "westerns." The powers of death are the bad guys, to be vanquished by the good guys, dressed in white coats rather than white hats. Every delay of death is a victory by the forces of good. Or, to change the analogy, the development and use of costly and dramatic end-stage therapies are seen as the "arms" to be used in a "crusade," a war fought over "holy places" because they were occupied by an alien, and therefore enemy, power. A "crusading mentality" comes into being; almost any means is justified when it will delay the enemy, death.

We do not wish our position to be construed as being obstructive to scientific and technologic research, but we do believe medical scientists should be reminded that death is as integral an aspect of human life as it is of all other biologic species. The development of technologies with the prime aim of prolonging life should be seriously questioned if the ultimate result is destined to be a grotesque, fragmented, or inordinately expensive existence. We were not privy to the discussions of the institutional review board at the University of Utah that led to the news report that the board had refused to approve continuing human experiments with the artificial heart, but it is possible that such considerations contributed to that decision.

Today's practicing physicians have accepted—often without knowing it—a far greater priestly role than any of their predecessors. In part this is attributable to the diminished impact of religion in our civilization. To a greater extent, this phenomenon is due to the immense power that medical science has placed in physicians' hands. However, given the frequent announcements of scientific "breakthroughs," the limitations of their power to diagnose and control diseases are not always appreciated by the public. The emphasis on mortality statistics as a measure of medical care effectiveness has tended to obscure the fact that most of the time and effort of practicing physicians is devoted to improving the life of their patients. The real enemies are disease, discomfort, disability, fear, and anxiety. Sensitive, perceptive physicians attempt to guide their patients, those who are relatively healthy as well as those who are seriously handicapped and ill, to a perspective in which the preservation of life is not their God.

Richard L. Landau and James M. Gustafson,
"Death is Not the Enemy,"
*Journal of the American Medical
Association*, Vol. 252(17), 1984, p. 2458.

technological developments. Racial hatred can induce terror only if a system exists whereby the hatred can be expressed. The terror of the holocaust, the transformation of evil attitudes into evil practices, was explicitly facilitated by the German trend toward rationalization, technocratization, bureaucratization, and mechanization. Understanding the Nazi world of the 1930s and 1940s can promote an understanding of the realities of the 1990s and the trends that are being established for the future. It is clear that the destructive capabilities of the Nazis in the death camps are child's play compared to the modern capacity for technological and nuclear destruction. The potential for world destruction, the possibility that there may not be a future, is an anxiety inherent in our nuclear age. For the first time in the history of human civilization, children are being born into and living in an ever-present shadow of genocidal death. The possibility of mass death from nuclear weapons and fleeting but haunting feelings of futurelessness have become an institutionalized part of everyday global life. The issue of genocide has evolved into new forms and new dangers, and fearful contingencies regularly occupy the attentions of modern civilizations.

It should also be noted that genocidal inclinations are not exclusively the domain of the holocaust or of nuclear destruction. Existence of savage inequalities in living arrangements and deeply felt hatred of others are prevalent throughout the world today. Hatred may assume an overt character, such as that currently found in Rwanda and Bosnia, or may be more subtly expressed, as in American racism. Inequality may result in the deaths of large numbers of people, such as found in the epidemic homicide rate of young black men in American inner cities, or in the starvation of masses of women, children, and men in various parts of Africa. Indeed, the seeming indifference of the modern world to these types of mass death bears a striking resemblance to the Nazis' indifference to the death of "subhuman" populations.

If the holocaust is the epitome of terror, destruction, and human cruelty, its study is precisely the opposite. It is an affirmation of the value of human life. To bear witness to the holocaust is to expose social realities and tendencies toward degradation, humiliation, and destruction. It represents a warning to all societies, especially technologically modern societies, of the potential for mass death and evil that inherently lies in their institutions, organizations, and ways of life. Thus, in facing genocide in its most grotesque and extreme form— the Nazi gas chambers—we develop deep appreciation and respect for life.

Similarly, thanatology itself stems not from a morbid preoccupation with mortality, but from a concern for and love of life. To ease death's sting is therefore to ease the sting of life! To be sure, one of the most important measures of the quality of living in society is the nature and quality of dying. It is for this reason, then, that despite much of the sadness, despair, and suffering that have

Thanatology Snapshot 9.2

Genocide Today

Baidoa, Somalia—This is the city of the walking dead. They move like an aimless army of skeletal zombies through dusty streets in desperate hope of finding a precious food scrap that will give them life.

For hundreds of Somalians each day, the price of failure is a ride on the death truck, stacked among a community of corpses on their way to mass burial grounds.

"These people look like they are from Auschwitz," said Rupert Lewis, a relief official in Baidoa. "Basically, they are human bones wrapped in skin."

Dorothy St. Germain, a U.S. member of the International Medical Corps, a Los Angeles-based charity organization, is the hospital chief in this southwestern city. Her job is to try to save those who have a chance. After all, scarce resources cannot be wasted.

"I would say 50 percent of the kids are dying. There is a measles epidemic and hepatitis is breaking out," she said last month.

More than 2,000 Somalians are dying daily from thirst, starvation, disease, and clan violence, according to Western relief agencies.

Relief officials estimate more than 350,000 people have died this year and 2 million more are on the verge of starving to death.

For those who die in Baidoa, there is a date on the death truck. The Red Crescent, the Muslim equivalent of the Red Cross, volunteered to collect and bury the increasing number of dead, but did not have a vehicle.

The International Committee of the Red Cross gave the Red Crescent a beat-up, red and green Fiat cattle hauler.

The death truck makes its rounds daily. On a typical day, it will collect about 200 shroud-wrapped bodies and deliver them for burial at a makeshift graveyard in the scrub.

To follow the death truck is to begin to understand the real human cost of this famine.

Remer Tyson, excerpt from "Pictures of Hell,"
The Indianapolis Star,
September 20, 1992, p. C1.

been dealt with in these pages, this study of death is devoted to life. From thanatology, we as individuals can better learn how to live. From thanatology, we can also assess how the organization of society fulfills the needs of living human beings. Thus, on both a societal and a personal level, the study of death is dedicated to the value of life.

L'chaim!

NOTES

Chapter 1

1. Phillipe Ariès, *The Hour of Our Death* (New York: Alfred A. Knopf, 1981).
2. Ariès, p. 64.
3. Phillipe Ariès, *Western Attitudes Towards Death* (Baltimore: Johns Hopkins University Press, 1974), Chapter 2.
4. Ivan Illich, *Medical Nemesis* (New York: Random House, 1976), p. 182.
5. Ariès, *The Hour of Our Death*, p. 380.
6. Ariès, p. 366.
7. David Stannard, *The Puritan Way of Death* (New York: Oxford University Press, 1979), p. 87.
8. Geddens, *A Directory for the Publique Worship of God* in Stannard, p. 101.
9. Stannard, p. 113.
10. Stannard, p. 171.
11. Leo Tolstoy, *The Death of Ivan Ilych* (New York: Signet, 1960).
12. Tolstoy, p. 98.
13. Tolstoy, p. 130.
14. Tolstoy, p. 131.
15. Tolstoy, p. 149.
16. Tolstoy, p. 151.
17. Edwin Schneidman, *Death: Current Perspectives* (Palo Alto, Calif.: Mayfield Publishing Co., 1980).
18. Ariès, *The Hour of Our Death*, Chapter 12.
19. Jessica Mitford, *The American Way of Death* (New York: Simon and Schuster, 1963).
20. David Wendell Moller, "On the Value of Suffering in the Shadow of Death," *Loss, Care and Grief: A Journal of Professional Practice*, Vol. 1(1/2), 1986–87, p. 127.
21. Geoffrey Gorer, *Death, Grief, and Mourning* (New York: Doubleday and Co., 1965).
22. Ernest Becker, *The Denial of Death* (New York: Free Press, 1973).
23. Gorer, op. cit.; Becker, op. cit.
24. Richard A. Kalish, *Death, Grief, and Caring Relationships* (Belmont, Calif.: Brooks/Cole, 1984), Chapter 4.
25. Allen Kellehear, "Are We a Death Denying Society?: A Sociological Review," (*Social Science and Medicine*, Vol. 18(9), 1984, pp. 713–723.

Chapter 2

Epigraph: Joseph Quinlan, as cited in *Readings in Aging and Death*, (New York: Harper and Row, 1978), p. 284.

1. Max Weber, "Bureaucracy," from *Max Weber*, translated by Hans Gerth and C. Wright Mills (New York: Free Press, 1968), p. 196.

2. Edward Shils, "Faith, Utility and Legitimacy of Science," *Daedalus*, Summer 1974, p. 3.

3. John Naisbitt, *Megatrends* (New York: Warren Books, 1984), pp. 35–36.

4. Lyn Lofland, *The Craft of Dying* (Beverly Hills, Calif.: Sage Press, 1978), Chapter 1.

5. David Barton, *Dying and Death: A Clinical Guide for Caregivers* (Baltimore: Williams and Wilkins Co., 1979), Chapter 6.

6. Robert Coombs and Pauline Powers, "Socialization for Death: The Physician's Role," in Lyn Lofland (ed.), *Toward a Sociology of Death and Dying* (Beverly Hills, Calif.: Sage Press, 1976), pp. 21–22.

7. Renée Fox, "The Autopsy: Its Place in the Attitude-Learning of Second-Year Medical Students," in Renée Fox, *Essays in Medical Sociology: Journeys Into the Field* (New York: John Wiley & Sons, 1979), Chapter three.

8. Fox, op. cit.

9. David Sudnow, *Passing On: The Social Organization of Dying* (Englewood Cliffs, N.J.: Prentice-Hall, 1967), p. 78.

10. Albert Camus, *The Myth of Sysyphus* (New York: Alfred A. Knopf, 1965).

11. Arthur Schopenhauer, *The World as Will and Idea*, Vol. 1, translated by R. Haldane and I. Kemp (London: Routledge and Kegan Paul, 1948); Leon Kass, "Thinking About the Body," *The Hastings Center Report*, Vol. 15(1), February 1985, pp. 20–30.

12. Ivan Illich, *Medical Nemesis* (New York: Random House, 1976), p. 205.

13. Herman Feifel et al., "Physicians Consider Death," *Proceedings of the 75th Annual Convention of American Psychological Association*, 1967.

14. Eric Cassell, "Dying in a Technological Society," in Peter Steinfels and Robert Veatch (eds.), *Death Inside Out* (New York: Harper and Row), p. 19.

15. David Wendell Moller, *On Death Without Dignity* (New York: Baywood Press, 1990), Chapter Three.

16. Coombs and Powers, op. cit.

17. Thomas Campbell et al., "Do Death Attitudes of Nurses and Physicians Differ?" *Omega*, Vol. 9(4), 1978–79, pp. 43–49.

18. Richard Schulz and David Alderman, "Physicians' Death Anxiety and Patient Outcomes," *Omega*, Vol. 9(4), 1978–79, pp. 327–332.

19. George Annas, "The Phoenix Heart: What We Have to Lose," *Hastings Center Report*, Vol. 15(3), June, 1985, pp. 15–16.

20. Joseph Fletcher, *Morals and Medicine* (Boston: Beacon Press, 1960); Richard Schulz and David Alderman, "How the Medical Staff Copes with Dying Patients: A Critical Review," *Omega*, Vol. 7(1), 1976, pp. 11–21.

21. Jay Katz, *The Silent World of Doctor and Patient* (New York: Free Press, 1984).

22. Anselm Strauss et al., *Social Organization of Medical Work* (Chicago: University of Chicago Press, 1985); Moller, op. cit..

23. Avery Weisman, "Misgivings and Misconceptions in the Psychiatric Care of Terminal Patients," in Charles Garfield (ed.), *Psychosocial Care of the Dying Patient* (New York: McGraw-Hill, 1978), pp. 162–172; P. B. Livingston and C. N. Zimet, "Death Anxiety, Authoritarianism and Choice of Specialty in Medical Students," *Journal of Nervous and Mental Diseases*, Vol. 140, 1965, pp. 222–230; R. Kastenbaum and R. Aisenberg, *The Psychology of Death* (New York: Springer Publishing Co., 1972).

24. Sudnow, op. cit.

25. Moller, op. cit.

26. Eleanor Polo Stoller, "The Impact of Death-Related Fears on Attitudes of Nurses in a Hospital Work Setting," *Omega*, Vol. 11(1), 1980–81, pp. 85–95.

27. Renée Fox, "Reflections and Opportunities in the Sociology of Medicine," *Journal of Health and Social Behavior*, Vol. 26, March 1985, pp. 6–14.

28. Stoller, op. cit.

29. Loretta Hoggatt and Bernard Spilka, "The Nurse and the Terminally Ill Patient: Some Perspectives and Projected Actions," *Omega*, Vol. 9(3), 1978–79, pp. 255–265.

30. Lawrence LeShan, "Psychotherapy and the Dying Patient," in Leonard Pearson (ed.), *Death and Dying* (Cleveland: Case Western Reserve University Press, 1969), Chapter 11.

31. Thomas O. Martin, "Death Anxiety and Social Desirability Among Nurses," *Omega*, Vol. 13, No. 1, 1982–83, pp. 51–58.

32. LeShan, op. cit.

33. Campbell, op. cit.

34. Campbell, op. cit.; Hoggatt and Spilka, op. cit.

35. Jean E. Kincade, "Attitudes of Physicians, House Staff and Nurses on Care for the Terminally Ill," *Omega*, Vol. 13(4), 1982–83, pp. 333–344; Mary Reardon and Patricia Keith, "Patient Concerns, Emotional Resources, and Perceptions of Nurse and Patient Roles," *Omega*, Vol. 10(1), 1979, pp. 27–33.

36. Barney G. Glaser and Anselm R. Strauss, *Awareness of Dying* (New York: Aldine Publishing Co., 1965).

37. Feifel et al., op. cit.; D. Oken, "What to Tell Cancer Patients: A Study of Medical Attitudes," *Journal of the American Medical Association*, Vol. 175, April 1961, pp. 1120–1128.

38. Moller, op. cit.

39. Glaser and Strauss, op. cit., p. 56.

40. Kathy Charmaz, *The Social Reality of Death* (Menlo Park, Calif.: Addison-Wesley Publishing Co., 1980), Chapter 5.

41. Elisabeth Kübler-Ross, *On Death and Dying* (New York: Macmillan Publishing Co., 1969); Kübler-Ross, *Death: The Final Stage of Growth* (Englewood Cliffs, N.J.: Prentice-Hall, 1975); Bernard Schoenbert et al., (eds.), *Loss and Grief* (New York: Columbia University Press, 1970).

42. David Wendell Moller, "Humanistic Care of the Dying," in Austin Kutscher (ed.) *Hospice U.S.A.* (New York: Columbia University Press, 1985).

43. C. B. Hatfield et al., "Attitudes About Death, Dying, and Terminal Care: Differences Among Groups at a University Teaching Hospital," *Omega*, Vol. 14(11), 1983–84, pp. 51–61.

44. Kincade, op. cit.

45. Naisbitt, op. cit.

46. Paul DuBois, *The Hospice Way of Death* (New York: Human Sciences Press, 1980), pp. 60–68.

47. Cicely Saunders, "The Moment of Truth: Care of the Dying Person," in Leonard Pearson (ed.) *Death and Dying* (Cleveland: Case Western Reserve University Press, 1969), p. 67.

48. Virginia Hine, "Dying at Home: Can Families Cope?" *Omega*, Vol. 10(2), 1979–80, pp. 175–187.

49. Robert Buckingham, *The Complete Hospice Guide* (New York: Harper and Row, 1983), p. 32.

Chapter 3

1. Leo Tolstoy, "Three Deaths," in *Leo Tolstoy: Short Stories*, selected by Ernest J. Simmons (New York: Random House, 1964).

2. Elisabeth Kübler-Ross, *On Death and Dying* (New York: Macmillan Publishing Co., 1969), pp. 5–6.

3. Kübler-Ross, p. 50.

4. Kübler-Ross, p. 54.

5. Kübler-Ross, p. 80.

6. Kübler-Ross, p. 81.

7. Kübler-Ross, p. 87.

8. Kübler-Ross, p. 113.

9. Kübler-Ross, p. 115.

10. Elisabeth Kübler-Ross, *To Live Until We Say Goodbye* (Englewood Cliffs, N.J.: Prentice-Hall, 1978).

11. Elisabeth Kübler-Ross, *Questions and Answers on Death and Dying* (New York: Macmillan Publishing Co., 1974).

12. Charmaz, op. cit.

13. Robert Bellah et al., *Habits of the Heart* (New York: Harper and Row, 1986), p. 22.

14. David Wendell Moller, *On Death Without Dignity* (New York: Baywood Press, 1990), Chapter two.

15. Dennis Klass, "Elisabeth Kübler-Ross and the Tradition of the Private Sphere: An Analysis of Symbols," *Omega*, Vol. 12(3), 1981, p. 256.

16. Kübler-Ross, *To Live Until We Say Goodbye*, p. 152.

17. Ron Rosenbaum, "Turn On, Tune In, Drop Dead," *Harper's*, Vol. 265, July 1982, p. 41.

18. Dennis Klass and Richard Hutch, "Elisabeth Kübler-Ross as a Religious Leader," *Omega*, Vol. 16(2), 1985, pp. 102–104.

19. Barney Glaser and Anselm Strauss, *Time for Dying* (New York: Aldine Publishing Co., 1968), p. 6.

20. Glaser and Strauss, pp. 99–102.

21. Glaser and Strauss, p. l00.

22. Glaser and Strauss, p. 102.

23. Anselm Strauss et al., *Social Organization of Medical Work* (Chicago: University of Chicago Press, 1985), p. 197.

24. Christina Mumma and Jeanne Quint Benoliel, "Care, Cure and Hospital Dying Trajectories, *Omega*, Vol. 15(3), 1984, pp. 285–286.

25. Moller, Chapter Seven.

26. Richard A. Hilbert, "The Acultural Dimensions of Chronic Pain: Flawed Reality Construction and the Problem of Meaning," *Social Problems*, Vol. 31(4), 1984, pp. 365–367.

27. Joseph A. Kortaka, "Perceptions of Death, Belief Systems and the Process of Coping with Chronic Pain," *Social Science and Medicine*, Vol. 17(10), 1983, p. 687.

28. Hilbert, op. cit., p. 375.

29. Norbert Elias, *The Loneliness of the Dying* (New York: Basil Blackwell, 1985); David Wendell Moller, "Humanistic Care of the Dying," in Austin Kutscher (ed.), *Hospice U.S.A.* (New York: Columbia University Press, 1985), pp. 37–49.

30. David Wendell Moller, "On the Value of Suffering in the Shadow of Death," in Robert DeBellis et al. (eds.), *Loss, Grief, and Care*, Vol. 1(1/2), 1986–1987, p. 131.

31. Leslie M. Thompson, "Cultural and Institutional Restrictions on Dying Styles in a Technological Society," *Death Education*, Vol. 8, 1984, p. 223.

Chapter 4

1. Thomas Lechford, as cited in David Stannard, *The Puritan Way of Death* (New York: Oxford University Press, 1979), p. 109.

2. Robert Habenstein and William Lamers, *The History of American Funeral Directory* (Milwaukee, Wisc.: Bulfin, 1962).

3. Habenstein and Lamers, Chapter 6.

4. Margaret Coffin, *Death in Early America* (New York: Thomas Nelson, 1976), p. 86.

5. Coffin, p. 104.

6. Vandelyn Pine, *Caretaker of the Dead: The American Funeral Director* (New York: Irvington Press, 1975).

7. Emile Durkheim, *The Elementary Forms of Religious Life* (New York: Free Press, 1965), p. 62.

8. Durkeim, p. 423.

9. Durkeim, p. 427.

10. Durkheim, p. 420.

11. Durkheim, p. 437.

12. Durkheim, pp. 445–446.

13. Robert Bellah et al., *Habits of the Heart* (Berkeley: University of California Press, 1985), pp. 76–83.

14. Randall Collins, *Sociological Insight: An Introduction to Non-Obvious Sociology* (New York: Oxford University Press, 1982), pp. 53–61.

15. Collins, p. 58 (emphasis added).

16. Howard Raether and Robert Slater, "Immediate Post-Death Activities in the United States," in Herman Feifel (ed.), *New Meanings of Death* (New York: McGraw-Hill Book Co., 1977), p. 244; Vandelyn Pine, "Social Meanings of the Funeral," in Vandelyn Pine et al. (eds.) *Acute Grief and the Funeral* (Springfield, Ill: Charles C. Thomas, Publisher, 1976), pp. 36–38.

17. Robert Fulton, "The Traditional Funeral and Contemporary Society," in Pine et al., p. 29.

18. Gene Hutchens, "Grief Therapy," in Pine et al., pp. 155–173.

19. John J. Schwab, "Funeral Behavior and Unresolved Grief," in Pine et al., pp. 241–249.

20. Ira O. Glick, Robert S. Weiss, and C. Murray Parkes, *The First Year of Bereavement* (New York: John Wiley & Sons, Inc., 1974).

21. Pine, op. cit.

22. Richard Kalish, *Death, Grief, and Caring Relationships* (Belmont, Calif.: Brooks/Cole, 1985), p. 216.

23. Vandelyn Pine, "The Cost of Dying: A Sociological Analysis of Funeral Expenditures," in Robert Fulton (ed.), *Death and Identity* (New York: John Wiley & Sons, 1966), p. 432.

24. Tillman Rodabaugh, "Funeral Roles: Ritualized Expectations," *Omega*, Vol. 12(3), 1981, pp. 227–240.

25. Rodabaugh, p. 237.

26. Leroy Bowman, *The American Funeral: A Study in Guilt, Extravagance, and Sublimity* (Westport, Conn.: Greenwood Press, 1959).

27. Paul Irion, *The Funeral: Vestige or Value* (Pearl River, Tenn.: Parthenon Press, 1966), p. 120.

28. Irion, p. 47.

29. Jessica Mitford, *The American Way of Death* (New York: Crest Books, 1964).

30. Irion, op. cit., p. 48.

31. Ruth Harmer, *The High Cost of Dying* (New York: Collier Books, 1963), p. 148.

32. Bowman, op. cit., Chapter 8.

33. Pine, *Caretaker of the Dead.*

34. Harmer, op. cit., p. 9.

35. Harmer, op. cit.

36. Vandelyn Pine, *A Statistical Abstract of Funeral Service Facts and Figures of the United States* (Milwaukee: National Funeral Directors Association, 1982).

37. Mitford, op. cit., p. 18.

38. Harmer, op. cit., p. 168.

39. Bowman, op. cit., p. 49.

40. Mitford, op. cit., p. 19.

41. Kenneth Doka, "Expectation of Death, Participation in Funeral Arrangements, and Grief Adjustment," *Omega*, Vol. 15(2), 1984, p. 119.

42. Helene Goldberg, "Funeral and Bereavement Rituals of Kota Indians and Orthodox Jews," *Omega*, Vol. 12(2), 1981, pp. 117–127.

43. Jean Masamba and Richard Kalish, "Death and Bereavement: The Role of the Black Church," *Omega*, Vol. 7(1), 1976, pp. 23–34.

44. Daniel David Cowell, "Funerals, Family, and Forefathers: A View of Italian-American Funeral Practices," *Omega*, Vol. 16(1), 1985, pp. 69–85.

45. Betty Bergen and Robert Williams, "Alternative Funerals: An Exploratory Study," *Omega*, Vol 12(1), 1981, pp. 71–77.

Chapter 5

1. Neil Postman, *Amusing Ourselves to Death: Public Discourse in the Age of Show Business* (New York: Viking Press, 1985).

2. David Wendell Moller, "On the Value of Suffering in the Shadow of Death," *Loss, Grief and Care*, Vol. 1(1/2), 1986–1987, pp. 127–136.

3. Mark Zborowski, *People in Pain* (San Francisco: Jossey-Bass, 1969).

4. B. Raphael, *The Anatomy of Bereavement* (New York: Basic Books, 1983).

5. Richard Kalish, *Death, Grief and Caring Relationships* (Belmont, Calif.: Brooks/Cole, 1985), p. 209.

6. Phillippe Ariès, *The Hour of Our Death* (New York: Alfred A. Knopf, 1981), p. 575.

7. Geoffrey Gorer, *Death, Grief, and Mourning* (New York: Doubleday and Co., 1965), pp. 126–127.

8. Robert Fulton, David Gottesman, and Greg Owen, "Loss, Social Change, and the Prospect of Mourning," *Death Education*, Vol. 6, 1982, p. 141.

9. Greg Owen, Robert Fulton, and Eric Markusen, "Death at a Distance: A Study of Family Survivors," *Omega*, Vol. 13(3), 1982, p. 216.

10. Owens et al., p. 215.

11. Larry Bugen, "Human Grief: A Model for Prediction and Intervention," *American Journal of Orthopsychiatry*, Vol. 47(2), April, 1977, pp. 196–206.

12. Bugen, p. 202.

13. Erich Lindemann. "Symptomatology and Management of Acute Grief," *American Journal of Psychiatry*, Vol. 101, 1944, pp. 141–148.

14. Catherine M. Sanders, "Effects of Sudden vs. Chronic Illness Death on Bereavement Outcome," *Omega*, Vol. 13(3), 1982, pp. 227–241.

15. Sanders, p. 236.

16. P. J. Clayton, et al., "The Bereavement of the Widowed," *Diseases of the Central Nervous System*, Vol. 32, 1971, pp. 597–604; P. Bornstein et al, "The Depression of Widowhood After Thirteen Months," *British Journal of Psychiatry*, Vol. 122, 1973, pp. 561–566.

17. Colin Murray Parkes, "Effects of Bereavement on Physical and Mental Health—A Study of the Medical Records of Widows," *British Medical Journal*, 2, 1964, pp. 274–279.

18. D. Maddison and W. Walker, "Factors Affecting the Outcome of Conjugal Bereavement," *British Journal of Psychiatry*, 113, 1967, pp. 297–300.

19. Vasnik Volkan, "Typical Findings in Pathological Grief," *Psychiatric Quarterly*, Vol. 44, 1970, pp. 231–250; Colin Murray Parkes, "Determinants of Outcome Following Bereavement," *Omega*, Vol. 6, 1975, pp. 303–323; Colin Murray Parkes and Robert Weiss, *Recovery From Bereavement* (New York: Basic Books, 1983).

20. Ira Glick, Robert Weiss, and Colin Murray Parkes, *The First Year of Bereavement* (New York: John Wiley and Sons, 1974), pp. 31–32.

21. Robert Fulton and David Gottesman, "Anticipatory Grief: A Psychosocial Concept Reconsidered," *British Journal of Psychiatry*, Vol. 137, 1980, pp. 45–54.

22. Fulton and Gottesman.

23. Robert Woodfield and Linda Viney, "A Personal Construct Approach to the Conjugally Bereaved Woman," *Omega*, Vol. 15(1), 1984, pp. 4–6.

24. Woodfield and Viney, p. 10.

25. Lindemann, op. cit.

26. Selby Jacobs and Adrian Ostfeld, "An Epidemiological Review of the Mortality of Bereavement," *Psychosomatic Medicine*, Vol. 39(5), 1977, p. 344.

27. Colin Murray Parkes et al., "Broken Heart: A Statistical Study of Increased Mortality Among Widowers," *British Medical Journal*, 1, 1969, pp. 740–743.

28. David Maddison and Agnes Viola, "The Health of Widows in the Year Following Bereavement," *Journal of Psychosomatic Research*, Vol. 12, 1969, p. 305.

29. David Maddison, "The Relevance of Conjugal Bereavement for Preventive Psychiatry," *British Journal of Medical Psychology*, Vol. 41, 1968, pp. 223–233.

30. Colin Murray Parkes, "The First Year of Bereavement," *Psychiatry*, Vol. 33, 1970, pp. 444–467.

31. D. Rees and S. Lutkins, "Mortality of Bereavement," *British Journal of Medicine*, Vol. 4, 1967, pp. 743–749.

32. Jacobs and Ostfeld. op. cit., pp. 349–352.

33. Margaret Stroebe et al., "The Broken Heart: Reality or Myth?" *Omega*, Vol 12(2), 1981, pp. 100–101.

34. Stroebe, p. 102.

35. Parkes, 1970, op. cit.

36. Parkes et al., 1969, op. cit., p. 743.

37. Jerome Frederick, "The Biochemistry of Bereavement: Possible Basis for Chemotherapy?" *Omega*, Vol. 13(4), 1983, p. 296.

38. Frederick, p. 299.

39. Frederick, p. 300.

40. Milton Greenblatt, "The Grieving Spouse," *American Journal of Psychiatry*, 135(1), 1978, p. 43; Richard Kalish, "Death and Survivorship: The Final Transition," *Annals of the American Academy of Political and Social Science*, Vol. 464, 1982, pp. 168–170.

41. Kalish, op. cit; Judith Brown and Christina Baldwin, *A Second Start: A Widow's Guide to Financial Survival at a Time of Emotional Crisis* (New York: Simon and Schuster, 1968).

42. Elizabeth A. Bankoff, "Social Support and Adaptation to Widowhood," *Journal of Marriage and the Family*, Vol. 45, 1983, p. 836.

43. Justine Ball, "Widow's Grief: The Impact of Age and Mode of Death," *Omega*, Vol. 7(4), 1977, p. 307.

44. Bankoff, op. cit., p. 836.

45. Phyllis Silverman, "The Widow-to-Widow Program: An Experiment in Preventive Intervention," *Mental Hygiene*, Vol. 53(3), 1969, p. 337.

46. Bankoff. op. cit., p. 831.

47. C. S. Lewis, *A Grief Observed* (New York: Seabury Press, 1963) p. 1.

Chapter 6

1. Norbert Elias, *The Loneliness of the Dying* (New York: Blackwell, 1985), pp. 23, 85.

2. Maria Nagy, "The Child's Theories Concerning Death," *Journal of Genetic Psychology*, Vol. 73, 1948, p. 12.

3. Nagy, op. cit., p. 13.

4. Nagy, op. cit., p. 25.

5. Richard Kalish, *Loss, Grief and Caring Relationships* (Belmont, Calif.: Brooks/Cole), 1985; Tillman Rodabough, "Helping Students Cope with Death," *Journal of Teacher Education*, Vol. 31(6), 1980, pp. 19–23; S. Salladay and M. Royal, "Children and Death: Guidelines for Grief Work," *Child Psychiatry and Human Development*, Vol. 11(4), 1981, pp. 203–212; Robert Kastenbaum, "We Covered Death Today," *Death Education*, Vol. 1, Spring 1977, pp. 86–92; Earl A. Grollman, *Talking About Death: A Dialogue Between Parent and Child* (Boston: Beacon Press, 1976).

6. Virginia Atwood, "Children's Concepts of Death: A Descriptive Study," *Child Study Journal*, Vol. 14(1), 1984, p. 13.

7. David Hicks, *Tetum Ghosts and Kin* (Palo Alto, Calif.: Mayfield Publishing Co., 1976).

8. G. Koocher, "Childhood, Death, and Cognitive Development," *Developmental Psychology*, Vol. 9, 1973, pp. 363–375.

9. F. Ferguson, "Children's Cognitive Discovery to Death," *Journal of the Association for the Care of Children in Hospitals*, Vol. 7, 1978, pp. 8–14; E. White, et al., "Children's Conceptions of Death," *Child Development*, Vol. 49, 1977, pp. 307–310.

10. B. Kane, "Children's Concepts of Death: A Descriptive Study," Journal of Genetic Psychology, Vol. 134, 1979, pp. 141–153; M. Speece and S. Brent, "Children's Understanding of Death," *Child Development*, Vol. 55, 1984, pp. 1671–1686.

11. Robert Kastenbaum and R. Aisenberg, *The Psychology of Death* (New York: Springer Publishing Co., 1972).

12. Speece and Brent, op. cit.

13. Israel Orbach et al., "Children's Perceptions of Death in Humans and Animals as a Function of Age, Anxiety and Cognitive Ability," *Journal of Child Psychological Psychiatry*, Vol. 26(3), 1985, pp. 453–463.

14. Orbach et al.; Atwood. op. cit.; Hannelore Wass et al., "Young Children's Death Concepts Revisited," *Death Education*, Vol. 7, 1983, pp. 385–394.

15. J. VanDovgen-Mehnow and J. Sanders-Woudstra, "Psychological Aspects of Childhood Cancer: A Review of the Literature," *Journal of Child Psychological Psychiatry*, Vol. 27(2), 1986, pp. 145, 149.

16. N. Alby, "Ending the Chemotherapy of Acute Leukemia: A Different Period of Meaning," in J. L. Schulman and M. J. Kupst (eds.), *The Child with Cancer* (Springfield, Ill.: Charles C. Thomas, Publisher, 1980), pp. 175–182; J. Spinetta and P. Deasy-Spinetta, *Living with Childhood Cancer* (St. Louis: C. V. Mosby Co., 1981).

17. Jean Comaroff and Peter Maguire, "Ambiguity and the Search for Meaning: Childhood Leukemia in the Modern Clinical Context," *Social Science and Medicine*, Vol. 15B, 1981, pp. 116–117.

18 Comaroff and Maguire, p. 118.

19. J. Vernick and M. Karon, "Who's Afraid of Death on a Leukemia Ward?" *American Journal of Diseases of Children*, Vol. 109, 1965, pp. 393–397; John Spinetta et al., "Anxiety and the Dying Child," *Pediatrics*, Vol. 52(6), 1973, pp. 841–845; Myra Bluebond-Langner, *The Private Worlds of Dying Children* (Princeton, N.J.: Princeton University Press, 1980; Elisabeth Kübler-Ross, *On Children and Death* (New York: Macmillan Publishing Co., 1983).

20. VanDovgen-Mehnow and Sanders-Woudstra, op. cit.

21. Bluebond-Langner, op. cit., pp. 166–168.

22. Bluebond-Langner, op. cit., p. 173.

23. Bluebond-Langner, op. cit.

24. Comaroff and Maguire. op. cit.; Bluebond-Langner, op. cit.

25. J. A. Cook, "Influence of Gender on the Problems of Parents of Fatally Ill Children," *Journal of Psychosocial Oncology*, Vol. 2(1), 1984, pp. 71–91.

26. Bluebond-Langner, op. cit.

27. Joseph M. Natterson and Alfred G. Knudson, "Observations Concerning Fear of Death in Fatally Ill Children and Their Mothers," *Psychosomatic Medicine*, Vol. 22(6), 1960, pp. 456–465.

28. Spinetta et al., op. cit; VanDovgen-Mehnow and Sanders-Woudstra, op. cit.

29. Eugenia Waechter, "Dying Children: Patterns of Coping," in Hannelore Wass and Charles Corr (eds.), *Childhood and Death* (New York: Hemisphere Publishing Corp., 1984), pp. 51–68.

30. J. Vernick, "Meaningful Communication with the Fatally Ill Child," in E. Anthony and C. Koupernik (eds.), *The Child in His Family* (New York: John Wiley and Sons, 1973); David W. Adams, "Helping the Dying Child: Practical Approaches for Nonphysicians," in Hannelore Wass and Charles Corr (eds.), *Childhood and Death* (New York: Hemisphere Publishing Corp., 1984), pp. 95–112.

31. Margaret Adams-Greenly, "Helping Children Communicate About Serious Illness and Death," *Journal of Psychological Oncology*, Vol. 2(2), Summer 1984, pp. 61–72.

32. Kübler-Ross, op. cit., pp. xvii, 2, 50.

33. Kastenbaum, op. cit.; David Carroll, *Living with Dying* (New York: Macmillan Publishing Co., 1985), pp. 168–189.

34. M. G. DeBruyn, *The Beaver Who Wouldn't Die* (Chicago: Follett Publishing Co., 1975); Judith Viorst, *The Tenth Good Thing About Barney* (New York: Athenaeum Press, 1971); David Carrick, *Accident* (New York: Seabury Press, 1976); Constance Greene, *Beat the Turtle Drum* (New York: Viking Press, 1976).

35. Earl A. Grollman, *Explaining Death to Children* (Boston: Beacon Press, 1967); Grollman, *Concerning Death: A Practical Guide for the Living* (Boston: Beacon Press, 1974), pp. 65–80; Grollman, *Talking About Death: A Dialogue Between Parent and Child* (Boston: Beacon Press, 1976); Sara Bonnett Stein, *About Dying* (New York: Walker and Company, 1974); Hannelore Wass et al., *Death Education: An Annotated Resource Guide* (Washington, D.C.: Hemisphere Publishing Co., 1980); Hannelore Wass and Charles Corr (eds.), *Helping Children Cope with Death: Guidelines and Resources* (Washington, D.C.: Hemisphere Publishing Co., 1984).

36. Joseph Weber and David Fournier, "Death in the Family: Children's Cognitive Understanding and Sculptures of Family Relationship Patterns," *Journal of Family Issues*, Vol. 7(3), 1986, pp. 277–296; Ester Gelcer, "Dealing with Loss in the Family Context," *Journal of Family Issues*, Vol. 7(3), 1986, pp. 315–335; Nancy Wedemeyer, "Transportation of Family Images Related to Death," *Journal of Family Issues*, Vol. 7(3), 1986, pp. 337–351; C. Fidley and H. McCublin (eds.), *Stress and the Family: Coping with Catastrophe* (New York: Brunner-Mazel, 1983).

37. Neil Postman, *The Disappearance of Childhood* (New York: Delacorte Press, 1983); Joshua Meyrowitz, *No Sense of Place* (New York: Oxford University Press, 1985); Vance Packard, *Our Endangered Children: Growing Up in a Changing World* (Boston: Little, Brown, and Co., 1983).

38. Cook, op. cit., p. 73.

39. Cook, op. cit., pp. 78–82.

40. Cook, op. cit., pp. 82–85.

41. Cook, op. cit., pp. 89–90.

42. Oscar A. Barbarim, Diane Hughes, and Mark A. Chesler, "Stress, Coping, and Marital Functioning Among Parents of Children with Cancer," *Journal of Marriage and the Family*, Vol. 47, 1985, p. 478.

43. Barbarim et al.

44. Judith A. Cook, "A Death in the Family: Parental Bereavement in the First Year," *Suicide and Life-Threatening Behavior*, Vol. 13, Spring 1983, pp. 42–61.

45. Cook.

46. Cook; Barbarim et al., op. cit.

47. C. M. Sanders, "A Comparison of Adult Bereavement in the Death of a Spouse, Child and Parent," *Omega*, Vol. 10(3), 1980, pp. 303–322.

48. J. A. Cook, "Influence," op. cit.

49. Richard A. Kalish and David K. Reynolds, *Death and Ethnicity* (Los Angeles: University of Southern California Press, 1976).

50. Richard A. Kalish, *Death, Grief, and Caring Relationships* (Belmont: Brooks/Cole Publishing Co., 1985), p. 247.

51. Joan Arnold and Penelope Gemma, *A Child Dies: A Portrait of Family Grief* (Rockville, Md.: Aspen Systems Corp., 1983).

52. Mary Smith and Darlene Francy, "When a Child Dies at Home," *Nursing*, Vol. 12, 1982, pp. 66–67.

53. Lynn Adkins, "Hospice Care for Terminally Ill Children," *Child Welfare*, Vol. 43(6), 1984, pp. 550–562.

54. Benjamin Siegel, "Helping Children Cope with Death," *American Funeral Director*, Vol. 31(3), pp. 175–180.

55. Barbara Soricelli and Carolyn Utech, "Mourning the Death of a Child: The Family and Group Process," *Social Work*, September-October, 1985, pp. 42–43 (emphasis added).

56. Jonathan Kotch and Susan Cohen, "SIDS Counselors' Reports of Own and Parents' Reactions to Reviewing the Autopsy Report," *Omega*, Vol. 16(2), 1985, pp. 129–139.

57. Peter De'Epiro, "When Sudden Infant Death Strikes," *Patient Care*, March 15, 1984, pp. 303–322.

58. Kübler-Ross, op. cit.; Roy Nichols and Jane Nichols, "Funerals: A Time for Grief and Growth," in Elisabeth Kübler-Ross (ed.), *Death: The Final Stage of Growth* (Englewood Cliffs, N.J.: Prentice-Hall, 1975), pp. 87–96.

59. Kübler-Ross, op. cit., pp. 197–202.

60. D. Klass and B. Shimers, "Professional Roles in a Self-Help Group for the Bereaved," *Omega*, Vol. 13(4), 1983, pp. 361–375.

61. Klass and Shimers, pp. 356–359.

62. Klass and Shimers, p. 363.

63. Klass and Shimers, pp. 363–367.

64. Klass and Shimers, pp. 368–371.

65. Klass and Shimers, pp. 369–370.

Chapter 7

1. Emile Durkheim, *Suicide: A Study in Sociology* (New York: Free Press, 1951), p. 299.

2. Durkheim, p. 335.

3. Durkheim, p. 335.

4. Durkheim, p. 209.

5. Durkheim, p. 338.

6. Durkheim, p. 338.

7. Durkheim, p. 220.

8. Whitney Pope, *Durkheim's Suicide: A Classic Analyzed* (Chicago: University of Chicago Press, 1976), p. 21 (emphasis added).

9. Durkheim, op. cit., pp. 238–239.

10. Joseph A. Blake, "Death by Hand Grenade: Altruistic Suicide in Combat," *Suicide and Life-Threatening Behavior*, Vol. 8(1), 1978, pp. 46–59.

11. James Selkins, "The Legacy of Emile Durkheim," *Suicide and Life-Threatening Behavior*, Vol. 13(1), 1983, p. 4.

12. Kathryn Johnson, "Durkheim Revisited: Why Do Women Kill Themselves?" *Suicide and Life-Threatening Behavior*, Vol. 9(3), 1979, pp. 145–153.

13. Durkheim, op. cit., p. 258.

14. Durkheim, pp. 248–250.

15. Durkheim, p. 259.

16. Durkheim, pp. 246–253.

17. Pope, op. cit., p. 28.

18. Durkheim, op. cit., p. 258.

19. Durkheim, op. cit., p. 276.

20. Durkheim, idem.

21. Durkheim, op. cit., p. 51.

22. Durkheim, idem. (emphasis added).

23. Sigmund Freud, "Mourning and Melancholia," Vol. 4, pp. 152–170, in Sigmund Freud, *Collected Papers* (New York: Basic Books, 1959); Sigmund Freud, *Beyond the Pleasure Principle* (New York: Bantam Books, 1959).

24. Karl Menninger, *Man Against Himself* (New York: Harcourt Brace Jovanovich, 1938).

25. Karl Menninger, *The Vital Balance* (New York: Viking Press, 1963), pp. 266–267.

26. Menninger, *Vital Balance*, pp. 267–269 (emphasis added).

27. Gregory Zilboorg, "Suicide Among Civilized and Primitive Races," *The American Journal of Psychiatry*, Vol. 92, 1936, pp. 1347–1369; Zilboorg, "Considerations in Suicide, with Particular Reference to that of the Young," *American Journal of Orthopsychiatry*, Vol. 17, 1937, pp. 15–31.

28. D. M. Palmer, "Factors in Suicidal Attempts: A Review of 25 Consecutive Cases," *The Journal of Mental and Nervous Disease*, Vol. 93(4), 1941, pp. 421–442.

29. Edmund Bergler, "Problems of Suicide," *Psychiatric Quarterly* (Supplement), Vol. 20, 1946, pp. 261–275.

30. Norman Farberow, "Summary" in Norman Farberow and Edwin Schneidman, *The Cry for Help* (New York: McGraw-Hill Book Co., 1961), pp. 290–321.

31. Pope, op. cit., p. 200.

32. "Suicide—Part I," *The Harvard Medical School Mental Health Letter*, Vol. 2(8), February 1986, p. 1.

33. Joseph Ross, moderator, "The Management of the Presuicidal, Suicidal and Post-suicidal Patient" (UCLA conference), *Annals of Internal Medicine*, Vol. 75(3), 1971, pp. 441–458; Rosemary Barnes, "The Recurrent Self-Harm Patient," *Suicide and Life-Threatening Behavior*, Vol. 16(4), Winter, 1986, pp. 399–408; Dean Schuyler, "Counseling Suicide Survivors: Issues and Answers," *Omega*, Vol. 4(4), 1973, pp. 313–321; Corrine Hutton and Sharon McBride Valente, "Bereavement Group for Parents Who Suffered a Suicidal Loss of a Child," *Suicide and Life-Threatening Behavior*, Vol. 11(3), Fall 1981, pp. 141–149.

34. "Suicide—Part 2," *The Harvard Medical School Mental Health Letter*, Vol. 2(9), 1986, p. 4.

35. Karen Siegel, "Rational Suicide: Considerations for the Clinician," *Psychiatric Quarterly*, Vol. 54(2), Summer 1982, p. 77.

36. Thomas Szasz, "The Ethics of Suicide," *The Antioch Review*, Vol. 31(1), Spring 1971, p. 13.

37. Thomas Szasz, *The Second Sin* (New York: Doubleday and Co., 1974), p. 75.

38. J. Minear and L. Brush, "The Correlations of Attitudes Toward Suicide with Death Anxiety, Religiosity, and Personal Closeness to Suicide," *Omega*, Vol. 11(4), 1981, p. 321.

39. Menno Boldt, "Normative Evaluations of Suicide and Death: A Cross-Generational Study," *Omega*, Vol. 13(2), 1982, p. 154.

40. B. K. Singh et al., "Public Approval of Suicide: A Situational Analysis," *Suicide and Life-Threatening Behavior*, Vol. 16(4), Winter 1986, pp. 409–417.

41. Thomas Widiger and Marie Rinaldi, "An Acceptance of Suicide," *Psychotherapy: Theory, Research and Practice*, Vol. 20(3), Fall 1983, p. 269.

42. Leslie Hansen and Charles McAleer, "Terminal Cancer and Suicide: The Health Care Professional's Dilemma," *Omega*, Vol. 14(3), 1983, pp. 241–247.

43. Betty Rollin, *Last Wish* (New York: Warner Books, 1985).

44. Siegel, op. cit., p. 78.

45. K. Rudestarn and D. Imbroll, "Societal Reactions to a Child's Death by Suicide," *Journal of Consulting and Clinical Psychology*, Vol. 51(3), 1983, pp. 461–462; B. Danto and A. H. Kutscher (eds.), *Suicide and Bereavement* (New York: Arno Press, 1977).

46. K. Siegel and P. Tucker, "Rational Suicide and the Terminally Ill Cancer Patient," *Omega*, Vol. 15(3), 1984, p. 267.

47. Siegel, op. cit., p. 80.

48. Francine Klagsbrun, *Too Young to Die* (New York: Pocket books, 1985).

49. Robert Litman and Julie Dieler, "Case Studies in Youth Suicide," in Michael Peck et al. (eds.), *Youth Suicide* (New York: Springer Publishing Co., 1985), p. 50.

50. Litman and Dieler, idem.

51. Brent Hafen and Kathryn Frandsen, *Youth Suicide: Depression and Loneliness* (Provo, Utah: Behavioral Health Associates, 1986), p. 39.

52. Litman and Dieler, op. cit., pp. 48–70.

53. Alan Berman and Teresa Carroll, "Adolescent Suicide: A Critical Review," *Death Education*, Vol. 8 (Supplement), 1984, pp. 58–59.

54. Carol Tishler et al., "Adolescent Suicide Attempts: Some Significant Factors," *Suicide and Life-Threatening Behavior*, Vol. 11(2), Summer 1981, pp. 86–92.

55. Hafen and Frandsen, op. cit., pp. 80–135.

56. James Selkin, in Hafen and Frandsen, op. cit., p. 80.

57. David Matza, *Delinquency and Drift* (New York: John Wiley and Sons, 1964), pp. 188–189.

58. Dennis L. Peck, "The Last Moments of Life: Learning to Cope," *Deviant Behavior*, Vol. 4, 1983, p. 314.

59. Phylis Topol and Marvin Reznikoff, "Perceived Peer and Family Relationships, Hopelessness and Locus of Control as Factors in Adolescent Suicide Attempts," *Suicide and Life-Threatening Behavior*, Vol. 12(3), 1982, pp. 148–149.

60. Topol and Reznikoff, p. 149.

61. Hafen and Frandsen, op. cit., p. 62.

62. Berman and Carroll, op. cit., p. 60.

63. Peter Sainsbury, *Suicide in London: An Ecological Study* (London: Chapman and Hall, 1955), p. 30.

64. Sainsbury, p. 76.

65. Michele Wilson, "Suicide Behavior: Toward an Explanation of Differences in Female and Male Rates," *Suicide and Life-Threatening Behavior*, Vol. 11(3), 1981, p. 132.

66. Norman Kreitman (ed.), *Parasuicide* (New York: John Wiley and Sons, 1977).

67. B. Joyce Stephens, "Suicidal Women and Their Relationships with Husbands, Boyfriends, and Lovers," *Suicide and Life-Threatening Behavior*, Vol. 15(2), Summer 1985, pp. 80–88.

68. Stephens, pp. 88–89.

69. Ronald Maris, "Deviance as Therapy: The Paradox of the Self-Destructive Female," *Journal of Health and Social Behavior*, Vol. 12, 1971, p. 120.

70. B. Joyce Stephens, "Suicidal Women and Their Relationships with Their Parents," *Omega*, Vol. 16(4), 1986, p. 291.

71. Stephens, "Suicidal Women . . . Parents," pp. 291–298.

72. Ronald Maris, "The Adolescent Suicide Problem," *Suicide and Life-Threatening Behavior*, Vol. 15(2), Summer 1985, p. 104.

73. Edwin Schneidman, "Postvention and the Survivor-Victim," in Schneidman, *Death: Current Perspective* (Palo Alto, Calif.: Mayfield Publishing Co., 1980), p. 234.

74. Earl Grollman, *Suicide: Prevention-Intervention-Postvention* (Boston: Beacon Press, 1971).

75. Erving Goffman, *Stigma: Notes on the Management of Spoiled Identity*, (Englewood Cliffs, N.J.: Prentice-Hall, 1963).

76. A. S. Demi, "Social Adjustment of Widows After a Sudden Death: Suicide and Non-Suicide Survivors Compared," *Death Education*, Vol. 8 (Supplement), 1984, pp. 91–111.

77. Albert C. Cain and Irene Fast, "The Legacy of Suicide: Observations on the Pathogenic Impact of Suicide upon Marital Partners," *Psychiatry*, Vol. 29, 1966, p. 408.

78. Hatton and Valente, op. cit., pp. 147–148.

79. Kjell Erik Rudestam, "Physical and Psychological Responses to Suicide in the Family," *Journal of Consulting and Clinical Psychology*, Vol. 45(2), 1977, p. 169.

80. Fady Hajal, "Post-Suicide Grief Work in Family Therapy," *Journal of Marriage and Family Counseling*, Vol. 3, 1977, pp. 35–42.

81. Hatton and Valente, op. cit.

82. Schuyler, op. cit., pp. 319–320.

Chapter 8

1. Martin Gilbert, *The Holocaust: A History of the Jews of Europe During the Second World War* (New York: Holt, Rinehart, and Winston, 1985), p. 129.

2. Donald J. Dietrich, "Holocaust as Public Policy: The Third Reich," *Human Relations*, Vol. 34(6), 1981, p. 450.

3. Gilbert, op. cit., p. 25.

4. Leni Yahil, *The Holocaust: The Fate of European Jewry* (New York: Oxford University Press, 1990), pp. 17–18.

5. Dietrich, op. cit. p. 456.

6. Robert Lifton, *The Nazi Doctors: Medical Killing and the Psychology of Genocide* (New York: Basic Books, 1986), p. 34.

7. Adolf Hitler, *Mein Kampf* (Boston: Houghton Mifflin Company, 1971), p. 398.

8. Hitler, p. 435.

9. Hitler, pp. 312–316.

10. Lifton, op. cit., p. 16.

11. Alexander Donat, *The Holocaust Kingdom* (New York: Holocaust Library, 1978), p. 7.

12. Donat, pp. 21–22.

13. Gilbert, op. cit., p. 129.

14. Nora Levin, *The Holocaust: The Destruction of European Jewry 1933–1945* (New York: Holocaust Library, 1968), pp. 207–208.

15. Gilbert, op. cit., p. 453.

16. Gilbert, p. 145.

17. Gilbert, idem.

18. Donat, op. cit., pp. 47–48.

19. Gilbert, op. cit., p. 368.

20. Donat, op. cit., p. 43.

21. Donat, op. cit., p. 53.

22. Gilbert, op. cit., p. 558.

23. Yahil, op. cit., p. 483.

24. Lucy Dawidowicz, *The War Against the Jews* (New York: Bantam Books, 1986), pp. 211–222.

25. Elie Weisel, *Night* (New York: Bantam Books, 1982), pp. 20–24.

26. Terence Des Pres, *The Survivor: An Anatomy of Life in the Death Camps* (New York: Oxford University Press, 1976), p. 75.

27. Des Pres, idem.

28. Lifton, op. cit., p. 163.

29. Lifton, op. cit., pp. 164–165.

30. Victor Frankl, *Man's Search for Meaning* (New York: Pocket Books, 1984), pp. 33–34.

31. Frankl, p. 51.

32. Sim Kessel, *Hanged at Auschwitz* (New York: Stein and Day, 1972), pp. 50–51.

33. Bruno Bettelheim, *The Informed Heart* (New York: Avon Books, 1971), pp. 133–134.

34. Bettelheim, idem.

35. Des Pres, op. cit., p. 57.

36. Des Pres, op. cit., pp. 53–71.

37. Kogon, op. cit., pp. 50–51.

38. Reska Weiss, *Journey Through Hell* (London: Vallentine, Mitchell, 1961), p. 211, as quoted in Des Pres, op. cit., p. 57.

39. Des Pres, op. cit., p. 59. This, of course, is also the central warning about power, that Orwell presents in *1984*.

40. Gilbert, op. cit., p. 171.

41. Gilbert, p. 167.

42. Gilbert, p. 178.

43. Gilbert, p. 380.

44. Gilbert, p. 403.

45. Raul Hilberg, *The Destruction of the European Jews* (Chicago: Quadrangle Books, 1967), p. 256.

46. Lifton, op. cit., p. 159.

47. Elie A. Cohen, *Human Behavior in the Concentration Camp* (Westport, Conn.: Greenwood Press, 1984), pp. 32–33.

48. Kogon, op. cit., p. 240.

49. Lifton, op. cit., p. 177.

50. Lifton, idem.

51. Lifton, p. 179.

52. Alvin Gouldner, *The Coming Crisis of Western Sociology* (New York: Avon Books, 1970), p. 500.

53. Max Weber, "Bureaucracy," in Hans Gerth and C. W. Mills, *From Max Weber* (New York: Oxford University Press, 1946), pp. 196–244.

54. Lifton, op. cit., p. 497.

55. Hannah Arendt, *The Origins of Totalitarianism* (New York: Harcourt Brace Jovanovich, 1973), p. 445.

56. Weisel, op. cit., p. 109.

57. Elie Weisel, "Foreword" to Sylvia Rothchild (ed.), *Voices from the Holocaust* (New York: New American Library, 1981), p. 4.

BIBLIOGRAPHY

Abram, Harry S. "Death and Denial in Conrad's 'Nigger of the Narcissus.'" *Omega*, Vol. 7(2), 1976, pp. 125–133.

Ad Hoc Committee of the Harvard Medical School to Examine the Definition of Brain Death. "A Definition of Irreversible Coma." *The Journal of the American Medical Association*, Vol. 205, No. 6, August 5, 1968, pp. 337–340.

Adams, Margaret. "A Hospital Play Program: Helping Children with Serious Illness." *American Journal of Orthopsychiatry*, 46(3), July 1976, pp. 416–424.

Adams-Greenly, Margaret. "Helping Children Communicate About Serious Illness and Death." *Journal of Psychosocial Oncology*, Vol. 2(2), Summer 1984, pp. 61–72.

Adkins, Lynn. "Hospice Care for Terminally Ill Children." *Child Welfare*, Vol. LXIII, Number 6, November–December 1984, pp. 559–562.

Ajemian, Ina and Balfour Mount. "The Adult Patient: Cultural Considerations in Palliative Care." In Cicely Saunders (ed.), *Hospice: The Living Idea* (Philadelphia: W. B. Saunders Co., 1981), p. 19.

Akiyama, Kiroko and Joseph M. Holtzman. "Pet Ownership and Health Status During Bereavement." *Omega*, Vol. 17(2), 1986–87, pp. 187–193.

Alexander, Charles P. "A Move to Ease Death's Sting." *Time*, May 14, 1984, p. 53.

Andersen, Barbara L. "Sexual Functioning Morbidity Among Cancer Survivors." *Cancer*, Vol. 55, April 15, 1985, pp. 1835–1842.

Andress, Vern, and David Corey. "Survivor-Victims: Who Discovers or Witnesses Suicide?" *Psychological Reports*, 1978, 42, 759–764.

Angier, Natalie. "Burying Bones of Contention." *Time*, September 10, 1984, p. 32.

Annas, George J. "At Law—Elizabeth Bouvia: Whose Space Is This Anyway?" *Hastings Center Report*, Vol. 16, No. 2, April, 1986, pp. 24–25.

Annas, George J. "At Law—The Phoenix Heart: What We Have to Lose." *Hastings Center Report*, Vol. 15, Number 3, June, 1985, pp. 15–16.

Annas, George J. "Consent to the Artificial Heart: The Lion and the Crocodiles." *Hastings Center Report*, Vol. 13, No. 2, April, 1983, pp. 20–22.

Annas, George J. "Defining Death: There Ought to Be a Law." *Hastings Center Report*, Vol. 13, No. 1, February, 1983, p. 20.

Annas, George J. "Nonfeeding: Lawful Killing in CA, Homicide in NJ." *Hastings Center Report*, Vol. 13, No. 6, December, 1983, pp. 19–20.

Anonymous (Sixteen-year-old high school student). "A Time of Hurt and Confusion." *Suicide and Life-Threatening Behavior*, Vol. 14, No. 4, Winter 1984, p. 284.

Ansel, Edward L. and Richard K. McGee. "Attitudes Toward Suicide Attempters." *National Institute of Mental Health Bulletin of Suicidology*, pp. 22–28.

Anthony, E. James and Cyrille Koupermilk (eds). *The Child in His Family: The Impact of Disease or Death, Vol. II* (New York: John Wiley & Sons, 1973).

Anthony, Sylvia. *The Discovery of Death in Childhood and After.* (London: Allen Lane, The Penguin Press, 1971).

Araki, Shunichi and Katsuyuki Murata. "Social Life Factors Affecting Suicide in Japanese Men and Women." *Suicide and Life-Threatening Behavior*, Vol. 16(4), Winter 1986, pp. 458–468.

Arens, Diana A. "Widowhood and Well-Being: An Examination of Sex Differences Within a Causal Model." *International Journal of Aging and Human Development*, Vol. 15(1), 1982–83, pp. 27–40.

Aries, Phillipe. *The Hour of Our Death* (New York: Alfred Knopf, 1981).

Arney, William, and Bernard Bergen. *Medicine and the Management of Living: Taming the Last Great Beast* (Chicago: University of Chicago Press, 1984).

"Art Therapy Helps Kids Cope with Death." *Hospitals*, Vol. 58, February 1, 1984, p. 44.

Atwood, Virginia A. "Children's Concepts of Death: A Descriptive Study." *Child Study Journal*, Vol. 14, No. 1, 1984, pp. 11–29.

Axelrod, Charles D. "Reflections on the Fear of Death." *Omega*, Vol. 17(1), 1986–87, pp. 51–64.

Backer, Barbara A., Natalie Hannon, and Noreen A. Russell. *Death and Dying: Individuals and Institutions* (New York: John Wiley & Sons, 1982).

Baldwin, David B. "What Dying Patients Fear Most." *Medical Economics*, Vol. 60, February 7, 1983, pp. 125–132.

Balk, David. "How Teenagers Cope with Sibling Death: Some Implications for School Counselors." *The School Counselor*, Vol. 31(2), November 1983, pp. 150–158.

Ball, Justine. "Widow's Grief: The Impact of Age and Mode of Death." *Omega*, Vol. 7(4), 1976–77, pp. 307–333.

Baltrusch, H. J. F. and Millard Waltz. "Cancer from a Biobehavioural and Social Epidemiological Perspective." *Social Science and Medicine*, Vol. 20, No. 8, 1985, pp. 789–794.

Bankoff, Elizabeth A. "Aged Parents and their Widowed Daughters: A Support Relationship." *Journal of Gerontology*, Vol. 38, No. 2, pp. 226–230.

Bankoff, Elizabeth A. "Social Support and Adaptation to Widowhood." *Journal of Marriage and the Family*, November, 1983, pp. 827–839.

Barbarin, Oscar A., Diane Hughes, and Mark A. Chesler. "Stress, Coping, and Marital Functioning Among Parents of Children with Cancer." *Journal of Marriage and the Family*, May, 1985, pp. 473–480.

Barinbaum, Lea. "Death of Young Sons and Husbands." *Omega*, Vol. 7(2), 1976, pp. 171–74.

Barnaby, Frank. "The Mounting Prospects of Nuclear War." *Bulletin of the Atomic Scientists*, June, 1977.

Barnard, Christian. "The Right to Die—Should a Doctor Decide?" (interview). *U.S. News & World Report*, November 3, 1975, p. 53ff.

Barnes, Gordon E. and Harry Prosen. "Parental Death and Depression." *Journal of Abnormal Psychology,* Vol. 94, No. 1, 1985, pp. 64–69.

Barnes, Rosemary A. "The Recurrent Self-Harm Patient." *Suicide and Life-Threatening Behavior*, Vol. 16(4), Winter 1986, pp. 399–408.

Barraclough, B. M. and D. M. Shepherd. "Impact of a Suicide Inquest," Editorial. *The Lancet*, 2(8091), 23 September 1978, pp. 666–667.

Barraclough, B. M. and D. M. Shepherd. "Impact of a Suicide Inquest," Reply to Editorial. *The Lancet*, 2(8093), 7 October 1978, p. 795.

Barrett, Carol J. and Karen M. Schneweis. "An Empirical Search for Stages of Widowhood." *Omega*, Vol. 11(2), 1980–81, pp. 97–103.

Barry, B. "Perceptions of Suicide." *Death Education*, Vol. 8, Supplement, 1984, pp. 17–25.

Barton, David (ed.). *Dying and Death: A Clinical Guide for Caregivers* (Baltimore: The Williams and Wilkins Co., 1977).

Bascue, Loy O., David J. Inman, and Wallace J. Kahn. "Recognition of Suicidal Lethality

Factors by Psychiatric Nursing Assistants." *Psychological Reports*, Vol. 51, 1982, pp. 197–198.

Bass, David M. "Response Bias in Studying Hospice Clients' Needs." *Omega*, Vol. 13(4), 1982–83, pp. 305–318.

Bass, David M., T. Neal Garland, and Melinda E. Otto. "Characteristics of Hospice Patients and Their Caregivers." *Omega*, Vol. 16(1), 1985–86, pp. 51–68.

Bass, David M., Fred P. Pestello, and T. Neal Garland. "Experiences with Home Hospice Care: Determinants of Place of Death." *Death Education*, Vol. 8, 1984, pp. 199–222.

Battin, Margaret Pabst. *Ethical Issues in Suicide* (Englewood Cliffs, NJ: Prentice-Hall, Inc., 1982).

Battin, Margaret Pabst. "The Least Worst Death." *Hastings Center Report*, Vol. 13, No. 2, April, 1983, p. 16.

Bayer, Ronald and Eric Feldman. "Hospice under the Medicine Wing." *Hastings Center Report*, Vol. 12, No. 6, December, 1982, pp. 5–6.

Becker, Ernest. *The Structure of Evil* (New York: The Free Press, 1968), p. 247.

Becker, Ernest. *Escape From Evil* (New York: The Free Press, 1975), p. xvii.

Becker, Ernest. *The Denial of Death* (New York: The Free Press, 1973), p. 11.

Beilin, Robert. "Social Functions of Denial of Death." *Omega*, Vol. 12(1), 1981–82, pp. 25–35.

Belk, Beverly, Larry Van De Creek, Norman W. Jankowski, and Samuel C. Klagsbrun. "When a Patient Dies Suddenly." *Patient Care*, September 15, 1985, pp. 78–83.

Bellah, Robert and Etzioni, Amatai. *An Immodest Agenda.* (NY: McGraw-Hill, 1983), Chapters 2 and 3.

Bellah, Robert et al. *Habits of the Heart* (Berkeley: University of California Press, 1985).

Bergen, M. Betsy and Robert R. Williams. "Alternative Funerals: An Exploratory Study." *Omega*, Vol. 12(1), 1981–82, pp. 71–75.

Berger, Peter. *Invitation to Sociology.* (New York: Doubleday and Co., 1963), p. 238.

Berger, Peter et al. *The Homeless Mind: Modernization and Consciousness* (New York: Random House, 1973), p. 57.

Bergesen, Albert. "Review Essay: Centuries of Death and Dying." *American Journal of Sociology*, Vol. 90, No. 2, pp. 435–439.

Berman, Alan L. "Dyadic Death: Murder-Suicide." *Suicide and Life-Threatening Behavior*, Vol. 9(1), Spring 1979, pp. 15–23.

Bertchnell, J. et al. *Effects of Early Parent Death* (New York: MSS Information Corp., 1973).

Betz, Michael and Lenahan O'Connell. "Changing Doctor-Patient Relationships and the Rise in Concern for Accountability." *Social Problems*, Vol. 31, No. 1, October, 1983, pp. 84–95.

Black, David. "What Gets Said about the Dead." *Harper's*, Vol. 226, September, pp. 66–67.

Blake, Joseph A. "Death by Hand Grenade: Altruistic Suicide in Combat." *Suicide and Life-Threatening Behavior*, Vol. 8(1), Spring 1978, pp. 46–59.

Bloom, Joan R. "Social Support, Accommodation to Stress and Adjustment to Breast Cancer." *Social Science and Medicine*, Vol. 16, 1982, pp. 4329–4338.

Bloom, Joan R. and David Spiegel. "The Relationship of Two Dimensions of Social Support to the Psychological Well-Being and Social Functioning of Women with Advanced Breast Cancer." *Social Science and Medicine*, Vol. 19, No. 8, 1984, pp. 831–837.

Bloom, Martin and James Halsema. "Survival in Extreme Conditions." *Suicide and Life-Threatening Behavior*, Vol. 13(3), Fall 1983, pp. 195–206.

Blues, Ann G. and Joyce V. Zerweth. *Hospice and Palliative Nursing Care* (Orlando, Fl: Grune & Stratton, 1984).

Boldt, Menno. "Normative Evaluations of Suicide and Death: A Cross-Generational Study." *Omega*, Vol. 13(2), 1982–83, pp. 145–157.

Boroson, Warren. "Protect Your Family from High-Pressure Undertakers." *Medical Economics*, Vol. 80, August 8, 1983, p. 118.

Bowie, N. E. "'Role' As a Moral Concept in Health Care." *The Journal of Medicine and Philosophy*, Vol. 7, 1982, pp. 57–63.

Bowlby, John. *Attachment and Loss, Vol. III: Loss* (New York: Basic Books, 1980).

Bowling, Ann and Ann Cartwright. *Life After Death* (New York: Tavistock Publications, 1982).

Bowman, Leroy. *The American Funeral* (Westport, CT: Greenwood Press Publishers, 1973).

Brearley, H. C. *Homicide in the United States* (Chapel Hill: The University of North Carolina Press, 1932).

Breault, K. D. and Karen Barkey. "A Comparative Analysis of Durkheim's Theory of Egoistic Suicide." *The Sociological Quarterly*, Vol. 23, Summer 1982, pp. 321–331.

Brennan, Paul. "Monetary Compensations of Death." *Society*, November–December, 1979, pp. 62–69.

Brescia, Frank J., Matthew Sadof, and Janice Barstow. "Retrospective Analysis of a Home Care Hospice Program." *Omega*, Vol. 15(1), 1984–85, pp. 37–44.

Brian, Orvill G., Jr. et al. (eds.). *The Dying Patient* (New York: Russell Sage Foundation, 1970).

Brown, James H. et al. "Is It Normal for Terminally Ill Patients to Desire Death?" *American Journal of Psychiatry*, Vol. 143, No. 2, February, 1986, pp. 208–211.

Brown, Mary H., Margaret E. Kiss, Sharon M. Glassman, and Barbara H. Popkin. "How the Quality of Care Improves the Quality of Life." *Cancer Nursing*, October, 1980, pp. 379–383.

Brown, Robert S., Nancy Massman, and Kathy Wornson. "Hospice Home Care: When Death Occurs." *Minnesota Medicine*, Vol. 64, 1981, pp. 368–369.

Bruhn, John G. "Broken Homes Among Attempted Suicides and Psychiatric Out-Patients: A Comparative Study." *The British Journal of Psychiatry*, Vol. 108, November, 1962, pp. 772–779.

Buckingham, Robert W. *The Complete Hospice Guide* (New York: Harper & Row Publishers, 1983).

Buckingham, Robert W. "Hospice Care in the United States: The Process Begins." *Omega*, Vol. 13(2), 1982–83, pp. 159–171.

Buckingham, Robert W. "Primary Care [for] the Terminally Ill." *Geriatrics*, December, 1979, pp. 73–75.

Buckingham, Robert W. et al. "Living with the Dying: Use of the Technique of Participant Observation." *CMA Journal*, Vol. 115, December 18, 1976, pp. 1211–1215.

Bugen, Larry A. "Human Grief: A Model for Prediction and Intervention." *American Journal of Orthopsychiatry*, Vol. 47(2), April, 1977, pp. 196–206.

Button, Sarah E. "Washington: Pricing Professional Services." *Money*, March, 1984, p. 200.

Cafferata, Gail L. "The Ideology of the American Medcial Profession: An Attribution Perspective." *Social Science and Medicine*, Vol. 15A, 1981, pp. 689–699.

Cain, Albert C. (ed.). *Survivors of Suicide* (Springfield, IL: Charles C. Thomas, Publishers, 1972).

Cain, Albert C. and Irene Fast. "The Legacy of Suicide." *Psychiatry*, Vol. 29, 1966, pp. 406–411.

Calhoun, Lawrence G., James W. Selby, and Carol B. Abernathy. "Suicidal Death: Social Reactions to Bereaved Survivors." *The Journal of Psychology*, Vol. 116, 1984, pp. 255–261.

Calhoun, Lawrence G., James W. Selby, and Candace M. Gribble. "Reactions to the Family of the Suicide." *American Journal of Community Psychology*, Vol. 7, No. 5, 1979, pp. 571–575.

Calhoun, Lawrence G., James W. Selby, and H. Elizabeth King. *Dealing with Crisis* (Englewood Cliffs, NJ: Prentice-Hall, Inc., 1976).

Calhoun, Lawrence G., James W. Selby, and Lisa E. Shelby. "The Psychological Aftermath of Suicide: An Analysis of Current Evidence." *Clinical Psychology Review*, Vol. 2, 1982, pp. 409–420.

Calhoun, Lawrence G., James W. Selby, and Peggy B. Walton. "Suicidal Death of a Spouse: The Social Perception of the Survivor." *Omega*, Vol. 16(4), 1985–86, pp. 283–288.

Callahan, Daniel. "Science: Limits and Prohibitions." Vol. 3, No. 6, *Hastings Center Report*, November, 1973, pp. 5–7.

Callen, Michael L. "Case Studies: 'If I Have AIDS, Then Let Me Die Now!'—Commentary." *Hastings Center Report*, Vol. 14, No. 1, February, 1984, p. 26.

Campbell, Sheila H. "The Meaning of the Breast Cancer/Mastectomy Experience." *Humane Medicine*, Vol. 2, No. 2, November, 1986, pp. 91–95.

Campbell, Thomas W., Virginia Abernathy, and Gloria J. Waterhouse. "Do Death Attitudes of Nurses and Physicians Differ?" *Omega*, Vol. 14(1), 1983–84, pp. 43–49.

Canny, Robert J. "I. Hospice: Family-Centered Care of the Dying." *Rhode Island Medical Journal*, Vol. 64, January, 1981, pp. 36–38.

Cantor, Norman L. *Legal Frontiers of Death and Dying* (Bloomington, IN: Indiana University Press, 1987).

Caplan, Arthur L. "Organ Transplants: The Costs of Success." *Hastings Center Report*, Vol. 13, No. 6, December, 1983, pp. 23–32.

Capron, Alexander M. "Determining Death: Do We Need a Statute?" *Hastings Center Report*, Vol. 3, No. 1, February, 1973, pp. 6–7.

Carey, Raymond G. "Weathering Widowhood: Problems and Adjustment of the Widowed During the First Year." *Omega*, Vol. 10(2), 1979–80, pp. 163–174.

"Case Conference: Where There's No Will There's No Way." *Journal of Medical Ethics*, Vol. 7, 1981, pp. 39–41.

Cassell, Eric F. *Talking with Patients, Vol. 1: The Theory of Doctor-Patient Communication* (Cambridge, MA: The M.I.T. Press, 1985.)

Castles, Mary R. and Patricia M. Keith. "Patient Concerns, Emotional Resources, and Perceptions of Nurse and Patient Roles." *Omega*, Vol. 10(1), 1979, pp. 27–33.

Chapman, C. Richard. "New Directions in the Understanding and Management of Pain." *Social Science and Medicine*, Vol. 19, No. 12, 1984, pp. 1261–1277.

Chesler, Mark, and Oscar Barbarin. "Difficulties of Providing Help in a Crisis: Relationships Between Parents of Children with Cancer and Their Friends." *Journal of Social Issues*, Vol. 40, No. 4, 1984, pp. 113–134.

Childress, James F. "Who Shall Live When Not All Can Live?" *Soundings*, Vol. 43, No. 4, Winter 1970, pp. 339–362.

Chiles, John A. et al. "The 24 Hours before Hospitalization: Factors Related to Suicide Attempting." *Suicide and Life-Threatening Behavior*, Vol. 16(3), Fall 1986, pp. 335–342.

Chng, Chwee L. and Michael K. Ramsey. "Volunteers and the Care of the Terminal Patient." *Omega*, Vol. 15(3), 1984–85, pp. 237–244.

Choron, Jacques. *Death and Modern Man* (New York: Collier Books, 1964).

Choron, Jacques. *Death and Western Thought* (New York: Macmillan and Co., 1963).

Clark, Elizabeth J. "Intervention for Cancer Patients: A Clinical Sociology Approach to Program Planning." *Journal of Applied Sociology*, Vol. 1, No. 1, January, 1984, pp. 83–96.

Clark, Matt, et al. "A Right to Die?" In Steven H. Zarit (ed.), *Readings in Aging and Death: Contemporary Perspectives* (New York: Harper & Row, Publishers, 1977).

Clark, Robert E. and Emily E. LaBeff. "Death Telling: Managing the Delivery of Bad News." *Journal of Health and Social Behavior*, Vol. 23, December, 1982, pp. 366–380.

Clark Grave Vault Company, The. *My Duty* (Columbus, OH: The Clark Grave Vault Company, 1968).

Code, Charles. "Determinants of Medical Care: A Plan for the Future." *The New England Journal of Medicine*, Vol. 283(133), 1979, p. 681.

Coffin, Margaret M. *Death in Early America* (New York: Thomas Nelson Inc., Publishers, 1976).

Cohen, Daniel. "Children's Hospices: Easing Early Deaths." *Esquire*, May, 1985, p. 78.

Cohen, John. "Death and the Danse Macabre." *History Today*, August, 1982, pp. 35–40.

Comaroff, Jean and Peter Maguire. "Ambiguity and the Search for Meaning: Childhood Leukaemia in the Modern Clinical Context." *Social Science and Medicine*, Vol. 15B, 1981, pp. 115–123.

Conroy, Robert W. and Kim Smith. "Family Loss and Hospital Suicide." *Suicide and Life-Threatening Behavior*, Vol. 13(3), Fall 1983, pp. 179–194.

Cook, Judith A. "A Death in the Family: Parental Bereavement in the First Year." *Suicide and Life-Threatening Behavior*, Vol. 13(1), Spring 1983, pp. 42–61.

Cook, Judith A. "If I Should Die Before I Wake: Religious Commitment and Adjustment to the Death of a Child." *Journal for the Scientific Study of Religion*, Vol. 22(3), 1983, pp. 222–238.

Cook, Judith A. "Influence of Gender on the Problems of Parents of Fatally Ill Children." *Journal of Psychosocial Oncology*, Vol. 2(1), Spring 1984, pp. 71–89.

Cooley, Charles Horton. *Human Nature and the Social Order* (New York: Scribner's, 1922).

Corr, Charles A. "A Model Syllabus for Children and Death Courses." *Death Education,* Vol. 8, 1984, pp. 11–28.

Corr, Charles A. and Donna M. Corr (eds.). *Hospice Care: Principles and Practice* (New York: Springer Publishing Co., 1983).

Covino, Susan. "To Die in Peace." *Progressive*, Vol. 49, February, 1984, p. 50.

Cowell, Daniel D. "Funerals, Family, and Forefathers: A View of Italian-American Funeral Practices." *Omega*, Vol. 16(1), 1985–86, pp. 69–86.

Cowgill, Robert. "Hospice Care for the Terminally Ill." *Journal of the Medical Association of Georgia*, Vol. 69, April, 1980, pp. 264–267.

Curl, James Steven. *A Celebration of Death* (London: Constable and Co., Ltd., 1980).

Curl, James Steven. *The Victorian Celebration of Death* (Detroit: The Partridge Press, 1972).

Curtis, Joy and Arthur L. Caplan. "When Patients Harm Themselves." *Hastings Center Report*, Vol. 14, No. 2, April, 1984, pp. 22–23.

Danto, Bruce L. *Jail House Blues: Studies of Suicidal Behavior in Jail and Prison* (Orchard Lake, MI: Epic Publications, Inc., 1973).

DaSilva, Anthony and M. Anthony Schork. "Gender Differences in Attitudes to Death Among a Group of Public Health Students." *Omega*, Vol. 15(1), 1984–85, pp. 77–84.

Davis, Gary. "A Content Analysis of Fifty-Seven Children's Books with Death Themes." *Child Study Journal*, Volume 16, No. 1, 1986, pp. 39–54.

Davis, Robert. "Black Suicide in the Seventies: Current Trends." *Suicide and Life-Threatening Behavior*, Vol. 9(3), Fall 1979, pp. 131–140.

Davis, Robert. "Suicide Among Young Blacks: Trends and Perspectives." *Phylon*, Vol. 41, pp. 223–229.

Davis, Robert and James F. Short. "Dimensions of Black Suicide: A Theoretical Model." *Suicide and Life-Threatening Behavior*, Vol. 8(3), Fall 1977, pp. 161–167.

Day, Lincoln H. "Death from Non-War Violence: An International Comparison." *Social Science and Medicine*, Vol. 19, No. 9, 1984, pp. 917–927.

DeBakey, Michael and Lois DeBakey. "The Ethics and Economics of High-Technology Medicine." *Comprehensive Therapy*, Vol. 9(12), 1983, pp. 6–16.

DeBois, Paul M. *The Hospice Way of Death* (New York: Herman Sciences Press, 1980).

de Haes, Johanna C. J. M. and Ferdinand C. E. van Knippenberg. "The Quality of Life of

Cancer Patients: A Review of the Literature." *Social Science and Medicine*, Vol. 20, No. 8, pp. 809–817.

Demi, Alice S. "Social Adjustment of Widows after a Sudden Death: Suicide and Non-Suicide Survivors Compared." *Death Education*, Vol. 8 (Supplement), 1984, pp. 91–111.

D'Epiro, Peter. "When Sudden Infant Death Strikes." *Patient Care*, March 15, 1984, pp. 18–42.

de Ramon, Pamela Babb. "The Final Task: Life Review for the Dying Patient." *Nursing*, Vol. 13, February, 1983, pp. 44–49.

Dietrich, David R. "Psychological Health of Young Adults Who Experienced Early Parent Death: MMPI Trends." *Journal of Clinical Psychology*, Vol. 40, No. 4, July, 1984, pp. 901–908.

Dietrich, Donald J. "Holocaust as Public Policy: The Third Reich." *Human Relations*, Vol. 34, No. 6, 1981, pp. 445–462.

Dobihal, Shirley V. "Hospice: Enabling a Patient to Die at Home." *American Journal of Nursing*, August, 1980, pp. 1448–1451.

Doka, Kenneth J. "Expectation of Death, Participation in Funeral Arrangements, and Grief Adjustment." *Omega*, Vol. 15(2), 1984–85, pp. 119–129.

Doka, Kenneth J. "Recent Bereavement and Registration for Death Studies Courses." *Omega*, Vol. 12(1), 1981–82, pp. 51–60.

Doka, Kenneth J. "The Social Organization of Terminal Care in Two Pediatric Hospitals." *Omega*, Vol. 12(4), 1981–82, pp. 345–353.

Domino, George, Valerie Domino, and Travis Berry. "Children's Attitudes Toward Suicide." *Omega*, Vol. 17(4), 1986–87, pp. 279–288.

Domino George and Barbara J. Swain. "Recognition of Suicide Lethality and Attitudes Toward Suicide in Mental Health Professionals." *Omega*, Vol. 16(4), 1985–86, pp. 301–308.

Donaldson, P. "Denying Death: A Note Regarding some Ambiguities in the Current Discussion." *Omega*, Vol. 3(3), 1972, pp. 285–290.

Drucker, Peter. *Technology, Management and Society* (New York: Harper and Row, 1970).

Dubos, René. "Health and Creative Adaptation." *Human Nature*, January, 1978, pp. 25–32.

Duda, Deborah. *A Guide to Dying at Home* (Santa Fe, NM: John Muir Publications, Inc., 1982).

Dunkel-Schetter, Christine. "Social Support and Cancer: Findings Based on Patient Interviews and Their Implications." *Journal of Social Issues*, Vol. 40, No. 4, 1984, pp. 77–98.

Dwyer, Philip M. "An Inquiry into the Psychological Dimensions of Cult Suicide." *Suicide and Life-Threatening Behavior*, Vol. 9(2), Summer 1979, pp. 120–127.

Earle, Ann M. et al. (eds.). *The Nurse as Caregiver for the Terminal Patient and His Family* (New York: Columbia University Press, 1976).

Edwards, Rem B. "Pain and the Ethics of Pain Management." *Social Science and Medicine*, Vol. 18, No. 6, 1984, pp. 515–523.

Eich, W. Foster. "How I Handled My 'Baby Doe' Case." *Medical Economics*, Vol. 86, July 9, 1984, pp. 100–110.

Elias, Norbert. *The Loneliness of the Dying* (New York: Basil Blackwell, 1985), p. 47.

Eliot, Thomas D. "The Bereaved Family." *Annals of the American Academy of Political and Social Science*, Vol. 160, 1932, pp. 184–190.

Ellul, Jacques. *The Technological Society* (New York: Vintage Books, 1974).

Elshtain, Jean Bethke. "On Humanness & Honoring the Dead." *Commonweal*, Vol. 110, March 11, 1983, pp. 144–146.

Engle, George L. "The Care of the Patient: Art or Science?" *The Johns Hopkins Medical Journal*, Vol. 140, 1977, pp. 222–232.

Engle, George L. "A Group Dynamic Approach to Teaching and Learning about Grief." *Omega*, Vol. 11(1), 1980–81, pp. 45–59.

Epley, Rita J. and Charles H. McCaghy. "The Stigma of Dying: Attitudes Toward the Terminally Ill." *Omega*, Vol. 8(4), 1977–78, pp. 379–393.

Eyler, John M. "Social Medicine: Early Efforts," (book review of William Coleman's *Death Is A Social Disease*). *Science*, Vol. 219, January 28, 1983, pp. 380–381.

Farrell, James J. *Inventing the American Way of Death, 1830–1920* (Philadelphia: Temple University Press, 1980).

Fedden, Henry Romilly. *Suicide: A Social and Historical Study* (London: Peter Davis Ltd., 1938).

Feifel, Herman and John Morgan. "Humanity Has to Be the Model." *Death Studies*, Vol. 10, 1986, pp. 1–9.

Fiefel, Herman and Daniel Schag. "Death Outlook and Social Issues." *Omega*, Vol. 11(3), 1980–81, pp. 201–215.

Feifel, Herman et al. "Physicians Consider Death." *The Proceedings, 75th Annual Convention, APA*, 1967, pp. 201–202.

Feinberg, Joel. "The Mistreatment of Dead Bodies." *The Hastings Center Report*, Vol. 15, No. 1, February, 1985, pp. 31–37.

Feldman, Marvin J. et al. *Fears Related to Death and Suicide* (NY: MSS Information Corp., 1974).

Fennell, F. V. "The Need for Hospices." *New Zealand Medical Journal*, August 27, 1980, pp. 158–161.

Fish, William C. and Edith Waldhart-Letzel. "Suicide and Children." *Death Education*, Vol. 5, 1981, pp. 215–222.

Fitts, W. T. and I. S. Ravin. "What Philadelphia Physicians Tell Patients With Cancer." *The Journal of the American Medical Association*, 153, 1953, pp. 901–904.

Fitzpatrick, Ray et al. *The Experience of Illness* (New York: Tavistock Publications, 1984).

Florian, Victor and Dov Har-Even. "Fear of Personal Death: The Effects of Sex and Religious Belief." *Omega*, Vol 14(1), 1983–84, pp. 83–94.

Flynn, Arthur and David E. Stewart. "Where Do Cancer Patients Die? A Review of Cancer Deaths in Cuyahoga County, Ohio, 1957–1974." *Journal of Community Health*, Vol. 5, No. 2, Winter 1979, pp. 126–130.

Fox, Renée C. "The Sting of Death in American Society." *Social Service Review*, March, 1981, pp. 42–59.

Frank, Jerome. "Nuclear Death: An Unprecedented Challenge to Psychiatry and Religion." *The American Journal of Psychiatry*, 141:11, November, 1984, 1343–48.

Frederick, Jerome F. "The Biochemistry of Bereavement: Possible Basis for Chemotherapy?" *Omega*, Vol. 13(4), 1982–83, pp. 295–303.

Frederick, Jerome F. "Grief As a Disease Process." *Omega*, Vol. 7(4), 1976–77, pp. 297–305.

Freese, Arthur. *Help for Your Grief* (NY: Schocken Books, 1977).

French, Laurence A. "Forensic Suicides and Attempted Suicides." *Omega*, Vol. 16(4), 1985–86, pp. 335–346.

Fromm, Erich. "Afterword to *1984*." In George Orwell, *1984* (New York: Harcourt Brace Jovanovich, Inc., 1949), p. 263.

Fromm, Erich. *The Art of Loving* (New York: Harper and Row, 1956). Chapter 1.

Fromm, Erich. *To Have Or To Be* (New York: Bantam Books, 1981), p. 15.

Fuller, Ruth L. and Sally Geis. "Communicating With the Grieving Family." *The Journal of Family Practice*, Vol. 21, No. 2, 1985, pp. 139–144.

Fulton, Robert and David J. Gottesman. "Anticipatory Grief: A Psychosocial Concept Reconsidered." *British Journal of Psychiatry*, Vol. 137, 1980, pp. 45–54.

Fulton, Robert, David J. Gottesman, and Greg M. Owen. "Loss, Social Change, and the Prospect of Mourning." *Death Education*, Vol. 6, 1982, pp. 137–153.

Fulton, Robert et al. (eds.) *Death and Dying: Challenge and Change* (Menlo Park, CA: Addison-Wesley Publishing Company, 1978).

Funch, Donna P., James R. Marshall, and Garren P. Gebhart. "Assessment of a Shorth Scale to Measure Social Support." *Social Science and Medicine*, Vol. 23, No. 3, 1986, pp. 337–344.

Garber, Benjamin. "Mourning in Adolescence: Normal and Pathological." *Adolescent Psychiatry*, 1985(12): 371–387.

Garfield, Charles A. and Gary J. Jenkins. "Stress and Coping of Volunteers Counseling the Dying and Bereaved." *Omega*, Vol. 12(1), 1981–82, pp. 1–13.

Garland, Robert. "Death Without Dishonour: Suicide in the Ancient World." *History Today*, January 1983, pp. 33–37.

Garman, E. Thomas and Charlotte A. Kidd. "Consumer Preparedness, Knowledge, and Opinions about Practices and Regulations of the Funeral Industry." *Death Education*, Vol. 6, 1983, pp. 341–352.

Garrity, Thomas P. "Medical Compliance and the Clinician-Patient Relationship: A Review." *Social Science and Medicine*, Vol. 15E, 1981, pp. 215–222.

Garvey, John. "The Last Solitude: Are We Honest About Death & Dying?" *Commonweal*, May 17, 1985, pp. 298–299.

Gastil, Raymond D. "Homicide and a Regional Culture of Violence." *American Sociological Review*, Vol. 36, June, 1971, pp. 412–427.

Gavey, Richard. *History of Mourning* (London: Jay's, Regent Street W., no date available).

Gaylin, Willard. "In Defense of the Dignity of Being Human." *Hastings Center Report*, Vol. 14, No. 4, August, 1984, pp. 18–22.

Gaylin, Willard et al. "Doctors Must Not Kill." *Journal of the American Medical Association*, Vol. 259, No. 14, April 8, 1988, pp. 2139–2140.

Gelman, David and Jerry Buckley. "Living with Cancer." *Newsweek*, April 8, 1985, pp. 64–71.

Gendron, Bernard. *Technology and the Human Condition* (New York: St. Martin's Press, 1977), Chapters 2–5.

Gill, W. Malcolm. "Subjective Well-Being: Properties of an Instrument for Measuring This (in the Chronically Ill)." *Social Science and Medicine*, Vol. 18, No. 8, 1984, pp. 683–691.

Ginsburg, G. P. "Public Conceptions and Attitudes About Suicide." *Journal of Health & Social Behavior*, Vol. 12, September, 1971, pp. 200–207.

Glaser, Barney and Anselm Strauss. *Awareness of Dying* (New York: Aldine, 1965).

Glaser, Barney and Anselm Strauss. *Discovery of Grounded Theory: Strategies for Qualitative Research* (Chicago: Aldine Publishing Co., 1967).

Glick, Ira O., Robert S. Weiss, and C. Murray Parkes. *The First Year of Bereavement* (New York: John Wiley & Sons, Inc., 1974).

Goffman, Erving. *Asylums* (New York: Doubleday, 1969), Chapter 1.

Goffman, Erving. *Stigma: Notes on the Management of Spoiled Identity* (Englewood Cliffs, NJ: Prentice-Hall, 1963), pp. 1–4.

Goldberg, Helene S. "Funeral and Bereavement Rituals of Kota Indians and Orthodox Jews." *Omega*, Vol. 12(2), 1981–82, pp. 117–128.

Goldberg, Richard J. and Leah O. Cullen. "Factors Important to Psychosocial Adjustment to Cancer: A Review of the Evidence." *Social Science and Medicine*, Vol. 20, No. 8, 1985, pp. 803–807.

Goldhagen, Daniel J. "Healers as Killers" (a book review of Robert J. Lifton's *The Nazi Doctors: Medical Killing and the Psychology of Genocide*). *Commentary*, Vol. 82, No. 6, December, 1986, pp. 77–80.

Goleman, Daniel. "Denial & Hope." *American Health*, December, 1984, pp. 55–61.

Gonda, Thomas Andrew and John Edward Ruark. *Dying Dignified: The Health Professional's Guide to Care* (California: Addison-Wesley Publishing Co., 1984), p. 207.

Gorer, Geoffrey. *Death, Grief, and Mourning* (New York: Doubleday & Co., Inc., 1965).

Gotay, Carolyn C. "Why Me? Attributions and Adjustment by Cancer Patients and Their Mates at Two Stages in the Disease Process." *Social Science and Medicine,* Vol. 20, No. 8, 1985, pp. 825–831.

Gouldner, Alvin. *The Dialectic of Technology and Ideology* (New York: Seabury Press, 1976), p. 7.

Graber, Richard F. "Helping Your Patient Face Death." *Patient Care,* May 30, 1985, pp. 40–69.

Grady, Denise. *American Health,* October, 1988, pp. 57–62.

Gray-Toft, Pamela A. and James G. Anderson. "Sources of Stress in Nursing Terminal Patients in a Hospice." *Omega,* Vol. 17, No. 1, 1986–87, pp. 27–40.

Greenblatt, Milton. "The Grieving Spouse." *American Journal of Psychiatry,* Vol. 135, No. 1, January, 1978, pp. 43–47.

Greer, Ann L. "Medical Technology and Professional Dominance Theory." *Social Science and Medicine,* Vol. 18, No. 10, 1984, pp. 809–817.

Greer, Steven. "The Psychological Dimension in Cancer Treatment." *Social Science and Medicine,* Vol. 18, No. 4, 1984, pp. 345–349.

Greer, Steven. "The Relationship Between Parental Loss and Attempted Suicide: A Control Study." *British Journal of Psychiatry,* Vol. 110, 1964, pp. 698–705.

Greer, Steven and Maggie Watson. "Towards a Psychobiological Model of Cancer: Psychological Considerations." *Social Science and Medicine,* Vol. 20, No. 8, 1985, pp. 773–777.

Grinsell, Leslie. *Barrow, Pyramid and Tomb* (London: Thomas and Hudson, 1975).

Grinspoon, Lester (ed.). "Suicide—Part I." *The Harvard Medical School Mental Health Letter,* Vol. 2, No. 8, February, 1986, pp. 1–4.

Grinspoon, Lester (ed.). "Suicide—Part II." *The Harvard Medical School Mental Health Letter,* Vol. 2, No. 9, March, 1986, pp. 1–4.

Grollman, Earl A. (ed.). *Concerning Death* (Boston: Beacon Press, 1974).

Grollman, Earl A. (ed.). *Explaining Death to Children* (Boston: Beacon Press, 1967).

Gruman, Gerald J. "Ethics of Death and Dying: Historical Perspective." *Omega,* Vol. 9(3), 1978–79, pp. 203–237.

Gyulay, Jo-Eileen. *The Dying Child* (New York: McGraw-Hill Book Co., 1978).

Habenstein, Robert W. and William M. Lamers. *Funeral Customs the World Over* (Milwaukee: Bulfin Printers Inc., 1960).

Habenstein, Robert W. and William M. Lamers. *The History of American Funeral Directing* (Milwaukee: Bulfin Printers Inc., 1955, 1962).

Haim, André. *Adolescent Suicide.* A. M. Sheridan Smith (trans.) (Great Britain: Tavistock Publications, 1974).

Hajal, Fady. "Post-Suicide Grief Work in Family Therapy." *Journal of Marriage and Family Counseling,* April, 1977, pp. 35–42.

Halbwachs, Maurice. *The Causes of Suicide.* Harold Goldblatt (trans.). (London: Routledge & Kegan Paul, 1978).

Hall, John R. "Apocalypse at Jonestown." *Society,* September–October, 1979, pp. 52–61.

Halpern, Werner. "Some Psychiatric Sequelae to Crib Death." *American Journal of Psychiatry,* 129:4, October 1972, pp. 398–402.

Hansen, Leslie C. and Charles A. McAleer. "Terminal Cancer and Suicide: The Health Care Professional's Dilemma." *Omega,* Vol. 14(3), 1983–84, pp. 241–248.

Harrison, Albert A. and Neal E. A. Kroll. "Variations in Death Rates in the Proximity of Christmas: An Opponent Process Interpretation." *Omega*, Vol. 16(3), 1985–86, pp. 181–192.

Hart, Kay. "Stress Encountered by Significant Others of Cancer Patients Receiving Chemotherapy." *Omega*, Vol. 17(2), 1986–87, pp. 151–168.

Hatfield, C. B. et al. "Attitudes About Death, Dying, and Terminal Care: Differences Among Groups at a University Teaching Hospital." *Omega*, Vol. 14(1), 1983–84, pp. 51–61.

Hatton, Corrine L. and Sharon M. Valente. "Bereavement Group for Parents Who Suffered a Suicidal Loss of a Child." *Suicide and Life-Threatening Behavior*, Vol. 11(3), Fall 1981, pp. 141–150.

Hawton, Keith and José Catàlan. *Attempted Suicide* (New York: Oxford University Press, 1982).

Hayslip, Bert, Jr. "The Measurement of Communication Apprehension Regarding the Terminally Ill." *Omega*, Vol. 17(3), 1986–87, pp. 251–262.

Hayslip, Bert, Jr. and Duke Stewart-Bussey. "Locus of Control-Levels of Death Anxiety Relationships." *Omega*, Vol. 17, No. 1, 1986–87, pp. 41–50.

Hayslip, Bert, Jr. and Mary L. Walling. "Impact of Hospice Volunteer Training on Death Anxiety and Locus of Control." *Omega*, Vol. 16(3), 1985–86, pp. 243–254.

Helgeland, John. "The Symbolism of Death in the Later Middle Ages." *Omega*, Vol. 15(2), 1984–85, pp. 145–160.

Henslin, James M. "Problems and Prospects in Studying Significant Others of Suicide." *National Institute of Mental Health Bulletin of Suicidology*, No. 8, Fall 1971.

Herzog, Alfred and H. L. P. Resnik. "A Clinical Study of Parental Response to Adolescent Death by Suicide with Recommendations for Approaching the Survivors." In Proceedings, 4th International Conference for Suicide Prevention, N. L. Farberow (ed.), Delmar Publishing, Los Angeles, 1968, pp. 381–390.

Hessing, Dick J. and Henk Elffers. "Attitude Toward Death, Fear of Being Declared Dead Too Soon, and Donation of Organs After Death." *Omega*, Vol. 17(2), 1986–87, pp. 115–126.

High, Dallas M. "Is 'Natural Death' an Illusion?" *Hastings Center Report*, August, 1978, pp. 37–42.

Hilbert, Richard A. "The Acultural Dimensions of Chronic Pain: Flawed Reality Construction and the Problems of Meaning." *Social Problems*, Vol. 31, No. 4, April, 1984, pp. 365–378.

Hine, Virginia. "Dying at Home: Can Families Cope?" *Omega*, Vol. 10(2), 1979–80, pp. 175–187.

Hinton, John. *Dying* (New York: Penguin Books Ltd., 1972).

Hirsh, Harold. "Should We Enact 'Death with Dignity' Legislation or 'Natural Death' Acts?" *Nursing Homes*, Sept/Oct 1985, pp. 10–16.

Hoffman, Sigal Ironi, and Sidney Strauss. "The Development of Children's Concepts of Death." *Death Studies*, 9:469–482, 1985.

Hoggatt, Loretta and Bernard Spilka. "The Nurse and the Terminally Ill Patient: Some Perspectives and Projected Actions." *Omega*, Vol. 9(3), 1978–79, pp. 255–265.

Hogman, Flora. "Role of Memories in Lives of World War II Orphans." *Journal of the American Academy of Child Psychiatry*, Vol. 24, No. 4, 1985, pp. 390–396.

Holinger, Paul C. "Violent Deaths as a Leading Cause of Mortality: An Epidemiologic Study of Suicide, Homicide, and Accidents." *American Journal of Psychiatry*, Vol. 137:4, April, 1980, pp. 472–476.

Howells, Kevin and David Field. "Fear of Death and Dying Among Medical Students." *Social Science and Medicine*, Vol. 16, 1982, pp. 1421–1424.

Hoyt, Michael F. "Clinical Notes Regarding the Experience of 'Presences' in Mourning." *Omega*, Vol. 11(2), 1980–81, pp. 105–111.

Humphrey, John A. and Stuart Palmer. "Stressful Life Events and Criminal Homicide." *Omega*, Vol. 17(4), 1986–87, pp. 299–308.

Iga, Mamoru. "A Concept of Anomie and Suicide of Japanese College Students." *Life-Threatening Behavior*, Vol. 1, No. 4, Winter 1971, pp. 232–244.

Iga, Mamoru. "Suicide of Japanese Youth." *Suicide and Life-Threatening Behavior*, Vol. 11(1), Spring 1981, pp. 17–30.

Illich, Ivan. *Medical Nemesis: The Expropriation of Death* (New York: Pantheon Books, 1976).

Irion, Paul. *The Funeral: Vestige or Value* (Tennessee: Parthenon Press, 1966).

Ishii, Kan'ichiro. "Adolescent Self-Destructive Behavior and Crisis Intervention in Japan." *Suicide and Life-Threatening Behavior*, Vol. 11(1), Spring 1981, pp. 51–61.

Ishii, Kan'ichiro. "Backgrounds of Higher Suicide Rates among 'Name University' Students: A Retrospective Study of the Past 25 Years." *Suicide and Life-Threatening Behavior*, Vol. 15, No. 1, Spring 1985, pp. 56–68.

Iverson, Kenneth V. and Charles M. Culver. "Using a Cadaver to Practice and Teach: Commentaries." *Hastings Center Report*, June, 1986, pp. 28–29.

Jackson, Edgar N. *The Many Faces of Grief* (Nashville: Parthenon Press, 1977).

Jacobs, Jerry, and Joseph Teicher. "Broken Homes and Social Isolation in Attempted Suicides of Adolescents." *International Journal of Social Psychiatry*, 13, 1967, 139–149.

Jacobs, Selby and Adrian Ostfeld. "An Epidemiological Review of the Mortality of Bereavement." *Psychosomatic Medicine*, Vol. 39, No. 5, September–October, 1977, pp. 344–357.

Jenkins, Richard A. and John C. Cavanaugh. "Examining the Relationship Between the Development of the Concept of Death and Overall Cognitive Development." *Omega*, Vol. 16(3), 1985–86, pp. 193–200.

Johnson, Kathryn K. "Durkheim Revisited: 'Why Do Women Kill Themselves?'" *Suicide and Life-Threatening Behavior*, Vol. 9(3), Fall 1979, pp. 145–153.

Johnson, Margaret. *Beyond Heartache* (Grand Rapids, MI: The Zondervan Corp., 1979).

Jonas, Hans. "The Right to Die." *Hastings Center Report*, August, 1978, pp. 31–33.

Jones, Barbara. *Design for Death* (Indianapolis: The Bobbs-Merrill Co., Inc., 1967).

Jordan, Terry G. *Texas Graveyards* (Austin: University of Texas Press, 1982).

Jorgenson, David E. and Ron C. Neubecker. "Euthanasia: A National Survey of Attitudes Toward Voluntary Termination of Life." *Omega*, Vol. 11(4), 1980–81, pp. 281–291.

Journal of the Medical Association of Georgia. Vol. 69, April, 1980, pp. 265–267.

Joyce, Christopher. "A Time for Grieving." *Psychology Today*, Vol. 18, November, 1984, pp. 42–46.

Kalish, Richard A. "Death and Survivorship: The Final Transition." *Annals of the American Academy of Political and Social Science*, Vol. 464, November, 1982, pp. 163–173.

Kalish, Richard A. *Death, Grief and Caring Relationships* (California: Brooks/Cole, 1985).

Kalish, Richard A. (ed.). *Death and Dying: Views from Many Cultures* (New York: Baywood Publishing Co., Inc., 1980).

Kalish, Richard A. and David K. Reynolds. *Death and Ethnicity: A Psychological Study* (University of Southern California, Ethel Percy Andrus Gerontology Center, 1976).

Kalish, Richard A., David K. Reynolds, and Norman L. Farberow. "Community Attitudes Toward Suicide." *Community Mental Health Journal*, Vol. 10(3), 1974, pp. 301–308.

Kane, Anne C. and John D. Hogan. "Death Anxiety in Physicians: Defensive Style, Medical Specialty, and Exposure to Death." *Omega*, Vol. 16(1), 1985–86, pp. 11–22.

Kane, Robert L. et al. "Terminal Care." *The Lancet*, April 21, 1984, pp. 890–894.

Karcher, Charles J. and Leonard L. Linden. "Is Work Conducive to Self-Destruction?" *Suicide and Life-Threatening Behavior*, Vol. 12(3), Fall 1982, pp. 151–175.

Karoly, Paul. "Cognitive Assessment in Behavioral Medicine." *Clinical Psychology Review*, Vol. 2, 1982, pp. 421–434.

Kass, Leon R. "The Case for Mortality." *American Scholar*, Vol. 52, Spring 1983, pp. 173–191.

Kastenbaum, Robert. "Book Review: *Suicide. The Will to Live vs. the Will to Die*, edited by Norman Linzer." *Omega*, Vol. 17(3), 1986–87, p. 277.

Kastenbaum, Robert. *Death, Society, and Human Experience* (St. Louis: The C. V. Mosby Co., 1981).

Kastenbaum, Robert. "Toward Standards of Care for the Terminally Ill." *Omega*, Vol. 7(3), 1976, pp. 191–193.

Katz, Jay. *The Silent World of Doctor and Patient* (New York: The Free Press, 1984).

Katz, Jay. "Why Doctors Don't Disclose Uncertainty." *Hastings Center Report*, Vol. 14, Number 1, February, 1984, pp. 35–44.

Kavanaugh, Robert E. *Facing Death* (New York: Penguin Books, 1972).

Kearl, Michael C. "Death as a Measure of Life: A Research Note on the Kastenbaum-Spilka Strategy of Obituary Analyses." *Omega*, Vol. 17, No. 1, 1986–87, pp. 65–78.

Kearl, Michael C. and Richard Harris. "Individualism and the Emerging 'Modern' Ideology of Death." *Omega*, Vol. 12(3), 1981–82, pp. 269–280.

Kearl, Michael C. and Anoel Rinaldi. "The Political Uses of the Dead as Symbols in Contemporary Civil Religions." *Social Forces*, Vol. 61:3, March, 1983, pp. 693–708.

Keith, Pat M. "Perceptions of Time Remaining and Distance from Death." *Omega*, Vol. 12(4), 1981–82, pp. 307–318.

Kellehear, Allan. "Are We a Death Defying Society? A Sociological Review." *Social Science and Medicine*, Vol. 18(9), 1984, p. 714.

Kenney, Edwin J., Jr. "Death's Other Kingdom." *Commonweal*, Vol. 109, November 19, 1982, pp. 627–632.

Kerr, Kathleen. "Reporting the Case of Baby Jane Doe." *Hastings Center Report*, Vol. 14, No. 4, August, 1984, pp. 7–9.

Killilea, Alfred. "Death and Social Consciousness." *Omega*, Vol. 11(30), 1980, p. 199.

Killilea, Alfred. "Death Consciousness and Social Consciousness: A Critique of Ernest Becker and Jacques Choron on Denying Death." *Omega*, Vol. 11(3), 1980–81, pp. 185–200.

Killilea, Alfred. "Nuclearism and the Denial of Death." *Death Studies*, 9:253–265, 1985.

Kincade, Jean E. "Attitudes of Physicians, Housestaff, and Nurses on Care for the Terminally Ill." *Omega*, Vol. 13(4), 1982–83, pp. 333–344.

Kirk, Alton R. and Robert A. Zucker. "Some Sociopsychological Factors in Attempted Suicide Among Urban Black Males." *Suicide and Life-Threatening Behavior*, Vol. 9(2), Summer, 1979, pp. 76–86.

Klapp, Orrin. *Collective Search for Identity* (New York: Holt, Rinehart and Winston, 1969), Chapter 1.

Klass, Dennis. "Bereaved Parents and the Compassionate Friends: Affiliation and Healing." *Omega*, Vol. 15(4), 1984–85, pp. 353–373.

Klass, Dennis. "Elizabeth Kübler-Ross and the Tradition of the Private Sphere: An Analysis of Symbols." *Omega*, Vol. 12(3), 1981–82, pp. 241–267.

Klass, Dennis. "Marriage and Divorce Among Bereaved Parents in a Self-Help Group." *Omega*, Vol. 17(3), 1986–87, pp. 237–250.

Klass, Dennis. "Professional Roles in a Self-Help Group for the Bereaved." *Omega*, Vol. 13(4), 1982–83, pp. 361–375.

Klass, Dennis. "To Work My Mind, When Body's Work's Expired, being a review of Dan Leviton's 'Thanatological Theory and My Dying Father.'" *Omega*, Vol. 17(2), 1986–87, pp. 145–150.

Klass, Dennis and Richard A. Hutch. "Elizabeth Kübler-Ross as a Religious Leader." *Omega*, Vol. 16(2), 1985–86, pp. 89–110.

Klass, Dennis and Beth Shinners. "Professional Roles in a Self-Help Group for the Bereaved." *Omega*, Vol. 13(4), 1982–83, pp. 361–375.

Kleiman, Dena. "In the Intensive-Care Unit, Doctors Find the Hippocratic Oath Is Redefined." *The New York Times*, January 16, 1985, Section Y, p. 11.

Klingman, Avigdor. "Simulation and Simulation Games as a Strategy for Death Education." *Death Education*, 7:339–352, 1983.

Kobasa, Suzanne C., Salvatore R. Maddi, and Mark C. Puccetti. "Personality and Exercise as Buffers in the Stress-Illness Relationship." *Journal of Behavioral Medicine*, Vol. 5, No. 4, 1982, pp. 391–402.

Kohn, Jane Burgess and Willard K. Kohn. *The Widower* (Boston: Beacon Press, 1978).

Kohut, Jeraldine Marasco and Sylvester Kohut, Jr. *Hospice: Caring for the Terminally Ill* (Springfield, IL: Charles C. Thomas, Publishers, 1984).

Kolb, Lawrence C. "Return of the Repressed: Delayed Stress Reaction to War." *Journal of The American Academy of Psychoanalysis*, Vol. 11, No. 4, 1983, pp. 531–545.

Kollar, Nathan. "Visions of the End of the World: Past and Present." *Death Education*, 7:9–24, 1983.

Kotarba, Joseph A. "Perceptions of Death, Belief Systems and the Process of Coping with Chronic Pain." *Social Science and Medicine*, Vol. 17, No. 10, 1983, pp. 681–689.

Kotarba, Joseph A. and John V. Seidel. "Managing the Problem Pain Patient: Compliance or Social Control?" *Social Science and Medicine*, Vol. 19, No. 12, 1984, pp. 1393–1400.

Kotch, Jonathan B. and Susan R. Cohen. "SIDS Counselors' Reports of Own and Parents' Reactions to Reviewing the Autopsy Report." *Omega*, Vol. 16(2), 1985–86, pp. 129–140.

Kramer, Bernard, Michael Kalick, and Michael Milburn. "Attitudes Toward Nuclear Weapons and Nuclear War: 1945–1982." *Journal of Social Issues*, Vol. 39, No. 1, 1983, pp. 7–24.

Krell, Robert. "Child Survivors of the Holocaust: 40 Years Later." *Journal of the American Academy of Child Psychiatry*, Vol. 24, No. 4, 1985, pp. 378–380.

Krell, Robert. "Therapeutic Value of Documenting Child Survivors." *Journal of the American Academy of Child Psychiatry*, Vol. 24, No. 4, 1985, pp. 397–400.

Kübler-Ross, Elisabeth. *To Live Until We Say Good-Bye* (Englewood Cliffs, NJ: Prentice-Hall, Inc. 1978).

Kübler-Ross, Elisabeth (interview). *Playboy*, Vol. 28(5), 1981, pp. 60–106.

Kuhse, Helga. "Extraordinary Means and the Intentional Termination of Life." *Social Science and Medicine*, Vol. 15F, pp. 117–121, 1981.

Kunkle, Sheryl and John A. Humphrey. "Murder of the Elderly: An Analysis of Increased Vulnerability." *Omega*, Vol. 13(1), 1982–83, pp. 27–33.

Kutner, Nancy G. "Issues in the Application of High Cost Medical Technology: The Case of Organ Transplantation." *Journal of Health and Social Behavior*, Vol. 28, No. 1, March, 1987, pp. 23–36.

Kuttner, Robert E. "The Genocidal Mentality: Philip II of Spain and Sultan Abdul Hamid II." *Omega*, Vol. 16(1), 1985–86, pp. 35–42.

Labus, Janet G. and Faye H. Dambrot. "A Comparative Study of Terminally Ill Hospice and Hospital Patients." *Omega*, Vol. 16(3), 1985–86, pp. 225–232.

Lamm, Richard D. "When 'Miracle Cures' Don't Cure." *The New Republic*, August 27, 1984, pp. 20–23.

Landsberg, Paul-Louis. *The Experience of Death* (New York: Arno Press, 1977).

Lane, Roger. "Suicide and the City." *Society*, Vol. 17, January/February, 1980, pp. 74–82.

Lasaga, Jose I. "Death in Jonestown: Techniques of Political Control by a Paranoid Leader." *Suicide and Life-Threatening Behavior*, Vol. 10(4), Winter, 1980, pp. 210–213.

Lattanzi, Marcia and Mary E. Hale. "Giving Grief Words: Writing During Bereavement." *Omega*, Vol. 15(1), 1984–85, pp. 45–52.

Lauter, H. and J. E. Meyer. "Active Euthanasia Without Consent: Historical Comments on a Current Debate." *Death Education*, Vol. 8, 1984, pp. 89–98.

Lauter, H. and J. E. Meyer. "Mercy Killing without Consent. Historical Comments on a Controversial Issue." *Acta Psychiat. Scand.*, Munksgaard, Copenhagen (1982) 65, pp. 134–141.

Leiderman, Deborah and Jean-Anne Grisso. "The Gomer Phenomenon." *Journal of Health and Social Behavior*, Vol. 25, September 1985, p. 222.

Leming, Michael R. and George E. Dickinson. *Understanding Dying, Death, and Bereavement* (New York: Holt, Rinehart and Winston, 1985).

LeShan, L. "Psychotherapy and the Dying Patient." In L. Pearson (ed.), *Death and Dying* (Cleveland: Case Western Reserve University Press, 1969).

Lester, David. "Depression and Fear of Death in a Normal Population." *Psychological Reports*, Vol. 56, 1985, p. 882.

Lester, David. "Regional Variation in Suicide and Homicide." *Suicide and Life-Threatening Behavior*, Vol. 15(2), Summer 1985, pp. 110–116.

Lester, David. "Southern Subculture, Personal Violence (Suicide and Homicide), and Firearms." *Omega*, Vol. 17(2), 1986–87, pp. 183–186.

Lester, David. *Why People Kill Themselves* (Springfield, IL: Charles C. Thomas Publishers, 1983).

Lester, David and Mary E. Murrell. "The Preventive Effect of Strict Gun Control Laws on Suicide and Homicide." *Suicide and Life-Threatening Behavior*, Vol. 12(3), Fall 1982, pp. 131–140.

Levine, Sol. "The Changing Terrains in Medical Sociology: Emergent Concern with Quality of Life." *Journal of Health and Social Behavior*, Vol. 28, No. 1, March, 1987, pp. 1–6.

Levinson, A-J Rock. "Case Studies: 'If I Have AIDS, Then Let Me Die Now!'—Commentary." *Hastings Center Report*, Vol. 14, No. 1, February, 1984, p. 25.

Leviton, Dan. "Thanatological Theory and My Dying Father." *Omega*, Vol. 17(2), 1986–87, pp. 127–144.

Lewis, Jeremy. "Death as a Party." *New Statesman*, Vol. 107, February 3, 1984, p. 32.

Liberman, Marla B. et al. "Development of a Behavior Rating Scale for Doctor-Patient Interactions and Its Implications for the Study of Death Anxiety." *Omega*, Vol. 14(3), 1983–84, pp. 231–239.

Lichtman, Helen. "Parental Communication of Holocaust Experiences and Personality Characteristics Among Second-Generation Survivors." *Journal of Clinical Psychology*, Vol. 40, No. 4, July, 1984, pp. 914–924.

Limbacher, Mary and George Domino. "Attitudes Toward Suicide Among Attempters, Contemplators, and Nonattempters." *Omega*, Vol. 16(4), 1985–86, pp. 325–334.

Lindemann, Erich. "Symptomatology and Management of Acute Grief." *American Journal of Psychiatry*, Vol. 101, 1944, pp. 141–148.

Linn, Margaret W., Bernard S. Linn, and Rachel Harris. "Effects of Counsel-ing for Late State Cancer Patients." *Cancer*, Vol. 49, March, 1982, pp. 1048–1055.

Litman, Robert E. "Hospital Suicides: Lawsuits and Standards." *Suicide and Life-Threatening Behavior*, Vol. 12(4), Winter 1982, pp. 212–233.

Litman, Robert E. "When Patients Commit Suicide." *American Journal of Psychotherapy*, Vol. 19, 1965, pp. 570–576.

Lofland, Lyn. *The Craft of Dying* (Beverly Hills: Sage Publications, 1978), p. 103.

Lonetto, Richard. *Children's Conceptions of Death* (New York: Springer Publishing Co., 1980).

Lund, Dale A. et al. "Identifying Elderly with Coping Difficulties after Two Years of Bereavement." *Omega*, Vol. 16(3), 1985–86, pp. 213–224.

Lundberg, George D. "'It's Over, Debbie' and the Euthanasia Debate." *Journal of the American Medical Association*, Vol. 259, No. 14, April 8, 1988, pp. 2142–2143.

Lundberg, Jean Pancner and Derek Lloyd Lundberg. *Teenage Suicide in America: A Handbook for Understanding* (Dayton: Pamphlet Publications, 1985).

Lunt, Barry and Richard Hillier. "Terminal Care: Present Services and Future Priorities." *British Medical Journal*, Vol. 283, August 29, 1981, pp. 595–597.

Lynn, Joanne. "Food and Water Can Be Withheld from Dying Patients: The Very Different Situations of Claire Conroy and Karen Quinlan." *Death Education*, Vol. 8, 1984, pp. 271–275.

Macon, Lillian B. "Help for Bereaved Parents." *Social Casework*, Vol. 60, November, 1979, pp. 558–561.

Maddison, David and Agnes Viola. "The Health of Widows in the Year Following Bereavement." *Journal of Psychosomatic Research*, Vol. 12, 1968, pp. 297–306.

Maguire, Peter. "Improving the Detection of Psychiatric Problems in Cancer Patients." *Social Science and Medicine*, Vol. 20, No. 8, 1985, pp. 819–823.

Maher, Ellen L. "Anomic Aspects of Recovery from Cancer." *Social Science and Medicine*, Vol. 16, 1982, pp. 907–912.

Malkinson, Ruth. "Helping and Being Helped: The Support Paradox." *Death Studies*, Vol. 11, 1987, pp. 205–219.

Maris, Ronald W. "The Adolescent Suicide Problem." *Suicide and Life-Threatening Behavior*, Vol. 15(2), Summer 1985, pp. 91–109.

Maris, Ronald W. "Deviance as Therapy: The Paradox of the Self-Destructive Female." *Journal of Health and Social Behavior*, Vol. 12, June, 1971, pp. 113–124.

Maris, Ronald W. *Pathways to Suicide* (Baltimore: The Johns Hopkins University Press, 1981).

Maris, Ronald W. *Special Forces in Urban Suicide* (Homewood, IL: The Dorsey Press, 1969).

Marks, Alan. "Race and Sex Differences and Fear of Dying: A Test of Two Hypotheses—High Risk or Social Loss?" *Omega*, Vol. 17(3), 1986–87, pp. 229–236.

Marks, Alan and Thomas Abernathy. "Toward a Sociocultural Perspective on Means of Self-Destruction." *Life-Threatening Behavior*, Vol. 4(1), Spring 1974, pp. 3–17.

Marks, Alan and Carolyn Riley. "Test of Goffman's Hypothesis of Familiarity and Deviance: Attempted Suicide and Tolerance of Deviant Behavior." *Psychological Reports*, Vol. 39, 1976, pp. 420–422.

Markusen, Eric and John Harris. "The Role of Education in Preventing Nuclear War." *Harvard Educational Review*, Vol. 54, No. 3, August 1984.

Marshall, George N. *Facing Death and Grief* (New York: Prometheus Books, 1981).

Marshall, James R., William Burnett, and John Brasure. "On Precipitating Factors: Cancer as a Cause of Suicide." *Suicide and Life-Threatening Behavior*, Vol. 13(1), Spring 1983, pp. 15–27.

Marshall, Karol A. "When a Patient Commits Suicide." *Suicide and Life-Threatening Behavior*, Vol. 10(1), Spring 1980, pp. 29–40.

Martin, Thomas O. "Death Anxiety and Social Desirability Among Nurses." *Omega*, Vol. 13(1), 1982–83, pp. 51–58.

Masamba, Jean and Richard A. Kalish. "Death and Bereavement: The Role of the Black Church." *Omega*, Vol. 7(1), 1976, pp. 23–34.

Maslow, Abraham. *Toward a Psychology of Being* (New York: Van Nostrand, 1968).

Matter, Darryl E. and Roxana M. Matter. "Developmental Sequences in Children's Understanding of Death with Implications for Counselors." *Elementary School Guidance and Counseling*, Vol. 17(2), December, 1982, pp. 112–118.

Mauksch, Hans O. "Ideology, Interaction and Patient Care in Hospitals." *Social Science and Medicine*, Vol. 7, 1973, pp. 817–830.

May, William F. "Religious Justifications for Donating Body Parts." *Hastings Center Report*, Vol. 15, No. 1, February, 1985, pp. 38–42.

McClain, Paula D. "Black Females and Lethal Violence: Has Time Changed the Circumstances Under Which They Kill?" *Omega*, Vol. 13(1), 1982–83, pp. 13–25.

McCown, Darlene E. "Funeral Attendance, Cremation, and Young Siblings." *Death Education*, Vol. 8, 1984, pp. 349–363.

McGee, Marsha. "Faith, Fantasy, and Flowers: A Content Analysis of the American Sympathy Card." *Omega*, Vol. 11(1), 1980–81, pp. 25–35.

McGuire, Donald J. and Margot Ely. "Childhood Suicide." *Child Welfare*, Vol. LXIII, Number 1, January–February, 1984, pp. 17–26.

McIntosh, John L. "Survivors of Suicide: A Comprehensive Bibliography." *Omega*, Vol. 16(4), 1985–86, pp. 355–370.

McIntosh, John L., Richard W. Hubbard, and John F. Santos. "Suicide Facts and Myths: A Study of Prevalence." *Death Studies*, Vol. 9, 1985, pp. 267–281.

McKeown, Thomas. "Determinants of Health." *Human Nature*, April 1978, pp. 6–14.

McKitrick, Daniel. "Counseling Dying Clients." *Omega*, Vol. 12(2), 1981–82, pp. 165–187.

McNeil, Joan N. "Young Mothers' Communication About Death with Their Children." *Death Education*, Vol. 6, 1983, pp. 323–339.

McQueen, David V. and Johannes Siegrist. "Social Factors in the Etiology of Chronic Disease: An Overview." *Social Science and Medicine*, Vol. 16, 1982, pp. 353–367.

Mechanic, David. *Medical Sociology* (New York: The Free Press, 1980), p. 171.

Meier, Diane E. and Christine K. Cassell. "Euthanasia in Old Age: A Case Study and Ethical Analysis." *Journal of the American Geriatrics Society*, Vol. 31, No. 5, May, 1983, pp. 294–298.

Mester, Roberto and Yoram Hazan. "Empathy and Death Expressions in a Therapy Group of Parents of Israeli Soldiers." *International Journal of Group Psychotherapy*, Vol. 34(4), October, 1984, pp. 627–637.

Meyrowitz, Joshua. *No Sense of Place: The Impact of Electronic Media on Social Behavior* (New York: Oxford University Press, 1985), pp. 116–118.

Michalowski, Raymond J. "The Social Meanings of Violent Death." *Omega*, Vol. 7(1), 1976, pp. 83–93.

Miles, Margaret S. and Alice S. Demi. "Toward the Development of a Theory of Bereavement Guilt: Sources of Guilt in Bereaved Parents." *Omega*, Vol. 14(4), 1983–84, pp. 299–314.

Mills, C. Wright. *The Sociological Imagination* (New York: Oxford University Press, 1959).

Milton, Isabel C. "Concerns of Final Year Baccalaureate Students About Nursing Dying Patients." *Journal of Nursing Education*, Vol. 23, No. 7, September, 1984, pp. 298–301.

Minear, Julianne D. and Lorelei R. Brush. "The Correlations of Attitudes toward Suicide with Death Anxiety, Religiosity, and Personal Closeness to Suicide." *Omega*, Vol. 11(4), 1980–81, pp. 317–324.

Mischel, Ellis. "Personal Reflections on the Holocaust and Holocaust Survivors." *The American Journal of Psychoanalysis*, Vol. 39, No. 4, 1979, pp. 369–376.

Mitchell, Marjorie Editha. *The Child's Attitude to Death* (New York: Schocken Books, 1967).

Mitford, Jessica. *The American Way of Death* (New York: Simon and Schuster, 1963).

Molin, Ronald S. "Covert Suicide and Families of Adolescents." *Adolescence*, Vol. XXI, No. 81, Spring 1986, pp. 177–184.

Moller, David Wendell. "On the Value of suffering in the Shadow of Death." *Loss, Grief, and Care: A Journal of Professional Practice*. Vol. 1, 1986, pp. 127–136.

Momeyer, Richard W. "Fearing Death and Caring for the Dying." *Omega*, Vol. 16(1), 1985–86, pp. 1–10.

Mor, Vincent and Jeffrey Hiris. "Determinants of Site of Death Among Hospice Cancer Patients." *Journal of Health and Social Behavior*, Vol. 24, December, 1983, pp. 375–385.

Morgan, Leslie A. "Economic Change at Mid-Life Widowhood: A Longitudinal Analysis." *Journal of Marriage and the Family*. November, 1981, pp. 899–907.

Morgan, Lucy G. "On Drinking the Hemlock." *Hastings Center Report*, Vol. 1, No. 3, December, 1971, pp. 4–5.

Morgenstern, Hal et al. "The Impact of a Psychosocial Support Program on Survival with Breast Cancer: The Importance of Selection Bias in Program Evaluation." *Journal of Chronic Disease*, Vol. 37, No. 4, 1984, pp. 273–282.

Moriarty, David M. (ed.). *The Loss of Loved Ones* (Springfield, IL: Chas. C. Thomas, Publishers, 1967).

Morley, John. *Death, Heaven and the Victorians* (Pittsburgh: University of Pittsburgh Press, 1971).

Morris, Tina, Susan Blake, and Miranda Buckley. "Development of a Method for Rating Cognitive Responses to a Diagnosis of Cancer." *Social Science and Medicine*, Vol. 20, No. 8, 1985, pp. 795–802.

Moskovitz, Sarah. "Longitudinal Follow-Up of Child Survivors of the Holocaust." *Journal of the American Academy of Child Psychiatry*, Vol. 24, No. 4, 1985, pp. 401–407.

Moss, Miriam S., Emerson L. Lesher, and Sidney Z. Moss. "Impact of the Death of an Adult Child on Elderly Parents: Some Observations." *Omega*, Vol. 17(3), 1986–87, pp. 209–218.

Moss, Miriam S. and Sidney Z. Moss. "The Impact of Parental Death on Middle-Aged Children." *Omega*, Vol. 14(1), 1983–84, pp. 65–75.

Moss, Miriam S. and Sidney Z. Moss. "Some Aspects of the Elderly Widow(er)'s Persistent Tie with the Deceased Spouse." *Omega*, Vol. 15(3), 1984–85, pp. 195–206.

Mount, Eric, Jr. "Individualism and Our Fears of Death." *Death Education*, Vol. 7, 1983, pp. 25–31.

Mullis, Marcia R. "Vietnam: The Human Fallout." *Journal of Psychosocial Nursing*, Vol. 22, No. 2, 1984, pp. 27–31.

Mumma, Christina M. and Jeanne Q. Benoliel. "Care, Cure, and Hospital Dying Trajectories." *Omega*, Vol. 15(3), 1984–85, pp. 275–287.

Munley, Anne. *The Hospice Alternative* (New York: Basic Books, Inc., Publishers, 1983).

Murphy, Patricia A. "Parental Death in Childhood and Loneliness in Young Adults." *Omega*, Vol. 17(3), 1986–87, pp. 219–228.

Murphy, Shirley A. and Barbara J. Stewart. "Linked Pairs of Bereaved Persons: A Method for Increasing the Sample Size in a Study of Bereavement." *Omega*, Vol. 16(2), 1985–86, pp. 141–154.

Murray, Thomas H., John D. Arras, Joseph F. Kett, and Leslie A. Fiedler. "On the Care of Imperiled Newborns." *Hastings Center Report*, Vol. 14, No. 2, April, 1984, pp. 24–42.

Nadler, Arie, Sophie Kav-Venaki, and Beny Gleitman. "Transgenerational Effects of the Holocaust: Externalization of Aggression in Second Generation Holocaust Survivors." *Journal of Consulting and Clinical Psychology*, Vol. 53, No. 3, 1985, pp. 365–369.

Nagy, Maria. "The Child's Theories Concerning Death." *The Journal of Genetic Psychology*, Vol. 73, 1948, pp. 3–27.

Naisbett, John. *Megatrends* (New York: Warren Books, 1984), p. 36.

Najman, Jackob and Sol Levine. "Evaluating the Impact of Medical Care and Technologies on the Quality of Life: A Review and Critique." *Social Science & Medicine*, Vol. 15F, 1981, pp. 107–115.

Natterson, Joseph M. and Alfred G. Knudson, Jr. "Observations Concerning Fear of Death in Fatally Ill Children and Their Mothers." *Psychosomatic Medicine*, Vol. XXII, No. 6, 1960, pp. 456–465.

Neimeyer, Robert. "Assessing Personal Meanings of Death." *Death Studies*, Vol. 13, No. 3, 1989, pp. 227–245.

Nelson, Franklyn L. "Suicide: Issues of Prevention, Intervention, and Facilitation." *Journal of Clinical Psychology*, Vol. 40, No. 6, November, 1984, pp. 1328–1333.

Nettler, Gwynn. *Killing One Another* (Cincinnati: Anderson Publishing Co., 1982).

Neuringer, Charles and Dan J. Lettieri. "Cognition, Attitude, and Affect in Suicidal Individuals." *Life-Threatening Behavior*, Vol. 1, No. 2, Summer 1971, pp. 106–124.

Newcomb, Michael. "Nuclear Attitudes and Reactions: Associations with Depression, Drug Use, and Quality of Life." *Journal of Personality and Social Psychology*, 1986, Vol. 50, No. 5, 906–920.

Newman, A. "Planetary Death." *Death Studies*, Vol. 11, 1987, pp. 131–135.

Nisbet, Robert. *The Quest for Community* (New York: Oxford University Press, 1981).

Nisbet, Robert. *Sociology as an Art Form* (New York: Oxford University Press, 1977), p. 42.

Noyes, Russell Jr. "The Human Experience of Death or, What Can We Learn from Near-Death Experiences?" *Omega*, Vol. 13(3), 1982–83, pp. 251–259.

O'Brien, Mary E. "An Identification of the Needs of Family Members of Terminally Ill Patients in a Hospital Setting." *Military Medicine*, Vol. 148, September, 1983, pp. 712–715.

O'Hare, Joseph A. "Of Many Things." *America*, Vol. 147, No. 8, September 25, 1982.

Oken, D. "What to Tell Cancer Patients." *The Journal of the American Medical Association*, 175, 1961, pp. 86–94.

Orbach, Israel. "Personality Characteristics, Life Circumstances, and Dynamics of Suicidal Children." *Death Education*, Vol. 8 (Supplement), 1984, pp. 37–52.

Orbach, Israel, Seymour Feshbach, Gabrielle Carlson, and Leah Ellenberg. "Attitudes Toward Life and Death in Suicidal, Normal, and Chronically Ill Children: An Extended Replication." *Journal of Counseling and Clinical Psychology*, Vol. 52, No. 6, 1984, pp. 1020–1027.

Orbach, Israel, Yigal Gross, Hananyah Glaubman, and Devora Berman. "Children's Perception of Death in Humans and Animals as a Function of Age, Anxiety and Cognitive Ability." *Journal of Child Psychol. Psychiat.*, Vol. 26, No. 3, 1985, pp. 453–463.

"Ordinary Care." *America*, April 30, 1983, p. 330.

O'Shea, John M. *Mortuary Variability: An Archaeological Investigation* (Orlando, FL: Academic Press, Inc., 1984).

Ostroff, Robert B. et al. "Adolescent Suicides Modeled After Television Movie." *American Journal of Psychiatry*, Vol. 142, No. 8, August, 1985, p. 989.

Owen, David. "Rest in Pieces." *Harper's*, Vol. 226, June, 1983, pp. 70–75.

Owen, Greg, Robert Fulton, and Eric Markusen. "Death at a Distance: A Study of Family Survivors." *Omega*, Vol. 13(3), 1982–83, pp. 191–225.

Page, Margot. "Go Gently . . ." *American Health*, October, 1988, pp. 60–61.

Palmer, Stuart. *The Violent Society* (New Haven, CT: College & University Press, 1972).

Parkes, Colin Murray, B. Benjamin, and R. G. Fitzgerald. "Broken Heart: A Statistical Study of Increased Mortality among Widowers." *British Medical Journal*, Vol. 1, March 22, 1969, pp. 740–743.

Parkes, Colin Murray and Robert S. Weiss. *Recovery from Bereavement* (New York: Basic Books, Inc., 1983).

Parson, Erwin Randolph. "Life After Death: Vietnam Veteran's Struggle for Meaning and Recovery." *Death Studies*, Vol. 10, 1986, pp. 11–26.

Parsons. *The Social System* (New York: The Free Press, 1964), p. 436.

Passuth, Patricia M. and Fay L. Cook. "Effects of Television Viewing on Knowledge and Attitudes about Older Adults: A Critical Reexamination." *The Gerontologist*, Vol. 25, No. 1, 1985, pp. 69–77.

Paulay, Dorothy. "Slow Death: One Survivor's Experience." *Omega*, Vol. 8(2), 1977–78, pp. 173–179.

Peach, Mary Rae and Dennis Klass. "Special Issues in the Grief of Parents of Murdered Children." *Death Studies*, Vol. 11, 1987, pp. 81–88.

Pearlman, Robert A. and Albert Jonsen. "The Use of Quality-of-Life Considerations in Medical Decision Making." *Journal of the American Geriatrics Society*, Vol. 33, No. 5, 1985, pp. 344–352.

Pearson, Leonard (ed.). *Death and Dying* (Cleveland: Case Western Reserve University Press, 1969).

Peck, Dennis L. "Complete Suicides: Correlates of Choice of Method." *Omega*, Vol. 16(4), 1985–86, pp. 309–324.

Peck, Dennis L. "The Last Moments of Life: Learning to Cope." *Deviant Behavior*, Vol. 4, 1983, pp. 313–332.

Peck, Dennis L. "'Official Documentation' of the Black Suicide Experience." *Omega*, Vol. 14(1), 1983–84, pp. 21–31.

Peck, Michael. "Youth Suicide." *Death Education*, Vol. 6, 1982, pp. 29–47.

Peck, Michael et al. (eds.). *Youth Suicide* (New York: Springer Publishing Co., 1985).

Pellegrino, Edmund. "The Right to Die—Should a Doctor Decide?" an interview. *U.S. News & World Report*, November 3, 1975, p. 53ff.

Pennebaker, James W. and Robin C. O'Heeron. "Confiding in Others and Illness Rate Among Spouses of Suicide and Accidental-Death Victims." *Journal of Abnormal Psychology*, Vol. 93, No. 4, 1984, pp. 473–476.

Peretti, Peter O. and Cedric Wilson. "Contemplated Suicide Among Voluntary and Involuntary Retirees." *Omega*, Vol. 9(2), 1978–79, pp. 193–201.

Perlin, Seymour (ed.). *A Handbook for the Study of Suicide* (New York: Oxford University Press, 1975).

Pernick, Martin S. "The Calculus of Suffering in Nineteenth-Century Surgery." *Hastings Center Report*, Vol. 13, No. 2, April, 1983, pp. 26–36.

Perris, C., S. Holmgren, L. Von Knorring, and H. Perris. "Parental Loss by Death in the Early Childhood of Depressed Patients and of their Healthy Siblings." *British Journal of Psychiatry*, Vol. 148, 1986, pp. 165–169.

Perrollaz, LaVerne and Margaret Mollica. "Public Knowledge of Hospice Care." *Nursing Outlook*, January, 1981, pp. 46–48.

Peters-Golden, Holly. "Breast Cancer: Varied Perceptions of Social Support in the Illness Experience." *Social Science and Medicine*, Vol. 16, 1982, pp. 483–491.

Peterson, James A. and Michael L. Briley. *Widows and Widowhood: A Creative Approach to Being Alone* (Chicago: Follett Publishing Co., 1977).

Pettingale, K. W. "Towards a Psychobiological Model of Cancer: Biological Considerations." *Social Science and Medicine*, Vol. 20, No. 8, 1985, pp. 779–787.

Pfeffer, Cynthia R. "Parental Suicide: An Organizing Event in the Development of Latency Age Children." *Suicide and Life-Threatening Behavior*, Vol. 11(1), Spring 1981, pp. 43–50.

Pfeffer, Cynthia R. et al. "Suicidal Behavior in Normal School Children: A Comparison with Child Psychiatric Inpatients." *Journal of the American Academy of Child Psychiatry*, Vol. 23, Number 4, 1984, pp. 416–423.

Piccione, Joseph J. *Last Rights: Treatment and Care Issues in Medical Ethics* (Washington, DC: Free Congress Research & Education Foundation, 1984).

Pine, Vanderlyn R. et al. (eds.). *Acute Grief and the Funeral* (Springfield, IL: Chas. C. Thomas, Publishers, 1976).

Pohlman, Joanne C. "Illness and Death of a Peer in a Group of Three-Year-Olds." *Death Education*, Vol. 8, 1984, pp. 123–136.

Pohlmeier, Hermann. "Suicide and Euthanasia—Special Types of Partner Relationships." *Suicide and Life-Threatening Behavior*, Vol. 15(2), Summer 1985, pp. 117–123.

Polito, Vincent. *An Historical Study of the American Funeral* (Boston: Suffolk University, 1964).

Pope, Whitney. *Durkheim's Suicide: A Classic Analyzed* (Chicago: The University of Chicago Press, 1976).

Porterfield, Austin L. *Cultures of Violence* (Fort Worth, TX: Les Potishman Foundation, 1965).

Postman, Neil. *The Disappearance of Childhood* (New York: Delacorte Press, 1983), p. 27.

Prentice, Ann E. *Suicide* (Metuchen, NJ: The Scarecrow Press Inc., 1974).

Prince, Robert M. "Second Generation Effects of Historical Trauma." *Psychoanalytic Review*, Vol. 72, No. 1, Spring 1985, pp. 9–29.

Puckle, Bertram S. *Funeral Customs: Their Origin and Development* (London: T. Werner Laurie Ltd., 1926).

Putnam, Sandra T., et al. "Home as a Place to Die." *American Journal of Nursing*, August, 1980, pp. 1451–1453.

Rachels, James. "Barney Clark's Key." *Hastings Center Report*, Vol. 13, No. 2, April, 1983, pp. 17–19.

Reynolds, David K. and Norman Farberow. *Suicide Inside and Out* (Berkley and Los Angeles: University of California Press, 1976).

Reynolds, David K., Richard A. Kalish, and Norman L. Farberow. "A Cross-Ethnic Study of Suicide Attitudes and Expectations in the United States." In *Suicide in Different Cultures*, Farberow, Normal L. (ed.) (Baltimore: University Park Press, 1975), pp. 35–50.

Rhodes, Colbert and Clyde B. Vedder. *An Introduction to Thanatology* (Springfield, IL: Chas. C. Thomas, Publishers, 1983).

Rigdon, Michael A. "Death Threat Before and After Attempted Suicide: A Clinical Investigation." *Death Education*, Vol. 7, (2–3), Summer, 1983, pp. 195–209.

Rodabough, Tillman. "Funeral Roles: Ritualized Expectations." *Omega*, Vol. 12(3), 1981–82, pp. 227–240.

Rofe, Yacov and Isaac Lewin. "Daydreaming in a War Environment." *Journal of Mental Imagery*, Vol. 4, 1980, pp. 59–75.

Rojcewicz, Stephen J., Jr. "War and Suicide." *Life-Threatening Behavior*, Vol. 1, Number 1, Spring 1971, pp. 45–54.

Rosenbaum, Ron. "Turn On, Tune In, Drop Dead." *Harper's*, Vol. 265, July, 1982, pp. 32–42.

Rosenbloom, Maria. "Implications of the Holocaust For Social Work." *Social Casework: The Journal of Contemporary Social Work*, Vol. 64, No. 4, April, 1983, pp. 205–213.

Rosenthal, Nina R. "Death Education and Suicide Potentiality." *Death Education*, Vol. 7, 1983, pp. 39–51.

Rosenthal, Perihan A. and Stuart Rosenthal. "Holocaust Effect in the Third Generation: Child of Another Time." *American Journal of Psychotherapy*, Vol. XXXIV, No. 4, October, 1980, pp. 572–580.

Ross, Eleanora. "Suicide and the Stages of Grief." *Death Education*, Vol. 2, 1979, pp. 407–411.

Ross, Joseph F. (moderator) "The Management of the Presuicidal, Suicidal, and Postsuicidal Patient." Interdepartmental Clinical Conference, UCLA School of Medicine. *Annals of Internal Medicine*, Vol. 75, 1971, pp. 441–458.

Rowland, Kay F. "Environmental Events Predicting Death for the Elderly." *Psychological Bulletin*, Vol. 84, No. 2, 1977, pp. 349–372.

Rozendal, Frederick G. "Halos vs. Stigmas: Long-Term Effects of Parent's Death or Divorce on College Students' Concepts of the Family." *Adolescence*, Vol. XVIII, No. 72, Winter 1983, pp. 947–955.

Rudestam, Kjell E. "Physical and Psychological Responses to Suicide in the Family." *Journal of Consulting and Clinical Psychology*, Vol. 45, No. 2, 1977, pp. 162–170.

Rudestam, Kjell E. "Some Notes on Conducting a Psychological Autopsy." *Suicide and Life-Threatening Behavior*, Vol. 9(3), Fall 1979, pp. 141–144.

Rudestam, Kjell E. and Doreen Imbroll. "Societal Reactions to a Child's Death by Suicide." *Journal of Consulting and Clinical Psychology*, Vol. 51, No. 3, 1983, pp. 461–462.

Rudolph, Marguerita. *Should the Children Know?* (New York: Schocken Books, 1978).

Rush, Alfred C. *Death and Burial in Christian Antiquity* (Washington, DC: The Catholic University of American Press, 1941).

Rushing, William A. "Alcoholism and Suicide Rates by Status Set and Occupation." *Quarterly Journal of Studies in Alcoholism*, Vol. 29, 1968, pp. 399–412.

Russell, Axel, Donna Plotkin, and Nelson Heapy. "Adaptive Abilities in Nonclinical Second-Generation Holocaust Survivors and Controls: A Comparison." *American Journal of Psychotherapy*, Vol. XXXIX, No. 4, October, 1985, pp. 564–579.

Ryan, Cornelius and Kathryn Morgan Ryan. *A Private Battle* (New York: Simon and Schuster, 1979), p. 84.

Ryn, Zdizislaw. "Suicides in the Nazi Concentration Camps." *Suicide and Life-Threatening Behavior*, Vol. 16(4), Winter 1986, pp. 419–433.

Rynearson, Edward K. "Bereavement After Homicide: A Descriptive Study." *American Journal of Psychiatry*, Vol. 141, No. 11, November, 1984, pp. 1452–1454.

Sabbath, Joseph C. "The Suicidal Adolescent—The Expendable Child." *American Academy of Child Psychiatry*, Vol. 8, 1969, pp. 272–289.

Sabo, Donald F., Jr. and L. Robert Paskoff. "Sharing Some of the Pain: Men's Response to Mastectomy." In Arthur B. Shostak and Gary McLouth (eds.), *Men and Abortion: Lessons, Losses, and Love* (New York: Praeger), pp. 285–291.

Sahler, Jane Z. (ed.). *The Child and Death* (St. Louis: The C. V. Mosby Co., 1978).

Sainsbury, Peter. *Suicide in London* (London: Chapman & Hall Ltd., 1955).

Sanders, Catherine M. "Effects of Sudden vs. Chronic Illness Death on Bereavement Outcome." *Omega*, Vol. 13(3), 1982–83, pp. 227–241.

Saubrier, J. P. and J. Vedrinne (eds.). *Depression and Suicide* (Paris: Pergamon Press, 1983).

Saunders, Judith M. and S. M. Valente. "Suicide Risk Among Gay Men and Lesbians: A Review." *Death Studies*, Vol. 11, 1987, pp. 1–23.

Saunders, Dame Cicely and Mary Baines. *Living with Dying: The Management of Terminal Disease* (New York: Oxford University Press, 1983).

Saunders, Dame Cicely et al. (eds.). *Hospice: The Living Idea* (Philadelphia: W. B. Saunders, Co., 1981).

Sawyer, Darwin O. "Public Attitudes Toward Life and Death." *Public Opinion Quarterly*, Vol. 46, 1982, pp. 521–533.

Schatzman, Leonard and Anselm Strauss. *Field Research: Strategies for a Natural Sociology* (Englewood Cliffs, NJ: Prentice-Hall, 1973).

Scher, Jordan M. "The Collapsing Perimeter." *American Journal of Psychotherapy*, Vol. XXX, Number 4, October, 1976, pp. 641–657.

Schindler, Ruben. "Reaction to Truth Telling Among Israeli Physicians." *Omega*, Vol. 13(3), 1982–83, pp. 277–286.

Schipper, H., J. Clinch, A. McMurray, and M. Levitt. "Measuring the Quality of Life of Cancer

Patients: The Functional Living Index—Cancer: Development and Validation." *Journal of Clinical Oncology*, Vol. 2, No. 5, 1984, pp. 472–483.

Schmitt, Raymond. "Symbolic Immortality in Ordinary Contexts: Impediments to the Nuclear Era." *Omega*, Vol. 13(2), 1982–83, pp. 95–116.

Schneidman, Edwin S. *Suicide Thoughts and Reflections, 1960–1980* (New York: Human Sciences Press, Inc., 1981).

Schneidman, Edwin S. (ed.). *Death and the College Student: A Collection of Brief Essays on Death and Suicide by Harvard Youth* (New York: Behavioral Publications, 1972).

Schneidman, Edwin S. (ed.). *On the Nature of Suicide* (San Francisco: Jossey-Bass, Inc., Publishers, 1969).

Schneidman, Edwin S. and Norman L. Farberow (eds.). *Clues to Suicide* (New York: McGraw-Hill, 1957).

Schneidman, Edwin S., Norman L. Farberow, and Robert E. Litman. *The Psychology of Suicide* (New York: Science House, Inc., 1970).

Schoenberg, Bernard, et al. (eds.). *Loss and Grief: Psychological Management in Medical Practice* (New York: Columbia University Press, 1970).

Schowalter, John E., et al. (eds.). *The Child and Death* (New York: Columbia University Press, 1983).

Schulz, Richard and David Aderman. "How the Medical Staff Copes with Dying Patients: A Critical Review." *Omega*, Vol. 7(1), 1976, pp. 11–21.

Schulz, Richard and David Aderman. "Physician's Death Anxiety and Patient Outcomes." *Omega*, Vol. 9(4), 1978–79, pp. 327–332.

Schuyler, Dean. "Counseling Suicide Survivors: Issues and Answers." *Omega*, Vol. 4(4), 1973, pp. 313–321.

Scott, Frances A. and Ruth M. Brewer (eds.). *Confrontations of Death* (Corvallis, OR: Continuing Education Publications, 1971).

Seeman, Melvin. "On the Meaning of Alienation." *American Sociological Review*, Vol. 24(6), 1959, pp. 784–785.

Seiden, Richard H. and Raymond P. Freitas. "Shifting Patterns of Deadly Violence." *Suicide and Life-Threatening Behavior*, Vol. 10(4), Winter 1980, pp. 195–207.

Selby, James W., III. "Situational Correlates of Death Anxiety: Reactions to Funeral Practices." *Omega*, Vol. 8(3), 1977, pp. 247–250.

Selkin, James. "The Legacy of Emile Durkheim." *Suicide and Life-Threatening Behavior*, Vol. 13(1), Spring, 1983, pp. 3–14.

Shanfield, Stephen B., G. Andrew H. Benjamin, and Barbara J. Swain. "Parents' Reactions to the Death of an Adult Child From Cancer." *American Journal of Psychiatry*, Vol. 141:9, September, 1984, pp. 1092–1094.

Shanfield, Stephen B. and Barbara J. Swain. "Death of Adult Children in Traffic Accidents." *The Journal of Nervous and Mental Disease*, Vol. 172, No. 9, 1984, pp. 533–538.

Shanfield, Stephen B., Barbara J. Swain, and G. Andrew H. Benjamin. "Parents' Responses to the Death of Adult Children from Accidents and Cancer: A Comparison." *Omega*, Vol. 17(4), 1986–87, pp. 289–298.

Shaw, Susan and Andrew Sims. "A Survey of Unexpected Deaths among Psychiatric In-Patients and Ex-Patients." *The British Journal of Psychiatry*, Vol. 108, 1962, pp. 473–476.

Shelp, Earl E. "Courage: A Neglected Virtue in the Patient-Physician Relationship." *Social Science and Medicine*, Vol. 18, No. 4, 1984, pp. 351–360.

Shepherd, Daphne and B. M. Barraclough. "The Aftermath of a Suicide." *British Medical Journal*, 15 June 1974, pp. 600–603.

Sherizen, Sanford and Lester Paul. "Dying in a Hospital Intensive Care Unit: The Social Significance for the Family of the Patient." *Omega*, 1977, pp. 29–39.

Sheskin, Arlene and Samuel E. Wallace. "Differing Bereavements: Suicide, Natural, and Accidential Death." *Omega*, Vol. 7(3), 1976, pp. 229–241.

Shils, Edward. "Faith, Utility and Legitimacy of Science." *Daedalus*. Summer, 1974, pp. 1–15.

Shneidman, Edwin S. "Postvention: The Care of the Bereaved." *Suicide and Life-Threatening Behavior*, Vol. 11, No. 4, Winter 1981, pp. 349–359.

Sickel, Ruth Z. "Children Who Experience Parental Suicide." *Pediatric Nursing*, May/June, 1979, pp. 37–39.

Siegel, Benjamin S. "Helping Children Cope with Death." *AFP*, Vol. 31, No. 3, March, 1985, pp. 175–180.

Siegel, Karolynn. "Rational Suicide: Considerations for the Clinician." *Psychiatric Quarterly*, Vol. 54(2), Summer, 1982, pp. 77–84.

Siegel, Karolynn and Peter Tuckel. "Rational Suicide and the Terminally Ill Cancer Patient." *Omega*, Vol. 15(3), 1984–85, pp. 263–269.

Sigal, J. J., M. Weinfeld, and W. W. Eaton. "Stability of Coping Style 33 Years After Prolonged Exposure to Extreme Stress." *Acta Psychiatr. Scand.*, Vol. 71, 1985, pp. 559–566.

Silberner, Joanne. "Biomedicine: Defining the End of the Road." *Science News*, December 15, 1984, p. 377.

Silberner, Joanne. "Psychological A-Bomb Wounds." *Science News*, Vol. 120, November 7, 1981, pp. 296–298.

Silverman, Phyllis R. "The Widow-to-Widow Program: An Experiment in Preventive Intervention." *Mental Hygiene*, Vol. 53, No. 3, July, 1969, pp. 333–337.

Silverman, Phyllis (interview). "Coping with Grief—It Can't Be Rushed." *U.S. News and World Report*, November 14, 1983, pp. 65–69.

Simpson, Michael A. "Brought in Dead." *Omega*, Vol. 7(3), 1976, pp. 243–248.

Simpson, Michael A. "The Lady Vanishes. . . . Or Getting Rid of the Body." *Omega*, Vol. 10(3), 1979–80, pp. 261–262.

Singer, Jerome E. "Some Issues in the Study of Coping." *Cancer*, Vol. 53, May 15, 1984, Supplement, pp. 2303–2315.

Singh, B. Krishna, J. Sherwood Williams, and Brenda J. Ryther. "Public Approval of Suicide: A Situational Analysis." *Suicide and Life-Threatening Behavior*, Vol. 16(4), Winter 1986, pp. 409–418.

Smith, George P., II, and Clare Hall. "Cryonic Suspension and the Law." *Omega*, Vol. 17, No. 1, 1986–87, pp. 1–8.

Smith, Mary L. and Darlene Francy. "When a Child Dies at Home." *Nursing*, Vol. 12, August, 1982, pp. 66–67.

Solkoff, Norman. "Children of Survivors of the Nazi Holocaust: A Critical Review of the Literature." *American Journal of Orthopsychiatry*, Vol. 51, No. 1, January, 1981, pp. 29–42.

Solomon, Mark I. "The Bereaved and the Stigma of Suicide." *Omega*, Vol. 13(4), 1982–83, pp. 377–387.

Sontag, Susan. *Illness as Metaphor* (New York: Vintage Books, 1979).

Soricelli, Barbara A. and Carolyn Lorenz Utech. "Mourning the Death of a Child: The Family and Group Process." *Social Work*, Sept.–Oct., 1985, pp. 429–434.

Speck, Peter W. *Loss and Grief in Medicine* (London: Brillière Tindall, 1978).

Speece, Mark W. and Sandor B. Brent. "Children's Understanding of Death: A Review of Three Components of a Death Concept." *Child Development*, Vol. 55, 1984, pp. 1671–1686.

Spero, Moshe H. "Psychophysiological Sequelae of Holocaust Trauma in a Jewish Child." *The American Journal of Psychoanalysis*, Vol. 40, No. 1, 1980, pp. 53–66.

Spinetta, John J. "The Dying Child's Awareness of Death: A Reviw." *Psychological Bulletin*, Vol. 81, No. 4, 1974, pp. 256–260.

Spinetta, John J. and Lorrie J. Maloney. "Death Anxiety in the Outpatient Leukemic Child." *Pediatrics*, Vol. 56, No. 6, December, 1975, pp. 1034–1037.

Spinetta, John J., David Rigler, and Myron Karon. "Anxiety in the Dying Child." *Pediatrics*, Vol. 52, No. 6, December, 1973, pp. 841–845.

Spiro, Jack D. *A Time to Mourn: Judaism and the Psychology of Bereavement* (New York: Bloch Publishing Co., 1967).

Spitzer, Walter O. et al. "Measuring the Quality of Life of Cancer Patients." *Journal of Chronic Illness*, Vol. 34, 1981, pp. 585–597.

Stack, Steven. "The Effect of Domestic/Religious Individualism on Suicide, 1954–1978." *Journal of Marriage and the Family*, May 1985, pp. 431–447.

Stack, Steven. "The Effects of Marital Dissolution on Suicide." *Journal of Marriage and the Family*, February, 1980, pp. 83–92.

Stannard, David. *The Puritan Way of Death* (New York: Oxford University Press, 1976).

Starker, Steven and Joan E. Starker. "A Group Awaiting Death: The Social Systems Perspective on a Naturally Occurring Group Situation." *Omega*, Vol. 13(1), 1982–83, pp. 79–89.

Start, Clarissa. *When You're a Widow* (St. Louis: Concordia Publishing House, 1968).

Steinbock, Bonnie. "The Removal of Mr. Herbert's Feeding Tube." *Hastings Center Report*, Vol. 13, No. 5, October, 1983, pp. 13–16.

Stephens, B. Joyce. "Suicidal Women and Their Relationships with Husbands, Boyfriends, and Lovers." *Suicide and Life-Threatening Behavior*, Vol. 15(2), Summer 1985, pp. 77–90.

Stephens, B. Joyce. "Suicidal Women and Their Relationships with Their Parents." *Omega*, Vol. 16(4), 1985–86, pp. 289–300.

Stephens, B. Joyce. "Vocabularies of Motive and Suicide." *Suicide and Life-Threatening Behavior*, Vol. 14(4), Winter 1984, pp. 243–253.

Still, Arthur and Chris Todd. "Differences Between Terminally Ill Patients Who Know, and Those Who Do Not Know, That They Are Dying." *Journal of Clinical Psychology*, Vol. 42, No. 2, March, 1986, pp. 287–296.

Still, Arthur and Chris Todd. "Role Ambiguity in General Practice: The Care of Patients Dying at Home." *Social Science and Medicine*, Vol. 23, No. 5, 1986, pp. 519–525.

Stillion, Judith. *Death and the Sexes* (New York: Hemisphere Publishing Corp., 1985).

Stillion, Judith M. "Perspectives on the Sex Differential in Death." *Death Education*, Vol. 8, 1984, pp. 237–256.

Stillion, Judith M., Eugene M. McDowell, and Jacque H. May. "Developmental Trends and Sex Differences in Adolescent Attitudes Toward Suicide." *Death Education*, Vol. 8 (Supplement), 1984, pp. 81–90.

Stillion, Judith M., Eugene M. McDowell, and Jane B. Shamblin. "The Suicide Attitude Vignette Experience: A Method for Measuring Adolescent Attitudes Toward Suicide." *Death Education*, Vol. 8 (Supplement), 1984, pp. 65–79.

Stoddard, Frederick J. and Sue S. Cahners. "Suicide Attempted by Self-Immolation during Adolescence II. Psychiatric Treatment and Outcome." *Adolescent Psychiatry*, Vol. 12, 1985, pp. 266–280.

Stoddard, Frederick J., Kambiz Pahlavan, and Sue S. Cahners. "Suicide Attempted by Self-Immolation during Adolescence I. Literature Review, Case Reports, and Personality Precursors." *Adolescent Psychiatry*, Vol. 12, 1985, pp. 251–265.

Stoesz, Willis. "Death and the Affirmation of Life: Robert Lifton's 'Sense of Immortality.'" *Soundings*, Vol. 62, No. 2, Summer 1979, pp. 187–208.

Stoker, Bram. *Dracula* (New York: Ballantine Books, 1959), p. 161.

Stoller, Eleanor P. "The Impact of Death-Related Fears on Attitudes of Nurses in a Hospital Work Setting." *Omega*, Vol. 11(1), 1980–81, pp. 85–95.

Strauss, Anselm. *Social Organization of Medical Work* (Chicago: University of Chicago Press, 1985).

Strauss, Anselm L. et al. "The Work of Hospitalized Patients." *Social Science and Medicine*, Vol. 16, 1982, pp. 977–986.

Stroebe, Margaret S. et al. "The Broken Heart: Reality or Myth?" *Omega*, Vol. 12(2), 1981–82, pp. 87–105.

Strong, Carson. "The Neonatologist's Duty to Patient and Parents." *Hastings Center Report*, Vol. 14, No. 4, August, 1984, pp. 10–16.

Sturma, Michael. "Death and Ritual on the Gallows: Public Executions in the Australian Penal Colonies." *Omega*, Vol. 17, No. 1, 1986–87, pp. 89–100.

Sudnow, David. *Passing On* (Englewood Cliffs, NJ: Prentice-Hall, 1967).

Sullivan, Michael T. "The Dying Person—His Plight and His Right." *New England Law Review*, Vol. 8, pp. 197–216.

Swain, Helen L. "Childhood Views of Death." *Death Education*, Vol. 2, 1979, pp. 341–358.

Szasz, Thomas S. "The Ethics of Suicide." *The Antioch Review*, Vol. XXXI, No. 1, Spring 1971, pp. 7–17.

Task Force on Death and Dying of the Institute of Society, Ethics, and the Life Sciences. "Refinements in Criteria for the Determination of Death: An Appraisal." *The Journal of the American Medical Association*, Vol. 221, No. 1, July 3, 1972, pp. 48–53.

Taylor, Lou. *Mourning Dress* (London: George Allen and Unwin, 1983).

Taylor, Maurice C. and Jerry W. Wicks. "The Choice of Weapons: A Study of Methods of Suicide by Sex, Race, and Religion." *Suicide and Life-Threatening Behavior*, Vol. 10(3), Fall 1980, pp. 142–149.

Taylor, Peggy and Rich Ingrasci. "Out of the Body: An Interview with Elisabeth Kübler-Ross." *Death and Dying: Challenge and Change*, Robert Fulton, et al. (eds.) (Menlo Park, CA: Addison-Wesley Publishing Company, 1978), pp. 144–153.

Taylor, Steve. *Durkheim and the Study of Suicide* (New York: St. Martin's Press, 1982).

Tegg, William. *The Last Act, Being the Funeral Rites of Nations and Individuals* (London: Wm. Tegg & Co., 1876; republished Detroit: Gale Research Co., 1973).

Teicher, Joseph D. and Jerry Jacobs. "Adolescents Who Attempt Suicide: Preliminary Findings." *American Journal of Psychiatry*, Vol. 122, 1966, pp. 1248–1257.

"The Cost of Dying." *Consumers' Research Magazine*, October, 1983, pp. 30–33.

"The Death Lobby." *The Nation*, September 11, 1982, pp. 196–197.

"The FTC Funeral Rule." *Consumer's Research*, June, 1984, pp. 28–29.

"The Limits to a Do-It-Yourself Will." *Changing Times*, November, 1984, pp. 82–85.

"The Long—and Unresolved—Goodbye." *Science News*, July 13, 1985, p. 25.

Theodoracoulos, Taki. "Dying Well Is the Best Revenge." *Esquire*, Vol. 97, February, 1982, p. 98.

Thompson, Edward H. Jr. "Palliative and Curative Care Nurses' Attitudes Toward Dying and Death in the Hospital Setting." *Omega*, Vol. 16(3), 1985–86, pp. 233–242.

Thompson, Leslie M. "Cultural and Institutional Restrictions on Dying Styles in a Technological Society." *Death Education*, Vol. 8, 1984, pp. 223–229.

Thorson, James A. "A Funny Thing Happened on the Way to the Morgue: Some Thoughts on Humor and Death, and a Taxonomy of the Humor Associated with Death." *Death Studies*, Vol. 9, 1985, pp. 201–216.

Thurlow, Setsuko. "Nuclear War in Human Perspective: A Survivor's Report." *American Journal of Orthopsychiatry*, 52(4), October 1982.

Tishler, Carl L., Patrick C. McKenry, and Karen Christman Morgan. "Adolescent Suicide Attempts: Some Significant Factors." *Suicide and Life-Threatening Behavior*, Vol. 11(2), Summer, 1981, pp. 86–92.

Toews, John, Robert Martin, and Harry Prosen. "Death Anxiety: The Prelude to Adolescence." *Adolescent Psychiatry*, Vol. 12, 1985, pp. 134–144.

Tokunaga, Howard T. "The Effect of Bereavement upon Death-Related Attitudes and Fears." *Omega*, Vol. 16(3), 1985–86, pp. 267–280.

Tolstoy, Leo. *The Death of Ivan Illych* (New York: The New American Library, A Signet Book, 1960), pp. 139–140.

Toolan, James M. "Suicide and Suicidal Attempts in Children and Adolescents." *American Journal of Psychiatry*, Vol. 118, Number 540, 1962, pp. 719–724.

Toolan, James M. "Suicide in Children and Adolescents." *American Journal of Psychotherapy*, Vol. XXIX, Number 3, July, 1975, pp. 339–344.

Topol, Phyllis and Marvin Reznikoff. "Perceived Peer and Family Relationships, Hopelessness and Locus of Control as Factors in Adolescent Suicide Attempts." *Suicide and Life-Threatening Behavior*, Vol. 12(3), Fall 1982, pp. 141–150.

Toynbee, Arnold, et al. *Man's Concern with Death* (New York: McGraw-Hill Book Co., 1968).

Trout, Deborah L. "The Role of Social Isolation in Suicide." *Suicide and Life-Threatening Behavior*, Vol. 10(1), Spring 1980, pp. 10–23.

Tuckman, Jacob and William F. Youngman. "Attempted Suicide and Family Disorganization." *The Journal of Genetic Psychology*, Vol. 105, 1964, pp. 187–193.

Turner, Ann Warren. *Houses for the Dead* (New York: David McKay Co., Inc., 1976).

"Undercutting the Undertakers." *Time*, August 8, 1984, p. 55.

U.S. Department of Health, Education and Welfare. "Trends Affecting the United States Health Care System." DHEW publication no. HRA-7614503 (January, 1976), p. 34.

Unruh, David R. "Death and Personal History: Strategies of Identity Preservation." *Social Problems*, Vol. 30, No. 3, February, 1983, pp. 340–351.

Van Der Zee, James, Owen Dobson, and Camille Billops. *The Harlem Book of the Dead* (New York: Morgan and Morgan, 1978).

Van Dongen-Melman, J. E. W. M. and J. A. R. Sanders-Woudstra. "Psychosocial Aspects of Childhood Cancer: A Review of the Literature." *Journal of Child Psychol. Psychiat.*, Vol. 27, No. 2, 1986, pp. 145–180.

Vasirub, Samuel. "Dying is Worked to Death." *The Journal of the American Medical Association*, 229(14), September 30, 1980, pp. 1909–1910.

Vaux, Kenneth L. "Debbie's Dying: Mercy Killing and the Good Death." *Journal of the American Medical Association*, Vol. 259, No. 14, April 8, 1988, pp. 2140–2141.

Veatch, Robert M. "Choosing Not To Prolong Dying." *Medical Dimensions*, December, 1972, pp. 8–10, 40.

Veatch, Robert M. "Models for Ethical Medicine in a Revolutionary Age," *Hastings Center Report*, Vol. 2, No. 3, June, 1972, pp. 5–7.

Venaki, Sophie K., Arie Nadler, and Hadas Gershoni. "Sharing the Holocaust Experience: Communication Behaviors and Their Consequences in Families of Ex-Partisans and Ex-Prisoners of Concentration Camps." *Family Process, Inc.*, Vol. 24, No. 2, June, 1985, pp. 273–280.

Vess, James, John Moreland, and Andrew I. Schwebel. "Understanding Family Role Reallocation Following a Death: A Theoretical Framework." *Omega*, Vol. 16(2), 1985–86, pp. 115–128.

Vincent, James. "When Laughter Is Okay at a Funeral." *Christianity Today*. August 9, 1985, p. 68.

Viney, Linda L. and Mary Westbrook. "Is There A Pattern of Psychological Reactions to Chronic Illness Which Is Associated with Death?" *Omega*, Vol. 17(2), 1986–87, pp. 169–182.

Vinogradov, Sophia and Joe E. Thornton. "Case Studies: 'If I Have AIDS, Then Let Me Die Now!'—Commentary." *Hastings Center Report*, Vol. 14, No. 1, February, 1984, pp. 24–25.

Vollman, Rita R. et al. "The Reactions of Family Systems to Sudden and Unexpected Death." *Omega*, Vol. 2, 1971, pp. 101–106.

Wagner, Karen D. and Raymond P. Lorion. "Correlates of Death Anxiety in Elderly Persons." *Journal of Clinical Psychology*, Vol. 40, No. 5, pp. 1234–1241.

Wagner, Richard V. "Psychology and the Threat of Nuclear War." *American Psychologist*, May, 1985, pp. 531–535.

Waller, Willard. "The Rating and Dating Complex." *American Sociological Review*, Vol. 2, 1937, pp. 727–734.

Wambach, Julie A. "The Grief Process as a Social Construct." *Omega*, Vol. 16 (3), 1985–86, pp. 201–212.

Ware, John E. "Conceptualizing Disease Impact and Treatment Outcomes." *Cancer*, Vol. 53, May 15, 1984, Supplement, pp. 2316–2326.

Wass, Hannelor and Charles A. Corr. *Helping Children Cope with Death* (New York: Hemisphere Publishing Co., 1984).

Wass, Hannelor and Charles A. Corr (eds.). *Childhood and Death* (New York: Hemisphere Publishing Co., 1984).

Wass, Hannelore et al. "Use of Play for Assessing Children's Death Concepts: A Reexamination." *Psychological Reports*, Vol. 53, 1983, pp. 799–803.

Wass, Hannelore et al. "Young Children's Death Concepts Revisited." *Death Education*, Vol. 7, 1983, pp. 385–394.

Wasserman, Ira M. "Political Business Cycles, Presidential Elections, and Suicide and Mortality Patterns." *American Sociological Review*, Vol. 48, October, 1983, pp. 711–720.

Watson, Maggie. "Psychosocial Intervention with Cancer Patients: A Review." *Psychological Medicine*, Vol. 13, 1983, pp. 839–846.

Wear, Delese. "Medical Students' Encounters with the Cadaver: A Poetic Response." *Death Studies*, Vol. 11, 1987, pp. 123–130.

Weber, Max. "Bureaucracy" in *From Max Weber*, translated by H. H. Gerth and C. W. Mills (New York: Oxford University Press, 1958), p. 196.

Weinfeld, Morton, John J. Sigal, and Sir Mortimer B. Davis. "The Effect of the Holocaust on Selected Socio-Political Attitudes of Adult Children of Survivors." *Canadian Review of Sociology and Anthropology*, Vol. 23, No. 3, 1986, pp. 365–382.

Weisman, Avery D. *The Coping Capacity: On the Nature of Being Mortal* (New York: Human Sciences Press, 1986).

Weisman, Avery D. "Is Suicide a Disease?" *Life-Threatening Behavior*, Vol. 1, No. 4, Winter 1971, pp. 219–231.

Weiss, Erwin, Agnes N. O'Connell, and Roland Siiter. "Comparisons of Second-Generation Holocaust Survivors, Immigrants, and Nonimmigrants on Measures of Mental Health." *Journal of Personality and Social Psychology*, Vol. 50, No. 4, 1986, pp. 828–831.

Weiss, James M. A. and Margaret E. Perry. "Transcultural Attitudes toward Homicide and Suicide." *Suicide*, Vol. 5(4), Winter 1975, pp. 223–227.

Weiss, Robert. "The Fund of Sociability." *Transaction*, July/August 1969, pp. 38–40.

Welch-McCaffrey, Deborah. "Cancer, Anxiety, and Quality of Life." *Cancer Nursing*, Vol. 8(3), June, 1985, pp. 151–158.

Wenestam, Claes-Göran. "Qualitative Age-Related Differences in the Meaning of the Word 'Death' to Children." *Death Education*, Vol. 8, 1984, pp. 333–347.

Werner-Beland, Jean A. (ed.). *Grief Responses to Long-Term Illness & Disability: Manifestations and Nursing Interventions* (Reston, Va: Reston Publishing Co., Inc., 1980).

Westin, Robert H. "*Ars Moriendi* Tradition and Visualization of Death in Roman Baroque Sculpture: Death Education in the Seventeenth Century." *Death Education*, Vol. 4, 1980, pp. 111–123.

Wharton, Robert and Frederick Mandell. "Violence on Television and Imitative Behavior: Impact on Parenting Practices." *Pediatrics*, Vol. 75, No. 6, 6 June 1985, pp. 1120–1123.

Whitis, Peter R. "The Legacy of a Child's Suicide." *Family Process*, Vol. 7, 1968, pp. 159–169.

Widiger, Thomas A. and Marie Rinaldi. "An Acceptance of Suicide." *Psychotherapy: Theory, Research and Practice*, Vol. 20, Number 3, Fall 1983, pp. 263–273.

Wilcox, Sandra Galdieri and Marilyn Sutton. *Understanding Death and Dying* (Calif: Mayfield Publishing Co., 1981).

Wilkes, Eric. *The Dying Patient* (NJ: George A. Bogden & Son, Inc., Publishers, 1982).

Wilkinson, H. Jean and John W. Wilkinson. "Evaluation of a Hospice Volunteer Training Program." *Omega*, Vol. 17(3), 1986–87, pp. 263–276.

Wilson, Ann L. and Douglas J. Soule. "The Role of a Self-Help Group in Working with Parents of a Stillborn Baby." *Death Education*, Vol. 5, 1981, pp. 175–186.

Wilson, Michele. "Suicidal Behavior: Toward an Explanation of Differences in Female and Male Rates." *Suicide and Life-Threatening Behavior*, Vol. 11(3), Fall 1981, pp. 131–140.

Wolfensberger, Wolf. "Holocaust II?" *Journal of Learning Disabilities*, Vol. 17, No. 7, August/September, 1984, pp. 439–440.

Wolfgang, Marvin and Franco Ferracuti. *The Subculture of Violence* (London: Tavistock Publications, 1967).

Woodfield, Robert L. and Linda L. Viney. "A Personal Construct Approach to the Conjugally Bereaved Woman." *Omega*, Vol. 15(1), 1984–85, pp. 1–13.

Wortman, Camille B. "Social Support and the Cancer Patient: Conceptual and Methodologic Issues." *Cancer*, Vol. 53, May 15, 1984, Supplement, pp. 2339–2360.

Wroblewski, Adina. "The Suicide Survivors Grief Group." *Omega*, Vol. 15(2), 1984–85, pp. 173–183.

Wuthnow, Robert, Kevin Christiano, and John Kuzlowski. "Religion and Bereavement: A Conceptual Framework." *Journal for the Scientific Study of Religion*, Vol. 19, No. 4, December, 1980, pp. 408–422.

"You're Named in the Will. Now What?" *Changing Times*, Vol. 37, November, 1983, pp. 56–60.

Zborowski, Mark. *People in Pain* (San Francisco: Jossey-Bass Inc., Publishers, 1969).

Zeldow, Peter B., et al. "Personality Indicators of Psychosocial Adjustment in First-Year Medical Students." *Social Science and Medicine*, Vol. 20, No. 1, 1985, pp. 95–100.

Zito, George. "Marx, Durkheim and Alienation." *Social Theory and Practice* (1974), Vol. 3, No. 2, Fall 1974, pp. 223–242.

CREDITS

Thanatology Snapshot 1.1: From *Hour of Our Death* by Philippe Aries. Copyright © 1980 by Alfred A. Knopf, Inc. Reprinted by permission of Random House, Inc.

1.3: Reprinted with permission from The New Physician, copyright 1985, American Medical Student Association.

2.1: Reprinted from *Human Medicine* by James Nelson, Copyright © 1973 Augsburg Publishing House. Used by permission of Augsburg Fortress.

2.2: Copyright 1985 by George J. Annas, reprinted with permission of the author.

2.3: Copyright © 1985 by The New York Times Company. Reprinted by permission.

3.1: Reprinted with permission of Simon & Schuster from *On Death and Dying* by Elisabeth Kübler-Ross. Copyright © 1969 by Elisabeth Kübler-Ross, M.D.

3.2: Reprinted by permission of the *New Age Journal*, 42 Pleasant Street, Watertown, MA 02172; $24/year.

3.3: Copyright © 1982 by *Harper's Magazine*. All rights reserved. Reproduced from the July issue by special permission.

3.5: From "Approaching Omega: The Roller Coaster of Dying" in *On Death Without Dignity* by Ernest Becker. Copyright © 1990 by Baywood Publishing Company Inc.

3.6: From "The Stigma of Dying" in *On Death Without Dignity: The Human Impact of technological Dying* by David Wendell Moller. Copyright © 1990 by Baywood Publishing Company, Inc.

4.2: Reprinted from *The American Way of Death* by Jessica Mitford, Crest Books, 1964.

4.3: From *Consumer's Research Magazine*, Vol. 67 June 1984.

5.3: Copyright © Commonweal Foundation 1984.

5.5: From *Death and Ethnicity: A Psychocultural Study*, V.4 edited by R. Kalish and D. Reynolds. Copyright © 1976 by Baywood Publishing Company, Inc.

6.1: From *The Lonliness of Dying* by Norbert Elias. Reprinted by permission of Blackwell Publishers.

6.2: Bluebond-Langner, Myra, *The Private Worlds of Dying Children*. Copyright © 1978 by Princeton University Press. Used with permission.

6.3: *Journal of Psychological Oncology* Vol. 2 (2) pgs. 61-72. © 1984 The Hadworth Press Inc. All rights reserved.

7.1: From "Suicide" in *The Harvard Medical School Mental Health Letter* 2(8) February 1986.

7. 3: Reprinted by permission of the William Morris Agency, Inc. on behalf of the Author. Copyright © 1986 Betty Rollin.

INDEX